Springer Monographs in Mathematics

For other titles published in this series, go to
http://www.springer.com/series/3733

Dmitry Treschev · Oleg Zubelevich

Introduction to the Perturbation Theory of Hamiltonian Systems

Springer

Dmitry Treschev
Department of Mechanics
Russian Academy of Sciences
Steklov Mathematical Institute
ul. Gubkina 8
Moscow 119991
Russia
treschev@mi.ras.ru

Oleg Zubelevich
Department of Mechanics
and Mathematics
Moscow State University
Leninskie Gory
Moscow 119899
Russia
ozubel@yandex.ru

ISSN 1439-7382
ISBN 978-3-642-03027-7 e-ISBN 978-3-642-03028-4
DOI 10.1007/978-3-642-03028-4
Springer Heidelberg Dordrecht London New York

Library of Congress Control Number: 2009937641

Mathematics Subject Classification (2000): 37JXX, 37J05, 37J10, 37J25, 37J30, 37J40, 37J45

Cover design: deblik

Printed on acid-free paper

Springer is part of Springer Science+Business Media (www.springer.com)

Preface

This book is an extended version of lectures given by the first author in 1995–1996 at the Department of Mechanics and Mathematics of Moscow State University. We believe that a major part of the book can be regarded as an additional material to the standard course of Hamiltonian mechanics. In comparison with the original Russian version[1] we have included new material, simplified some proofs and corrected misprints.

Hamiltonian equations first appeared in connection with problems of geometric optics and celestial mechanics. Later it became clear that these equations describe a large class of systems in classical mechanics, physics, chemistry, and other domains. Hamiltonian systems and their discrete analogs play a basic role in such problems as rigid body dynamics, geodesics on Riemann surfaces, quasi-classic approximation in quantum mechanics, cosmological models, dynamics of particles in an accelerator, billiards and other systems with elastic reflections, many infinite-dimensional models in mathematical physics, etc.

In this book we study Hamiltonian systems assuming that they depend on some parameter (usually ε), where for $\varepsilon = 0$ the dynamics is in a sense simple (as a rule, integrable). Frequently such a parameter appears naturally. For example, in celestial mechanics it is accepted to take ε equal to the ratio: the mass of Jupiter over the mass of the Sun. In other cases it is possible to introduce the small parameter artificially. For example, if we are interested in trajectories near an invariant manifold, where the dynamics is known (in the simplest case the manifold is an equilibrium of the system), we can take as such a parameter the distance to the manifold. In some systems it is possible to use as a small parameter the quantity inverse to the total energy. In particular, the problem of the motion of a particle on a compact Riemannian manifold M in a potential force field for large values of the energy is close to the problem of geodesic lines on M.

A standard example of a Hamiltonian system depending on a small parameter is presented by the equations

[1] D. Treschev. Introduction to the perturbation theory of Hamiltonian systems. Moscow, Phasis 1998 (in Russian).

$$\dot{x} = \partial H/\partial y, \quad \dot{y} = -\partial H/\partial x, \tag{1}$$

$$H(x, y, \varepsilon) = H_0(y) + \varepsilon H_1(x, y) + \varepsilon^2 H_2(x, y) + \cdots. \tag{2}$$

Here y belongs to some domain in $\mathbb{R}^m$ and x mod 2π is a point of the m-dimensional torus $\mathbb{T}^m$. The Hamiltonian H can depend explicitly on time t. The dependence on time is usually periodic. We assume that the function H is infinitely smooth in all arguments. Systems of form (1)–(2) are below called *close to integrable* or *near-integrable*.

Poincaré called the studying of near-integrable systems *the basic problem of the dynamics*. At present Poincaré's "basic problem of the dynamics" continues to occupy one of the most important places in the theory of dynamical systems.

Frequently it is more convenient to consider instead of Hamiltonian systems their discrete analogs, *symplectic maps*, i.e., self-maps of the phase space that preserve the symplectic structure (in the classical situation the form $dy \wedge dx$). Near-integrable symplectic maps have the form

$$\begin{pmatrix} x \\ y \end{pmatrix} \mapsto \begin{pmatrix} x + \partial f/\partial y \\ y \end{pmatrix} + O(\varepsilon), \tag{3}$$

where $f = f(y)$ is a smooth function. The perturbation theory for such systems is parallel to the corresponding theory for Hamiltonian systems (1)–(2).

The basic problems considered in the book are integrability of systems (1)–(2) and (3), existence of invariant tori, and the description of chaos for small values of the parameter ε. Chapter 7 is unusual from this point of view since there we study systems which are, on the contrary, very far from integrable ones.

Naturally, we are not able to discuss all methods and problems of the Hamiltonian perturbation theory. In particular, Hamiltonian systems with an infinite number of degrees of freedom, systems with a degenerate Poisson bracket, the theory of normal forms, and the theory of singular perturbations are not presented here.

The first chapter can be regarded as a technical introduction. A reader familiar with such concepts as symplectic structures, Poisson brackets, Hamiltonian vector fields, the Poincaré maps, and Liouville integrability, can skip the first two sections. In Sect. 1.3 the problem of the representation of a given diffeomorphism as the Poincaré map for some non-autonomous system of ordinary differential equations is discussed. In Sect. 1.4 we present the procedure of the classical perturbation theory and the Poincaré theorem on nonintegrability.

In Chap. 2 we discuss basic ideas and results of the Kolmogorov–Arnold–Moser (KAM) theory. KAM theory is a collection of theorems on the preservation of quasi-periodic motions after a perturbation of the system.[2] Such motions lie on invariant tori in the phase space. We present results about the preservation of Lagrangian and lower-dimensional tori. Special attention is devoted to hyperbolic tori which play an important role in the perturbed dynamics. We prove convergence of the KAM procedure in one of the simplest model examples, where the problem of small divisors

[2] KAM-theory also contains reversible, volume-preserving, and dissipative versions which are not discussed here.

appears. KAM theory explains many dynamical effects in perturbed systems. Some of these effects are shortly discussed in Sect. 2.6.

In Chap. 3 we present the Poincaré–Melnikov theory of the splitting of asymptotic manifolds (separatrices), including both traditional and multidimensional cases. Separatrix splitting is known as one of fundamental dynamical phenomena, generating chaos in perturbed systems.

In Chap. 4 we construct the separatrix map for a two-dimensional symplectic map and for a Hamiltonian system with one and a half degrees of freedom. The separatrix map is the main tool which is used in Chap. 5, where we analyze the stochastic layer appearing in the vicinity of separatrices, generated by a hyperbolic fixed point of a two-dimensional symplectic diffeomorphism. We present several asymptotic formulas describing quantitatively stochastic layers in systems close to integrable.

In Chap. 6 a special averaging procedure is described. This procedure was originally invented for the analysis of exponentially small effects in near-integrable Hamiltonian systems. Later it turned out to be effective in other problems, and even outside the Hamiltonian perturbation theory. We apply the procedure in the proof of the theorem on an inclusion of a diffeomorphism into a flow and in the analysis of one-frequency slow-fast systems.

In Chap. 7 we consider systems far from integrable. We prove a generalized version of the Aubry–Abramovici theorem on the anti-integrable limit. Probably, the anti-integrable limit can be regarded as methodically the simplest introduction to symbolic dynamics.

In Chap. 8, written jointly with Serge Bolotin, we discuss various versions of the Hill formula. This formula connects geometric and dynamical properties of a periodic solution of a Lagrangian system. The first ones are determined by the characteristic polynomial of the monodromy matrix while the second by the determinant of the corresponding Hesse matrix, its generalizations and regularizations. As a corollary we obtain sufficient conditions for dynamical stability or instability of the solution in terms of its Morse index.

We present also several number-theoretic statements concerning the Diophantine properties of the frequency vectors. The corresponding results are gathered in Sect. 9.1. In Sect. 9.2 we prove two theorems on the coincidence of closures of the stable and unstable separatrix in a two-dimensional area-preserving map. Section 9.3 contains results on the difference of frequency vectors on two KAM-tori, situated from two different sides with respect to a resonance. In Sect. 9.4 we show that separatrix lobes of near-integrable systems can contain stability islands. In Sect. 9.5 we gather some facts from functional analysis necessary for proofs dealing with the continuous averaging method. The main technical tools we need are the Schauder and Nirenberg–Nishida theorems.

Chapters 2–8 of the book can be regarded independent with the following exception: methods used in Chap. 5 are based on results of Chap. 4.

Now several words about the notation. The sets of complex, real, rational, integer, and natural numbers are as usual denoted $\mathbb{C}$, $\mathbb{R}$, $\mathbb{Q}$, $\mathbb{Z}$, and $\mathbb{N}$ respectively. The one-dimensional torus $\mathbb{T} = \mathbb{R}/(2\pi\mathbb{Z})$ (sometimes we assume that $\mathbb{T} = \mathbb{R}/\mathbb{Z}$). If we want to indicate the coordinate used on a given set, we write this coordinate as a subscript. For example $\mathbb{R}_y$ is the real line with the coordinate y.

Vectors appearing below are regarded as columns. To obtain a row we use the operation of matrix transposition T. For example, $v \in \mathbb{R}^m$ is a column vector and v^T is a row vector.

The standard inner product in $\mathbb{R}^m$ is denoted $\langle\, ,\rangle$: for any two vectors $u = (u_1, \dots, u_m)^T$, $v = (v_1, \dots, v_m)^T$ we have $\langle u, v\rangle = \sum_{j=1}^m u_j v_j$. Note that a more careful consideration shows that u and v in expressions like $\langle u, v\rangle$ are usually elements of dual vector spaces, so $\langle\, ,\rangle$ can be regarded as the operation of pairing of a vector and a covector.

Functions of order n with respect to the variable y for small $|y|$ are denoted $O_n(y)$. A dot denotes differentiation with respect to time t. For example, $\dot{y} = dy/dt$.

We usually reserve the letter i to denote of the imaginary unit and the symbol $\square$ to denote the end of a proof.

Section n of Chapter m is called Section $m.n$. Theorems and other statements are numerated as follows:

the number of the chapter . the number of the statement.

Formulas also have double numbers:

(the number of the chapter . the number of the formula).

During the work on the book we benefited much from communications with our friends and colleagues Valery Kozlov, Serge Bolotin, Anatoly Neishtadt, and Gena Piftankin. We are grateful to them and to many other people, forming the dynamical systems community, inside which the results presented below were born. The work was supported by Russian Foundation of Basic Research grant 08-01-00681-a.

Moscow

Dmitry Treschev
Oleg Zubelevich

Contents

Chapter 1
Hamiltonian Equations

1.1 Hamiltonian Systems: Geometric Point of View

It is well known that the form

$$\dot{q} = \partial H/\partial p, \quad \dot{p} = -\partial H/\partial q \tag{1.1}$$

of the Hamiltonian equations is preserved only by canonical (symplectic) changes of variables. On the other hand it is clear that specific properties of Hamiltonian systems should be presented independently of local coordinates on the phase space. In this section we discuss the invariant nature of Hamiltonian equations. Since the corresponding material is essentially standard, see for example [1, 9]; we usually skip proofs.

Hamiltonian vector fields. Recall that a smooth manifold M, $\dim M = 2m$, endowed with a nondegenerate closed differential 2-form ω (a symplectic structure) is called a *symplectic manifold*. In local coordinates

$$\omega = \sum_{1\le j<k\le 2m} a_{jk}(z)\, dz_j \wedge dz_k, \qquad a_{jk} = -a_{kj}, \quad d\omega = 0. \tag{1.2}$$

According to the definition of the symplectic structure, the matrix $A = (a_{jk})$ is nondegenerate and skew-symmetric. The condition $d\omega = 0$ is equivalent to the equations

$$\frac{\partial a_{jk}}{\partial z_l} + \frac{\partial a_{kl}}{\partial z_j} + \frac{\partial a_{lj}}{\partial z_k} = 0$$

for any three indices $1 \le j < k < l \le 2m$.

According to the Darboux theorem (see for example [10]), in a neighborhood of any point of a symplectic manifold there exist local coordinates (q, p) such that $\omega = \sum_{j=1}^{m} dp_j \wedge dq_j$. The matrix A takes the form $A = \begin{pmatrix} 0 & -I \\ I & 0 \end{pmatrix}$, where I is the

D. Treschev, O. Zubelevich, *Introduction to the Perturbation Theory of Hamiltonian Systems*, Springer Monographs in Mathematics,
DOI 10.1007/978-3-642-03028-4_1,

identity $(m \times m)$ matrix. Such coordinates are called *canonical* or *symplectic*.[1] The coordinates p and q are called *canonically conjugate* to one another.

The diffeomorphism $f : M_1 \to M_2$ of two symplectic manifolds (M_1, ω_1) and (M_2, ω_2) is called symplectic (or a *symplectomorphism*) if it maps one symplectic structure to another: $f^*\omega_2 = \omega_1$. Hence, the Darboux theorem asserts that any two symplectic manifolds of the same dimension are locally symplectomorphic.

Let N be an m-dimensional manifold and T^*N the corresponding cotangent bundle. In local coordinates the symplectic structure has the form $\omega = dp \wedge dq$, where $q = (q_1, \dots, q_m)$ are coordinates on N and $p = (p_1, \dots, p_m)$ are coordinates on the fiber T_q^*N dual to the coordinates $dq_1, \dots, dq_m$ on the fiber T_qN of the tangent bundle TN. Such symplectic manifolds play an important role in classical mechanics.

Recall that the Legendre transform of the Lagrangian system

$$\frac{d}{dt}\frac{\partial \mathscr{L}}{\partial \dot{q}} - \frac{\partial \mathscr{L}}{\partial q} = 0, \qquad \mathscr{L} = \mathscr{L}(q, \dot{q}, t),$$

where the Lagrangian function $\mathscr{L}$ is convex in the velocities $\dot{q}$ and the configurational space is N, is a Hamiltonian system whose phase space is the symplectic manifold $(T^*N, dp \wedge dq)$.

It is generally accepted to regard the Hamiltonian presentation of the dynamics as more perfect in comparison with the Lagrangian presentation. The reason is as follows. In the Lagrangian presentation we can simplify the coordinate form of the Lagrangian function $\mathscr{L}$ by using only changes of coordinates on the configurational space N: the corresponding transformation of the velocities $\dot{q}$ is then uniquely determined. In the Hamiltonian presentation we usually want to preserve the canonical form of the symplectic structure $dp \wedge dq$. However even this class of admissible (canonical) coordinate changes is much wider than the extension to the phase space of coordinate changes on the configurational space. This circumstance is especially important in the theory of integrable systems as well as in the perturbation theory. However it is necessary to keep in mind one important advantage of the Lagrangian formalism: convenient variational principles, primarily the Hamilton's principle $\delta \int_{t_0}^{t_1} \mathscr{L}(q, \dot{q}, t)\, dt = 0$. Due to this in some situations the Lagrangian presentation is more adequate. We will see this in Chaps. 7 and 8.

Now let us return to the geometry of the phase space in the Hamiltonian formalism. The submanifold L in a symplectic manifold is called *symplectic* if the restriction of the form ω to L is nondegenerate. If $\omega|_L = 0$ then L is called *isotropic*. An isotropic m-dimensional submanifold is called *Lagrangian*. For example, for any smooth $S : N \to \mathbb{R}$ the manifold

$$L_S = \{(q, p) \in T^*N : p = \partial S/\partial q\}$$

[1] Note that in the literature there is no unique tradition concerning definitions of canonical form of the symplectic structure and of the Poisson bracket: in different textbooks these objects can differ by sign.

is Lagrangian. If $S \equiv 0$, L_S is the zero section of the cotangent bundle T^*N. Below we need two generalizations of the Darboux theorem (see for example [10]).

Theorem 1.1. *A small neighborhood of the Lagrangian submanifold L in the symplectic manifold (M, ω) is symplectomorphic to a neighborhood of the zero section in the bundle T^*L.*

Theorem 1.2. *Let ω_0, ω_1 be two symplectic structures on M and let their restrictions to a submanifold $N \subset M$ coincide. Suppose that ω_0 can be continuously deformed to ω_1 in the class of symplectic structures on M coinciding on N with $\omega_0|_N = \omega_1|_N$. Then there exist neighborhoods U_0 and U_1 of N and a diffeomorphism $g : U_0 \to U_1$ identical on N which transforms $\omega_1|_{U_1}$ to $\omega_0|_{U_0}$: $g^*\omega_1 = \omega_0$.*

The condition about the possibility to deform ω_0 to ω_1 is satisfied for example if ω_0 and ω_1 coincide on any pair of vectors from TM issuing from a point of N (this is a stronger condition than $\omega_0|_N = \omega_1|_N$). In this case the forms $\omega_0 + (1 - s)\omega_1$, $0 \le s \le 1$, do not degenerate in a neighborhood of N.

By (1.2) for any point $z \in M$ and any two vectors $u, v \in T_zM$

$$\omega(u, v) = \langle u, Av\rangle, \qquad A = (a_{jk}(z)) = -A^T.$$

Then A can be regarded as a linear operator acting from the tangent space T_zM at an arbitrary point $z \in M$ to the cotangent space T_z^*M:

$$v \mapsto Av = \omega(\cdot, v), \qquad v \in T_zM,$$

where the covector $\omega(\cdot, v)$, applied to the vector u, yields $\omega(u, v)$. Since A is nondegenerate, it is a linear isomorphism of the spaces T_zM and T_z^*M. Let J be the inverse isomorphism:

$$\langle f, u\rangle = \omega(u, Jf) \quad \text{for any } u \in T_zM, \quad f \in T_z^*M,$$

where $\langle f, u\rangle$ is the action of the covector f on the vector v.

Let H be a function on M. Then the 1-form dH defines $v_H = JdH$, which is called the *Hamiltonian vector field* associated with the Hamiltonian function H. Then

$$\langle dH, \cdot\rangle = \omega(\cdot, v_H).$$

Thus a Hamiltonian system is determined by the triple (M, ω, H) and m is said to be the number of degrees of freedom. It is easy to check that in canonical coordinates p, q the Hamiltonian equations $\dot z = v_H(z)$ have the usual form (1.1).

For any system (M, ω, H) the function H is a *first integral*, i.e., it preserves its value on any solution. Indeed,

$$\dot H = \langle dH, v_H\rangle = \omega(v_H, v_H) = 0.$$

Usually H is called the *energy integral*.

Below we frequently consider nonautonomous systems. Their Hamiltonians depend explicitly on time. The extended phase space of such a system is the direct product of the manifold M and the real time axis $\mathbb{R}$ or $M \times \mathbb{T}$ if the dependence on time is periodic. In the latter case we say that the system has $m + 1/2$ degrees of freedom.

Any autonomous system (M, ω, H) can be locally reduced on the level of the energy integral $H = h$ to a nonautonomous system (*isoenergetic reduction*). The main idea of this procedure is to decrease the dimension of the phase space. In canonical local coordinates (q, p) near a nonsingular point (q^0, p^0) this reduction looks as follows. Let p_j be the coordinate such that $\partial H/\partial p_j \neq 0$. Without loss of generality $j = m$. Taking q_m as a new time $q_m = \tau$, $d/d\tau = (\cdot)'$ and putting $\tilde{q} = (q_1, \dots, q_{m-1})$, $\tilde{p} = (p_1, \dots, p_{m-1})$, we define the new Hamiltonian function $K(\tilde{q}, \tilde{p}, \tau, h)$ as a solution $p_m = -K$ of the equation $H = h$ with respect to p_m. Then the system (M, ω, H), $H = h$, is locally equivalent to the nonautonomous system

$$\dot{\tilde{q}} = \frac{\partial K}{\partial \tilde{p}}, \quad \dot{\tilde{p}} = -\frac{\partial K}{\partial \tilde{q}}.$$

The nonautonomous case can be reduced to the autonomous one with the help of the following standard trick, called the *autonomization*. Introduce the new phase variable E and the new independent variable $\tau = t + \text{const}$, $d/d\tau = (\cdot)'$. We assume that E is the momentum canonically conjugate to the time variable t. The value of the constant in the definition of τ plays no role. In particular, it is possible to take this constant equal to zero and to identify τ and t. The system $(M \times \mathbb{R}_t \times \mathbb{R}_E, \omega + dE \wedge dt, H + E)$ has the form

$$q' = \frac{\partial H}{\partial p}, \quad p' = -\frac{\partial H}{\partial q}, \quad t' = 1, \quad E' = -\frac{\partial H}{\partial t}, \quad (\cdot)' = \frac{d}{d\tau}. \tag{1.3}$$

Therefore it implies (M, ω, H) in the sense that trajectories of system (1.3), being projected to the space $M \times \mathbb{R}_t$, give solutions of the system (M, ω, H).

We see that it is possible to consider only the autonomous case. However some formulas and statements look simpler for nonautonomous systems.

Theorem 1.3. *For any pair of real numbers τ_1, τ_2 the shift $g_{\tau_1}^{\tau_2}$ along solutions of a (may be, nonautonomous) Hamiltonian system, when the time passes from $t = \tau_1$ to $t = \tau_2$, preserves the form ω on the phase space.*

In the autonomous case the maps $g_{\tau_1}^{\tau_2} = g_0^{\tau_2 - \tau_1}$ form a one-parameter transformation group of M. This group is called the *phase flow* g^t or g_H^t.

Corollary 1.1. *The phase flow g_H^t of an autonomous Hamiltonian system preserves the symplectic structure: $(g_H^t)^* \omega = \omega$.*

Corollary 1.2. *The transformations $g_{\tau_1}^{\tau_2}$ preserve the forms $\omega^{(k)} = \underbrace{\omega \wedge \dots \wedge \omega}_{k \text{ times}}$.*

In particular, the maps $g_{\tau_1}^{\tau_2}$ preserve the canonical volume form $dp_1 \wedge \dots \wedge dp_m \wedge dq_1 \wedge \dots \wedge dq_m$, since it is proportional to $\omega^{(m)}$.

In general not all closed 1-forms are exact. Primitives of such forms are multi-valued functions. If we take such a function as a Hamiltonian, the corresponding Hamiltonian vector field is obviously single-valued. Vector fields obtained in this way are called *locally Hamiltonian*. It is easy to notice that Theorem 1.3 remains true if we assume that the system is locally Hamiltonian. Due to this observation the following theorem can be regarded as inverse to Theorem 1.3.

Theorem 1.4. *Suppose that for any finite time interval the corresponding shift along solutions of the systems $\dot{z} = v(z, t)$, $z \in M$, preserves the symplectic structure. Then the vector field v is locally Hamiltonian.*

The Poisson bracket. Recall that the vector space L together with the bilinear skew-symmetric operation (the commutator) $[\,,\,] : L \times L \to L$ satisfying the Jacobi identity

$$[[A, B], C] + [[B, C], A] + [[C, A], B] = 0, \qquad A, B, C \in L,$$

is called a *Lie algebra*.

The simplest example of a Lie algebra is the space of square matrices of a fixed order, where $[A, B] = AB - BA$.

The Lie algebra of smooth vector fields on M is a more important example for us. Here the commutator is introduced as follows. Any vector field v determines on the space $C^\infty(M)$ the differential operator ∂_v. The coordinate definition of ∂_v is $\sum v_j \partial/\partial z_j$. Conversely, any differential operator of the first order uniquely determines a vector field.

For any two smooth vector fields u, v the differential operator $[\partial_u, \partial_v] = \partial_u \partial_v - \partial_v \partial_u$ is of first order. Therefore, it determines a vector field. This vector field is called the *commutator* of u and v.

Theorem 1.5. *Let u, v be smooth vector fields on M and g_u^t, g_v^s their phase flows. Then g_u^t and g_v^s commute*[2] *if and only if $[u, v] = 0$.*

If the vector fields u and v commute, v is called the symmetry field for the system $\dot{z} = u(z)$ and vice versa. As a corollary we obtain that the phase flow of a symmetry field maps solutions of a system of ordinary differential equations to solutions of the same system.

For any smooth function H on the symplectic manifold M we put $\partial_H = \partial_{v_H}$, where v_H is the Hamiltonian vector field with Hamiltonian H.

Let F and H be two smooth functions on M. The Poisson bracket of these functions is defined as follows:

$$\{H, F\} = \partial_H F = (dF, v_H).$$

(The first equation is a definition while the second one is an identity.) The functions H and F, satisfying the equality $\{H, F\} = 0$, are said to commute or to be in involution.

[2] I.e., $g_u^t \circ g_v^s = g_v^s \circ g_u^t$.

There are several properties of the Poisson bracket which follow directly from the definition.

1. The smooth function F is a first integral of the system with Hamiltonian H if and only if $\{H, F\} = 0$.
2. $\{H, F\} = \omega(v_H, v_F)$. Indeed, from the definition of v_H we have: $(dF, v_H) = \omega(v_H, v_F)$.
3. The operation $\{\,,\}$ is bilinear and skew-symmetric.
4. In canonical coordinates $\partial_H = \frac{\partial H}{\partial p}\frac{\partial}{\partial q} - \frac{\partial H}{\partial q}\frac{\partial}{\partial p}$. Therefore

$$\{H, F\} = \sum_j \left(\frac{\partial F}{\partial q_j}\frac{\partial H}{\partial p_j} - \frac{\partial F}{\partial p_j}\frac{\partial H}{\partial q_j} \right).$$

5. The Poisson bracket satisfies the Jacobi identity

$$\{\{F, G\}, H\} + \{\{G, H\}, F\} + \{\{H, F\}, G\} = 0.$$

Hence, the space $(C^\infty(M), \{\,,\})$ is a Lie algebra.

Note that any Poisson bracket generated by a symplectic structure is non-degenerate, i.e., for any $z \in M$ and any function F such that $dF(z) \neq 0$ there exists a function G such that $\{F, G\}(z) \neq 0$. In some physical models degenerate Poisson brackets appear, but below we do not consider such cases.

Theorem 1.6. *For any two smooth functions F and G on M*

$$[v_F, v_G] = v_{\{F,G\}}.$$

Proof. Let φ be an arbitrary function on M. Then

$$\begin{aligned} \partial_{\{F,G\}}\varphi &= \{\{F, G\}, \varphi\} = -\{\{G, \varphi\}, F\} - \{\{\varphi, F\}, G\} \\ &= \{F\{G, \varphi\}\} - \{G\{F, \varphi\}\} \\ &= \partial_F \partial_G \varphi - \partial_G \partial_F \varphi = [\partial_F, \partial_G]\varphi. \quad \square \end{aligned}$$

The Liouville theorem on integrability. Let the system (M, ω, H) have m first integrals in involution:

$$F_1, \ldots, F_m, \quad \{F_j, F_k\} = 0, \qquad 1 \le j, k \le m.$$

Usually it is assummed that $F_1 = H$. We assume that flows of the vector fields v_{F_j} are complete.

Consider the manifold

$$M_f = \{z \in M : F_j(z) = f_j = \text{const}, \; j = 1, \ldots, m\}. \tag{1.4}$$

Theorem 1.7 (Liouville-Arnold). *Suppose that on M_f the functions F_j are independent. Then*

(1) *M_f is a smooth manifold invariant with respect to the system $\dot{z} = v_H$.*
(2) *Each compact connected component of M_f is diffeomorphic to the m-dimensional torus $\mathbb{T}^m$.*
(3) *In some coordinates $(\varphi_1, \dots, \varphi_m) \bmod 2\pi$ on $\mathbb{T}^m$ the Hamiltonian equations have the form $\dot{\varphi} = \nu = \nu(f)$.*

Systems satisfying the conditions of Theorem 1.7 almost everywhere on M_f are called *Liouville integrable.*

Remark 1.1. Any Liouville integrable system is integrable by quadratures in a neighborhood of the invariant manifold M_f. Initially it is this fact which was called the Liouville theorem. Arnold has discovered that Liouville integrable systems possess the geometric structure described in Theorem 1.7.

Proof. Assertion (1) obviously follows from the implicit function theorem. Consider the Hamiltonian vector fields $v_j = v_{F_j}$, $j = 1, \dots, m$. They are tangent to the manifold M_f since for any $j, k \in \{1, \dots, m\}$ we have: $\partial_{v_j} F_k = \{F_j, F_k\} = 0$. Moreover, these vector fields are linearly independent at any point of M_f and commute. The independence follows from the independence of the covectors dF_j and from the nondegeneracy of the operator J. Commutators of the fields $v_1, \dots, v_m$ satisfy the equations

$$[v_j, v_k] = v_{\{F_j, F_k\}} = 0.$$

To finish the proof of Theorem 1.7 we use the following geometric lemma (see for example [9]). □

Lemma 1.1. *Suppose that on a compact connected m-dimensional manifold there are m everywhere linearly independent commuting vector fields with complete flows. Then the manifold is diffeomorphic to the m-dimensional torus $\mathbb{T}^m$. Moreover, there are angular coordinates $(\varphi_1, \dots, \varphi_m) \bmod 2\pi$ on the manifold such that all these vector fields have constant components.*

Remark 1.2. The restriction of the form ω to M_f vanishes, i.e., M_f is a Lagrangian manifold. Indeed, the vector fields $v_1, \dots, v_m$ are pairwise skew-orthogonal ($\omega(v_j, v_k) = \{F_j, F_k\} = 0$) and form bases in tangent spaces to M_f.

Invariant Lagrangian tori of Liouville integrable systems are called *Liouville tori.* In a neighborhood of a Liouville torus on the phase space it is possible to introduce convenient canonical coordinates: the *action-angle variables* (I, φ). The coordinates φ are the same as in Lemma 1.1, and $I = (I_1, \dots, I_m)$ are functions of the first integrals $F_1, \dots, F_m$. In the new variables the Hamiltonian function $H = H(I)$ does not depend on the angles φ. As a result the Hamiltonian equations can be easily solved explicitly. Action-angle variables play an essential role in the perturbation theory. Below we use the notation x for angles and y for actions.

1.2 Symplectic Maps

Systems with discrete time occupy an important place in the world of dynamical systems. Such systems are generated by self-maps of a manifold which is usually called the phase space. The discrete analog of a Hamiltonian system is a *symplectic map* i.e., the phase space is a symplectic manifold and the map preserves ω.

One of the simplest examples is the standard Chirikov map:

$$\begin{pmatrix} x \bmod 2\pi \\ y \end{pmatrix} \mapsto \begin{pmatrix} X \bmod 2\pi \\ Y \end{pmatrix} = \begin{pmatrix} x + Y \\ y + \varepsilon \sin x \end{pmatrix}. \tag{1.5}$$

In this case the phase space is $(M, \omega) = (\mathbb{T}_x \times \mathbb{R}_y, dy \wedge dx)$ while ε is a parameter.

According to Theorem 1.3 the shift $g_{\tau_1}^{\tau_2}$ along trajectories of a Hamiltonian system is a symplectic map. In particular, the time-τ map g_0^τ in a nonautonomous system with τ-periodic in time Hamiltonian is symplectic. Below we call it the Poincaré map.

Poincaré maps corresponding to autonomous Hamiltonian systems form another important class of symplectic maps. The construction is as follows. Let (M, ω, H) be an autonomous Hamiltonian system, g^t its phase flow, $M_h = \{z \in M : H(z) = h\}$ an energy level, and let $S \subset M_h$ be a $(2m-2)$-dimensional surface transversal to the vector field v_H. The Poincaré map P acts on the surface S in the following way. It associates to the point $z \in S$ the point $g^t(z) \in S$ with the minimal positive t. In other words P associates to $z \in S$ the point at which the positive semitrajectory of the system with the initial condition z intersects S for the first time (Fig. 1.1).

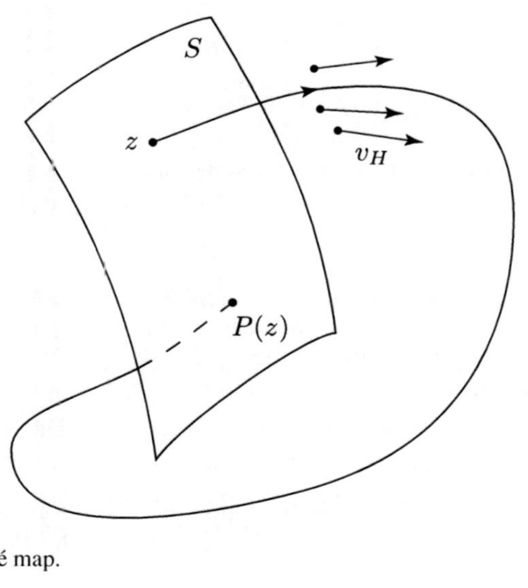

Fig. 1.1 The Poincaré map.

Note that in general the map P is not defined on the whole surface S. Usually it is considered in a neighborhood of some invariant set (a fixed point, an invariant torus, etc).

Theorem 1.8.

(1) *The vector field* v_H *is an annihilator of the form* $\omega|_{M_H}$, *i.e.,* $\omega(u, v_H) = 0$ *for any vector* $u \in TM_h$.
(2) *The form* $\omega|_S$ *is nondegenerate.*
(3) *The map* P *preserves the symplectic structure* $\omega|_S$.

Proof. Assertion (1) follows from the equations

$$\omega(u, v_H) = \partial_u H = 0 \quad \text{for any } u \in TM_h.$$

(2) Let the vector $w \in T_z S$ be an annihilator of the form $\omega|_S$ (i.e. for any vector $u \in T_z S$ we have $\omega(u, w) = 0$). Let $L(v_H) \subset T_z M_h$ be the line spanned by the vector $v_H(z)$. Then the tangent space $T_z M_h$ is the direct sum $T_z S \oplus L(v_H)$ and therefore, w is an annihilator of the form $\omega|_{M_h}$.

The form ω is nondegenerate on M. Hence for some $\alpha \in T_z M$ the quantity $\omega(\alpha, v_H)$ does not vanish. Since $\alpha \notin T_z M_h$, we have $T_z M = T_z M_h \oplus L(\alpha)$, where $L(\alpha)$ is the line spanned by the vector α. The vector

$$\beta = w - \frac{\omega(\alpha, w)}{\omega(\alpha, v_H)} v_H$$

is an annihilator of $\omega|_{M_h}$, because it is a linear combination of w and v_H. Furthermore, since $\omega(\alpha, \beta) = 0$, the vector β is an annihilator of ω. Nondegeneracy of the symplectic structure implies $\beta = 0$. Therefore $w \parallel v_H$. Now we get $w = 0$ because S is transversal to the field v_H.

(3) Let $\Lambda_0 \subset S$ be an arbitrary 2-dimensional disk in S and $\Lambda_1 = P(\Lambda_0) \subset S$. The boundaries $\partial \Lambda_0$ and $\partial \Lambda_1$ of the disks are joined by the tube of trajectories Σ. The trajectories, beginning on Λ_0 and finishing on Λ_1 form a 3-dimensional manifold $\Gamma \subset M$ with the boundary $\Lambda_0 \cup \Lambda_1 \cup \Sigma$. By the Stokes formula

$$\int_\Gamma d\omega = \int_{\Lambda_0} \omega - \int_{\Lambda_1} \omega + \int_\Sigma \omega. \tag{1.6}$$

(The minus sign is taken for compatibility of the orientations.) The left-hand side of (1.6) vanishes because $d\omega = 0$. The third term on the right-hand side vanishes because $\omega|_\Sigma = 0$.[3] Finally we obtain: $\int_{\Lambda_0} \omega = \int_{P(\Lambda_0)} \omega$. This implies that $P^* \omega = \omega$. □

Note that the Poincaré map in a nonautonomous Hamiltonian system (M, ω, H) with periodic in time Hamiltonian can be regarded as the Poincaré map for the

[3] Indeed, the vector field v_H is tangent to $\Sigma \subset M_h$. It remains to use the fact that v_H is an annihilator of ω_{M_h} and Σ is 2-dimensional.

corresponding autonomous system $(M \times \mathbb{T}_t \times \mathbb{R}_E, \omega + dE \wedge dt, H + E)$ (see Sect. 1.1). Indeed, consider the energy level $M_0 = \{H + E = 0\}$ and the transversal $S = M_0 \cap \{t = \text{const}\}$ to the Hamiltonian flow. In this case $(S, (\omega + dE \wedge dt)|_S)$ is symplectomorphic to (M, ω), and the map, corresponding to the surface S, is conjugated to the Poincaré map in the original nonautonomous system.

If the symplectic structure ω is an exact form: $\omega = d\omega^1$, then for any symplectic diffeomorphism the 1-form $\tilde{\omega} = \omega^1 - T^*\omega^1$ is closed, because its differential equals $d\tilde{\omega} = \omega - T^*\omega = 0$. If $\tilde{\omega}$ is exact, T is called an exact symplectic map.

Proposition 1.1. *Suppose that the form ω is exact and T is the Poincaré map corresponding to the system (M, ω, H). Then T is an exact symplectic map.*

Corollary 1.3. *The Poincaré map in a system with exact symplectic structure and periodic in time Hamiltonian is exact symplectic.*

Proof (of Proposition 1.1). Let the map T be defined on the surface $S \subset M$. Let l be an arbitrary closed curve on S. Then by the definition of the Poincaré map and the Stokes formula

$$\int_l (T^*\omega^1 - \omega^1) = \int_{T(l)} \omega^1 - \int_l \omega^1 = \int_\Sigma \omega,$$

where Σ is the two-dimensional surface formed by segments of trajectories of the Hamiltonian system which begin on the curve l and finish on $T(l)$. As in (1.6) we have: $\int_\Sigma \omega = 0$. Therefore $\int_l \tilde{\omega} = 0$ for any l. □

A coordinate interpretation of a symplectic map is a canonical change of variables. If (q, p) are canonical local coordinates on (M, ω) and $T : M \to M$ is a symplectic map, the coordinates $(Q, P) = T(q, p)$ are also canonical: $\omega = dp \wedge dq = dP \wedge dQ$. For any $H : M \to \mathbb{R}$ the Hamiltonian vector field v_H has the form

$$\frac{\partial H}{\partial p}\frac{\partial}{\partial q} - \frac{\partial H}{\partial q}\frac{\partial}{\partial p} = \frac{\partial \mathcal{H}}{\partial P}\frac{\partial}{\partial Q} - \frac{\partial \mathcal{H}}{\partial Q}\frac{\partial}{\partial P},$$

where $\mathcal{H}(Q, P) = H(q, p)$, i.e., $\mathcal{H} \circ T = H$.

If the system is non-autonomous ($H = H(q, p, t)$), it is natural to deal with non-autonomous changes of variables, i.e., symplectic maps parametrized by time: $T : M \times \mathbb{R}_t \to M$. In this case the Hamiltonian vector field on the extended phase space $M \times \mathbb{R}_t$ again preserves its canonical form:

$$\frac{\partial H}{\partial p}\frac{\partial}{\partial q} - \frac{\partial H}{\partial q}\frac{\partial}{\partial p} + \frac{\partial}{\partial t} = \frac{\partial \mathcal{H}}{\partial P}\frac{\partial}{\partial Q} - \frac{\partial \mathcal{H}}{\partial Q}\frac{\partial}{\partial P} + \frac{\partial}{\partial t}$$

in the new coordinates $(Q, P) = (Q(q, p, t), P(q, p, t)) = T(q, p, t)$, but the connection between H and $\mathcal{H}$ becomes more complicated.

Proposition 1.2. *The form $\nu = -\frac{\partial P}{\partial t} dQ + \frac{\partial Q}{\partial t} dP$ on M is closed for any value of the parameter t.*

If $\nu = d\Phi$ then $\mathcal{H} \circ T = H + \Phi$.

Proof. The first statement of Proposition 1.2 follows from the equation

$$dv = -\frac{\partial}{\partial t}(dP \wedge dQ) = -\frac{\partial}{\partial t}(dp \wedge dq) = 0.$$

By using the autonomization (Sect. 1.1) we reduce the situation to the autonomous case. To this end we put $\mathscr{E} = E - \Phi$. Then

$$\begin{aligned} & dP \wedge dQ + d\mathscr{E} \wedge dt \\ &= dp \wedge dq + \frac{\partial P}{\partial t} dt \wedge dQ + \frac{\partial Q}{\partial t} dP \wedge dt - d\Phi \wedge dt + dE \wedge dt \\ &= dp \wedge dq + dE \wedge dt. \end{aligned}$$

Therefore the change of coordinates

$$M \times \mathbb{R}^2 \to M \times \mathbb{R}^2, \quad (q, p, t, E) \mapsto (Q, P, t, \mathscr{E}),$$

is canonical and the new Hamiltonian $\mathscr{H} \circ T + \mathscr{E} = H + E = H + \Phi + \mathscr{E}$. □

Symplectic maps can be locally determined with the help of generating functions. For example the map $(q, p) \mapsto (Q, P) = T(q, p, t)$,

$$p = \frac{\partial W}{\partial q}, \quad Q = \frac{\partial W}{\partial P}, \qquad W = W(q, P, t),$$

is locally defined if $\det(\frac{\partial^2 W}{\partial q_j \partial P_k}) \neq 0$. In this case the Hamiltonian is transformed as follows:

$$\mathscr{H}(Q, P, t) = H(q, p, t) + \frac{\partial W}{\partial t}(q, P, t).$$

Usually the same map T can be obtained from another generating function S:

$$\begin{aligned} & p = -\frac{\partial S}{\partial q}, \quad P = \frac{\partial S}{\partial Q}, \qquad S(q, Q, t) = QP - W(q, P, t), \\ & \mathscr{H}(Q, P, t) = H(q, p, t) - \frac{\partial S}{\partial t}(q, Q, t). \end{aligned} \tag{1.7}$$

If T is considered from the viewpoint of the dynamics (in this case T does not depend on time), the functions $W(q, P)$ and $S(q, Q)$ are called the discrete Hamiltonian and the discrete Lagrangian respectively. They are connected by the Legendre transform (1.7). For example, for the standard map (1.5) we have:

$$W(x, Y) = xY + \frac{1}{2}Y^2 + \varepsilon \cos x, \quad S(x, X) = \frac{1}{2}(X - x)^2 - \varepsilon \cos x.$$

If the dynamics is determined by a discrete Lagrangian S, the system is called a *discrete Lagrangian system*. Discrete Lagrangian systems admit a variational principle:

the sequence $(\hat{q}_k, \hat{p}_k)$, $k \in \mathbb{Z}$, *is a trajectory (i.e.,* $(\hat{q}_{k+1}, \hat{p}_{k+1}) = T(\hat{q}_k, \hat{p}_k)$ *for all* k*) if and only if the sequence* $\hat{q} = (\ldots, \hat{q}_{-1}, \hat{q}_0, \hat{q}_1, \ldots)$ *is an extremal of the formal functional*

$$A(q) = \sum_{k=-\infty}^{\infty} S(q_k, q_{k+1}). \tag{1.8}$$

Series (1.8) in general diverge. The statement "$\hat{q}$ is an extremal of the formal functional $A(q)$" by definition means that, for any sequence $\lambda_k \in \mathbb{R}^m$ such that $\lambda_k \neq 0$ only for a finite number of indices k,

$$\frac{d}{d\varepsilon}\bigg|_{\varepsilon=0} \sum_{k=-\infty}^{\infty} S(q_k + \varepsilon\lambda_k, q_{k+1} + \varepsilon\lambda_{k+1}) = 0.$$

The left-hand side is the sum of only a finite number of nonzero terms. Chapters 7 and 8 contain more information on discrete Lagrangian systems.

Let the symplectic self-map T of the $2m$-dimensional manifold M possess m commuting first integrals $F_1, \ldots, F_m$.[4]

Theorem 1.9. *Suppose that on the common integral level* M_f, *see* (1.4), *the functions* F_j *are independent. Then the following assertions hold:*

(1) M_f *is a smooth manifold invariant with respect to the map* T.
(2) *Each compact connected component of* M_f *is diffeomorphic to* $\mathbb{T}^m$.
(3) *In some canonical coordinates* (I, φ) *in a neighborhood of the torus* $\mathbb{T}^m \subset M_f$

$$T(I, \varphi) = (I, \varphi + \nu(I)), \qquad \nu(I) = \partial\Phi(I)/\partial I, \tag{1.9}$$

where Φ *is a smooth function.*

Theorem 1.9 follows from the ordinary Liouville theorem and from the theorem on the existence of action-angle coordinates in a Liouville integrable Hamiltonian system. Indeed, assertion (1) of Theorem 1.9 is obvious, and assertion (2) can be proved in the same way as in Theorem 1.7. The action-angle variables are determined by a complete involutive set of first integrals rather than by the Hamiltonian system itself. Therefore they can be determined in the discrete situation as well. We just need to prove formulas (1.9).

By construction the coordinates (I, φ) are canonical. The variables I are functions of the first integrals $F_1, \ldots, F_m$ and therefore they are also first integrals. Let $T(I, \varphi) = (J, \psi)$. Then $J = I$ and

$$dI \wedge d\varphi = dJ \wedge d\psi = dI \wedge d\psi(I, \varphi).$$

We put $\psi(I, \varphi) = \varphi + \nu(I, \varphi)$. Then the 2-form $dI \wedge d\nu$ vanishes. This implies that the functions ν do not depend on the angles φ and the 1-form $\nu(I)dI$ is closed. The function Φ is its (local) primitive.

[4] The function $F : M \to \mathbb{R}$ is called a first integral for T if it is preserved by the map: $F \circ T = F$.

1.3 Inclusion of a Diffeomorphism into a Flow

In the previous section we noted that the Poincaré map in a nonautonomous Hamiltonian system, where the Hamiltonian is periodic in time, is symplectic. Naturally the inverse problem appears: to present a given symplectic self-map of a manifold as the map for a period (the Poincaré map) in some Hamiltonian system with periodic in time Hamiltonian function.

This problem is called the problem of the inclusion of a diffeomorphism into a flow in the symplectic set up. The problem can be formulated for other classes of maps and the corresponding vector fields. For example, it is possible to consider generic maps and vector fields, reversible ones with respect to some involution, preserving a volume, etc. One should distinguish also smooth and analytic situations. Here we mean that for a smooth map it is natural to search for an inclusion into a smooth flow, and for analytic into an analytic flow.

The following construction is well-known. Given a diffeomorphism T of a smooth manifold M onto itself, consider the direct product $M \times [0, 1]$ with the vector field $\partial/\partial t$, where t is the coordinate on $[0, 1]$. The map T generates the identification

$$M \times \{0\} \sim M \times \{1\}, \qquad (z, 0) \sim (T(z), 1).$$

This identification converts $M \times [0, 1]$ into a smooth manifold $\mathscr{M}$ (the class of smoothness remains the same). Let $\pi : M \times [0, 1] \to \mathscr{M}$ be the natural projection. The smooth vector field $\partial/\partial t$ generates on the section $\pi(M \times \{0\}) \subset \mathscr{M}$ the Poincaré map coinciding with T.

This construction, traditionally called a *suspension*, does not solve the above problem because it is not clear if $\mathscr{M}$ is diffeomorphic to $M \times \mathbb{T}^1$. Nevertheless, sometimes this can be proven [74, 76]. Below we use another method.

There are obvious topological obstacles to the inclusion of a map into a flow. Consider for example, the diffeomorphism

$$T : \mathbb{T}^2 \to \mathbb{T}^2, \qquad T(\varphi_1, \varphi_2) = (-\varphi_2, \varphi_1).$$

This map can not be included into any flow. Indeed, T transforms the cycle $\{\varphi_1 = 0\}$ onto $\{\varphi_2 = 0\}$. These cycles are not homotopic (i.e., they can not be continuously deformed to one another). But any shift along solutions of differential equations transforms a cycle onto a homotopic cycle. Generalizing this simple observation we see that any map which is not isotopic to the identity[5] cannot be included into a flow.

Douady [39] constructed inclusions of smooth symplectic maps defined by generating functions into smooth flows.

[5] Two smooth maps $T_j : M' \to M''$, $j = 0, 1$ (M' and M'' are manifolds) are called isotopic if there exists a family of maps $\hat{T}_s : M' \to M''$ of the same smoothness class continuous in the parameter $s \in [0, 1]$, such that $\hat{T}_0 = T_0$ and $\hat{T}_1 = T_1$. In other words, if T_0 can be continuously deformed into T_1.

The case of smooth general maps is almost trivial. Indeed, let $T : M \to M$ be a diffeomorphism, where the manifold M is compact. If T is isotopic to the identity, the isotopy $\hat{T}_t$ can be chosen to be smooth in t. Moreover, it is easy to take $\hat{T}_t$ so that the vector field $v(x,t) = (\frac{d}{dt}\hat{T}_t) \circ \hat{T}_t^{-1}$ is 1-periodic and smooth (including the points $t \in \mathbb{Z}$).

The analytic case turns out to be more complicated. Douady, Kuksin, and Pöschel [40, 74, 76] solved the problem in the symplectic set up for maps which are close to integrable. Trifonov [143] solved the problem for generic maps. The proofs in [40, 76, 143] use the Grauert theorem on the inclusion of an analytic manifold into Euclidean space. In [74] KAM-technique is used.

Theorem 1.10, formulated below, solves the problem of the inclusion of a map into a flow in an analytic set up. Here instead of the Grauert theorem the main tool is a special averaging method.

We begin with some definitions and notation. Let M be an m-dimensional real-analytic manifold. Then M is a real part of a complex analytic manifold $\tilde{M}_s$. Below we assume that M is compact (otherwise one should deal with a compact domain in M).

We also assume that $\overline{\tilde{M}_s}$, the closure of $\tilde{M}_s$ is a compact manifold (with a boundary). Then $\tilde{M}_s$ can be covered with a finite collection of charts $\{U_i\}$ with coordinate bijections $\varphi_i : U_i \to \tilde{D}_i^s$. The domains $\tilde{D}_i^s \subset \mathbb{C}^m$ have the following structure. There exist domains $D_i \subset \mathbb{R}^m$ such that

$$\tilde{D}_i^s = \{x + w : x \in D_i,\ w \in \mathbb{C}^m,\ \|w\| < sr\} \quad \tilde{D}_i = \tilde{D}_i^1$$

for some positive r and $0 < s \leq 1$. Moreover the manifold M can be covered with a collection of charts $V_i \subseteq U_i \cap M$ such that the mappings $\varphi_i|_{V_i} : V_i \to D_i$ are the coordinate bijections. We shall say that $\tilde{M}_s$ is a complex neighborhood of the real-analytic manifold M, $\tilde{M}_1 = \tilde{M}$.

Remark 1.3. Sometimes when convenient we cover the domains $\tilde{D}_i$ with polydiscs

$$B_{w_0}(r) = \{w \in \mathbb{C}^m \mid \|w - w_0\| < r\}.$$

These polydiscs form another atlas of $\tilde{M}$ with the charts $A = \{\varphi_i^{-1}(B_{w_0}(r)) \mid w_0 \in D_i\}$. In the sequel we develop local theory in each chart of A, where without loss of generality we assume that $w_0 = 0$.

Denote by $\mathcal{O}(\tilde{M})$ the space of real analytic functions $f : \tilde{M} \to \mathbb{C},\ f(M) \subseteq \mathbb{R}$. Being equipped with the collection of norms

$$\|f\|_s = \sup_{z \in \tilde{M}_s} |f(z)|, \qquad 0 < s < 1,$$

the space $\mathcal{O}(\tilde{M})$ becomes a locally convex space (see Sect. 9.5).

Let $g = (g_{i,j}(z, \bar{z}))$ be a Hermitian metric on $\overline{\tilde{M}}$ with continuous components. Then, for the space of real-analytic vector-fields $v(z) = (v^1, \ldots, v^m)(z) \in T\tilde{M}_z$, we shall use the following collection of norms:

$$\|v\|_s = \sup_{z,\bar{z}\in\tilde{M}_s} \sqrt{g_{i,j}(z,\bar{z})\, v^i(z)\, \overline{v^j(z)}}.$$

The space of real-analytic vector-fields $\mathscr{V}(\tilde{M})$ is also a locally convex space with respect to this collection of norms. It is not hard to show that another metric tensor g generates the same topology in $\mathscr{V}(\tilde{M})$.

Let χ be a closed (with respect to the topology of $\mathscr{V}(\tilde{M})$) subalgebra in the Lie algebra $(\mathscr{L}, [\,,\,])$. The basic examples are as follows.

(a) $\chi = \mathscr{L}$,
(b) M is a symplectic manifold and χ is the algebra of Hamiltonian (or locally Hamiltonian) vector fields,
(c) χ is the algebra of volume-preserving vector fields.

Let X be the subset of all analytic diffeomorphisms of M obtained as a result of the time-2π shift along solutions of the system

$$\dot{z} = u(z,t), \qquad u(\cdot,t) \in \chi, \quad t \in [0,2\pi], \quad z \in M. \tag{1.10}$$

We assume that the vector field u is C^2-smooth in time. This smoothness condition is technical. It can be weakened in the Hamiltonian and general cases with the help of the following lemmas.

Lemma 1.2. *If a real-analytic map T is continuously isotopic to the identity map inside the set of real-analytic maps, the isotopy g^t can be chosen to be smooth in time.*

Lemma 1.3. *If a real-analytic symplectic map T is continuously isotopic to the identity map inside the set of real-analytic simplectic maps, the isotopy g^t can be chosen to be smooth in time.*

We present proofs of these lemmas at the end of this section.

Below it is convenient to continue the vector fields u by 2π-periodicity to the whole the real axis of time. The obtained vector fields are in general discontinuous in t for $t = 2\pi n$, $n \in \mathbb{Z}$. However, we can achieve smoothness by changing time. Indeed, let us perform the change $t \mapsto t'$. The function $t(t')$ (Fig. 1.2) is smooth, odd, strictly monotone, $t(0) = 0$, $t(2\pi + t') = 2\pi + t(t')$, and the derivatives $d^l t/(dt')^l$, $l = 1,2,3$, vanish at the points $t' = 2\pi k$, $k \in \mathbb{Z}$. We put $\hat{u}(z,t') = u(z,t(t'))\,dt/dt'$. Then $\hat{u}(z,\cdot)$ is C^2-smooth in time on all the real axis and the time-2π shifts corresponding to u and $\hat{u}$ coincide. Below we deal with the function $\hat{u}$ and drop the prime in t' for brevity.

Reversible vector fields. Consider the space $C(\mathbb{T}, \mathscr{V}(\tilde{M}))$ of nonautonomous vector fields $u(z,t)$, $t \in \mathbb{T}$. Now we give a definition of a reversible vector field. Let $I : \tilde{M} \to \tilde{M}$, $I(M) \subseteq M$ be a real-analytic involution (I^2 is the identity map). The vector field $u(z,t) \in C(\mathbb{T}, \mathscr{V}(\tilde{M}))$ is said to be reversible with respect to I (or I-reversible) if

$$u(z,t) = -dI\, u(Iz,-t). \tag{1.11}$$

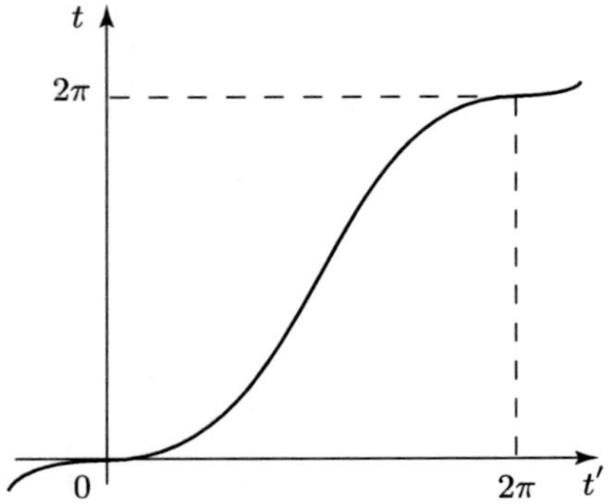

Fig. 1.2 Graph of the function $t(t')$.

In particular, in the autonomous case the involution I transforms the vector field u to $-u$.[6] The space of reversible vector fields is a closed subspace of $C(\mathbb{T}, \mathscr{V}(\tilde{M}))$.

We define the set X_I as a subset of X generated by I-reversible vector fields u. Obviously, all diffeomorphisms from X (X_I) are isotopic to the identity inside X (X_I).

Theorem 1.10 ([110]). *Suppose that the map T belongs to X (respectively, to X_I). Then there exists a vector field (respectively, an I-reversible vector field)*

$$U = U(z, t), \qquad U(\cdot, t) \in \chi, \quad t \in \mathbb{R}, \quad z \in M,$$

which is analytic in z and t, 2π-periodic in t, and such that the time-2π shift along its trajectories coincides with T.

As a corollary we obtain the possibility of the inclusion of analytic maps into analytic flows in the general, symplectic, volume-preserving and reversible cases.

Remark 1.4. The vector field U is not uniquely defined.

Remark 1.5. In the symplectic case U is in general locally Hamiltonian. The system can be made globally Hamiltonian provided the vector field u that defines T is globally Hamiltonian.

Remark 1.6. Suppose that the map T is close to T_0 ($\text{dist}(T, T_0) = \varepsilon$ in $\tilde{M}$),[7] where T_0 has already been included into the flow generated by a periodic analytic vector

[6] The Hamiltonian system with Hamiltonian $H(q, p)$, which is even in the momenta p is a standard example of a reversible system of classical mechanics. The corresponding involution has the form $(q, p) \mapsto (q, -p)$.

[7] The distance can be defined for example as follows:

$$\text{dist}(T, T_0) = \sup_{z \in \tilde{M}} \rho(T(z), T_0(z)),$$

where ρ is some metric on the closure of $\tilde{M}$. The choice of the metric plays no role.

field U_0. Then the vector field U can be chosen close to U_0 ($|U - U_0| = O(\varepsilon)$ in a smaller complex neighborhood of M).

In particular, in the symplectic case if T is close to an integrable map, a Hamiltonian system associated with T also can be chosen close to an integrable one and the orders of closeness are the same.

We prove Theorem 1.10 in Chap. 6. Now let us prove Lemmas 1.2 and 1.3.

Proposition 1.3. *Lemmas* 1.2 *and* 1.3 *hold if* T *is close to the identity.*

This proposition easily implies the lemmas. Indeed, suppose that, Proposition 1.3 is valid. Let g^t be the initial (continuous) isotopy. We divide the interval $(0, 2\pi)$ by the points $0 = t_0 < t_1 < \cdots < t_K = 2\pi$ into a large amount of small intervals (t_{k-1}, t_k), $1 \le k \le K$. We have:

$$T = Q_K \circ \cdots \circ Q_1, \qquad Q_k = g^{t_k} \circ g^{-t_{k-1}}, \qquad 1 \le k \le K, \qquad Q_0 = \mathrm{id}.$$

All the maps Q_k are close to the identity. Hence, there exist smooth isotopies ϕ_k^t which link Q_k with the identity map. The isotopy $T_t = \phi_k^t \circ \cdots \circ \phi_1^t$ belongs to the desired smoothness class.

Proof (of Proposition 1.3). First, consider the case of general maps. Fix some Riemannian metric on M. Since M is compact and T is close to the identity, for any $z \in M$ the points z and $T(z)$ are joined by a unique shortest geodesic γ_z. We can parameterize γ_z by a natural parameter s so that $\gamma_z(0) = z$ and $\gamma_z(1) = T(z)$. Then the maps $z \mapsto \gamma_z(s)$ form the required smooth isotopy.

Now turn to the symplectic case. Consider the graph $\mathscr{G}$ of T in the direct product $M \times M$. The set $\mathscr{G}$ consists of the pairs $(z, T(z)) \in M \times M$. Since M is compact and T is close to the identity, $\mathscr{G}$ lies in a small neighborhood $\mathscr{N}$ of the diagonal

$$\mathscr{D} = \{(z_1, z_2) \in M \times M : z_1 = z_2\}.$$

The manifold $M \times M$ is endowed with the natural symplectic structure $\Omega = \pi_1^*\omega - \pi_2^*\omega$, where $\pi_j : M \times M \to M$, $j = 1, 2$, is the projection on the j-th multiplier. The manifolds $\mathscr{D}$ and $\mathscr{G}$ are Lagrangian with respect to Ω because the identity map and T are symplectic.

According to Theorem 1.1, the neighborhood $\mathscr{N}$ is symplectomorphic to a neighborhood of the zero section M_0 of the cotangent bundle $(T^*M, dp \wedge dz)$. Let $f : (\mathscr{N}, \Omega) \to (T^*M, dp \wedge dz)$ be this symplectomorphism and $\Lambda = f(\mathscr{G}) \subset T^*M$. We can assume that $f(\mathscr{D}) = M_0$. The manifold Λ projects one-to-one onto M_0 under the natural projection $(p, z) \mapsto (0, z)$. Therefore, Λ can be determined by the equation $p = p(z)$. Since Λ is Lagrangian, $p(z) = \partial S(z)/\partial z$, where S is a (in general multivalued) real-valued function on M.

Consider the family Λ_t, $t \in [0, 1]$ of Lagrangian manifolds in T^*M determined by the equation $p = (2\pi)^{-1} t\, \partial S/\partial z$. Obviously, $\Lambda_0 = M_0$, $\Lambda_{2\pi} = \Lambda$. The Lagrangian manifolds $\mathscr{G}_t = f^{-1}(\Lambda_t) \subset \mathscr{N}$ can be presented in the form $\mathscr{G}_t = \{(z_1, z_2) : z_2 = T_t(z_1)\}$, where T_t are some symplectic maps. Obviously, T_t are smooth with respect to t and

$$\mathscr{G}_0 = \mathscr{D}, \qquad \mathscr{G}_{2\pi} = \mathscr{G}, \qquad T_0 = \mathrm{id}, \qquad T_{2\pi} = T.$$

Hence, T_t is the isotopy we are looking for. □

1.4 The Classical Perturbation Theory

Recall that Hamiltonian systems (1)–(2) are called close to integrable or near-integrable. The unperturbed ($\varepsilon = 0$) equations are

$$\dot{x} = \nu(y) = \partial H_0/\partial y, \quad \dot{y} = 0.$$

They can be easily solved:

$$x = x^0 + t\nu(y^0), \quad y = y^0.$$

Hence, the phase space of the unperturbed system is foliated by invariant m-dimensional tori $N_y = \{(x, y) : y = \text{const}\}$. The motion on the torus N_y is quasiperiodic with the frequencies $\nu(y)$. Its properties depend on arithmetic properties of the frequency vector ν. The vector $\nu \in \mathbb{R}^m$ (as well as the corresponding torus N_y) is called *nonresonant* if for any nonzero vector $k \in \mathbb{Z}^m$ the quantity $\langle k, \nu\rangle$ does not vanish. Otherwise the frequency vector (and the corresponding torus) is called resonant. If the vector $\nu(y)$ is nonresonant, any trajectory lying on N_y fills the torus densely. If the frequencies are resonant, a trajectory lying on the invariant torus fills densely some torus, which lies on N_y and has a smaller dimension. In particular, if the vector $\nu(y)$ is collinear to an integer one, all trajectories on N_y are periodic.

Proposition 1.4. *Suppose that the unperturbed system is nondegenerate, i.e.,* $\det(\partial^2 H_0/\partial y^2) \neq 0$ *almost everywhere in D. Then nonresonant tori are dense in the phase space of the unperturbed system.*

Proof. Obviously, it is sufficient to check that the set

$$\{y \in D : \nu(y) \text{ is a nonresonant vector}\}$$

is dense in D. According to the nondegeneracy assumption, the map $y \mapsto \nu(y)$ is a local diffeomorphism almost everywhere in D. Since the set of nonresonant frequencies ν is dense in $\mathbb{R}^m_\nu$, its preimage in the domain D is also dense. □

According to the main idea of the classical perturbation theory, let us try to eliminate the angular variables from the Hamiltonian by the canonical change

$$(x \bmod 2\pi, y) \mapsto (X \bmod 2\pi, Y).$$

In the new variables the system will be easily integrated. The change will be determined by the generating function

$$S(x, Y) = \langle x, Y \rangle + \varepsilon S_1(x, Y) + \cdots,$$
$$X = \partial S/\partial Y, \quad y = \partial S/\partial x.$$

Let $\mathcal{H} = \mathcal{H}_0(Y) + \varepsilon \mathcal{H}_1(Y) + \cdots$ be the Hamiltonian function in the new variables. Then

$$\mathcal{H}(Y, \varepsilon) = H_0(Y + \varepsilon \partial S_1/\partial x + \cdots) + \varepsilon H_1(x, Y + \cdots) + \cdots.$$

In the zero approximation with respect to the small parameter we have: $H_0 = \mathcal{H}_0$. The first approximation is

$$\langle \nu, \partial S_1/\partial x \rangle + H_1(x, Y) = \mathcal{H}_1(Y). \tag{1.12}$$

To solve this equation we expand the functions S_1 and H_1 into the Fourier series

$$S_1(x, Y) = \sum_{k \in \mathbb{Z}^m} S_1^k(Y) e^{i\langle k, x \rangle}, \qquad H_1(x, Y) = \sum_{k \in \mathbb{Z}^m} H_1^k(Y) e^{i\langle k, x \rangle}.$$

Equating coefficients of the Fourier series in (1.12), we get

$$\mathcal{H}_1 = H_1^0, \quad S_1^k = -\frac{H_1^k}{i\langle k, \nu \rangle}, \qquad k \neq 0.$$

The function S_1^0 is arbitrary. It is usually assumed that $S_1^0 = 0$.

The next approximations are essentially analogous. The functions S_j and $\mathcal{H}_j$ satisfy equations of the form

$$\langle \nu, \partial S_j/\partial x \rangle - \mathcal{H}_j(Y) = \Phi_j(x, Y),$$

where the function Φ_j is known from previous steps of the procedure. Therefore, S_j and $\mathcal{H}_j$ are well-defined provided that the mean value in x of any function S_j vanishes.

Thus, on a formal level the change of variables is determined. However, the series in the classical perturbation theory as a rule diverge. The reason for this phenomenon is that systems (1)–(2) in general are nonintegrable. Formally the series for the functions S and $\mathcal{H}$ diverge because of the presence of "*small divisors*" $\langle k, \nu \rangle$. Here we mean the following. The functions S_j can be presented as fractions with denominators in the form of products

$$\langle k_1, \nu(Y) \rangle \cdots \langle k_l, \nu(Y) \rangle, \qquad l \in \mathbb{N},$$

with integer vectors $k_1, \ldots, k_l$. Any multiplier $\langle k_r, \nu(Y) \rangle$ vanishes on a certain hypersurface in the space of the variables Y. The function S is not defined on such hypersurfaces even formally (provided there is no cancelation in the fractions S). It remains to note that as a rule, the union of these surfaces is dense in the domain D.

To present a formal version of this argument, we introduce the following definition.

Definition 1.1. The set $\mathscr{B}$ is said to be a *secular set* if it consists of points $y \in D$ satisfying the following conditions. For some nonzero vector $k \in \mathbb{Z}^m$

(1) $\langle k, \nu(y)\rangle = 0$;
(2) $H_1^k(y) \neq 0$.

Poincaré has proved that, if the secular set is dense in D, the perturbed system is nonintegrable [105]. The conditions of the following theorem are slightly less restrictive.

Theorem 1.11 ([105]). *Suppose that the following assumptions hold.*

(1) *The Hamiltonian H is real-analytic.*
(2) *The unperturbed system is nondegenerate.*
(3) *The set $\mathscr{B}$ is such that any real-analytic function vanishing identically on $\mathscr{B}$ also vanishes everywhere on D.*[8]

Then the perturbed system does not have first integrals

$$F^{(j)} = F_0^{(j)}(x, y) + \varepsilon F_1^{(j)}(x, y) + \varepsilon^2 F_2^{(j)}(x, y) + \cdots, \qquad j = 1, \dots, m, \quad (1.13)$$

independent for $\varepsilon = 0$, where the coefficients $F_l^{(j)}$ are real-analytic.

Remark 1.7. The series (1.13) are not supposed to converge, i.e., if assumptions of the theorem hold, there is no complete set of first integrals even presented by formal series (1.13).

Remark 1.8. If the series (1.13) converge then it would be more natural to deal with integrals independent for small $\varepsilon \neq 0$. This condition is weaker than the independence for $\varepsilon = 0$ which we have in the theorem. A theorem with such an assumption has been proved only for $m = 2$ (see [70, 105]).

Remark 1.9. In Theorem 1.11 we do not assume that the first integrals are in involution. However, it is possible to show [126] that if a nondegenerate system (1)–(2) has m first integrals (1.13) independent for $\varepsilon = 0$ then they are in involution.

The proof of Theorem 1.11 is based on the following auxiliary statement.

Lemma 1.4. *Suppose that conditions of Theorem* 1.11 *hold. Then*

(1) *The functions $F_0^{(j)}$ do not depend on x.*
(2) *For any $y \in \mathscr{B}$ we have*

$$\det \frac{\partial(F_0^{(1)}, \dots, F_0^{(m)})}{\partial(y_1, \dots, y_m)} = 0.$$

[8] For example, $\mathscr{B}$ is dense in an open subset of D. It is easy to present much weaker sufficient conditions.

Proof (of Lemma 1.4). (1) The functions $F_0^{(j)}$ are first integrals of the unperturbed system. Let the torus N_{y^0} be nonresonant. Then any trajectory fills this torus densely. The functions $F_0^{(j)}(y^0, x)$ do not depend on x, because the integral is constant on trajectories. According to Proposition 1.4, nonresonant tori are dense in the phase space. Hence $F_0^{(j)}$ do not depend on x.

(2) The functions $F^{(j)}$ satisfy the equations

$$\{H, F^{(j)}\} = \{H_0, F_0^{(j)}\} + \varepsilon(\{H_1, F_0^{(j)}\} + \{H_0, F_1^{(j)}\}) + \cdots = 0.$$

Therefore, $\{H_0, F_1^{(j)}\} = \{F_0^{(j)}, H_1\}$. Let us expand this equation into Fourier series. Putting

$$F_1^{(j)}(x, y) = \sum_{k\in\mathbb{Z}^m} F_1^{(j)k}(y)\, e^{i\langle k,x\rangle},$$

we obtain the equations

$$\langle k, \nu\rangle F_1^{(j)k} = \langle k, \partial F_0^{(j)}/\partial y\rangle H_1^k, \qquad k \in \mathbb{Z}^m. \tag{1.14}$$

If $y \in \mathscr{B}$, for some $k \neq 0$ we have $\langle k, \nu\rangle = 0$ and $H_1^k \neq 0$. Then by (1.14)

$$\langle k, \partial F_0^{(j)}/\partial y\rangle = 0, \qquad j = 1, \ldots, m.$$

Since the m vectors $\partial F_0^{(j)}/\partial y$ are orthogonal to some vector $k \neq 0$, they are linearly dependent. The lemma is proved. □

According to condition (3) of Theorem 1.11, the Jacobian $\det(\partial F_0^{(j)}/\partial y)$ vanishes identically on the domain D. This concludes the proof of Theorem 1.11.

Poincaré used Theorem 1.11 to prove nonintegrability of the restricted three-body problem. Later this theorem turned out to be one of the most convenient tools for the proof of nonintegrability of near-integrable Hamiltonian systems. Sometimes the set $\mathscr{B}$ is not big enough for application of Theorem 1.11. Then obstacles to integrability can be constructed, by considering secular sets which appear on higher steps of the perturbation theory (details can be found in [72]).

Unfortunately, the statement on nonintegrability presented in Theorem 1.11 is in a sense formal, i.e., it gives almost no information about the dynamics.[9] In this sense nonintegrability proofs based on a complicated behavior of solutions of the perturbed system are more interesting. Here basic ideas also belong to Poincaré. Dynamical obstacles to integrability include the existence of a sufficiently large set of nondegenerate periodic solutions (or lower-dimensional invariant tori) and the splitting of separatrices (or of invariant surfaces asymptotic to hyperbolic tori in a general situation).[10] The problems of integrability and nonintegrability in Hamiltonian dynamics are discussed in detail in [72] (see also [70]).

---

[9] It is just possible to assert that resonant tori of the unperturbed system with the frequencies $\nu(y)$, $y \in \mathscr{B}$ are destroyed by the perturbation.

[10] Some multidimensional results are presented in [44].

Chapter 2
Introduction to the KAM Theory

The Kolmogorov–Arnold–Moser theory showed that quasi-periodic motions are generic in Hamiltonian systems. Moreover, they usually form a set of a positive measure in the phase space. This changed considerably the generally accepted idea of the dynamics in Hamiltonian systems close to integrable. Earlier such systems were supposed to be as a rule ergodic on compact energy levels.[1] In the present chapter we discuss basic facts and ideas of the KAM theory and prove one of the simplest theorems of this type.

2.1 The Kolmogorov Theorem

Consider the Hamiltonian system with real-analytic Hamiltonian

$$H(x, y, \varepsilon) = H_0(y) + \varepsilon H_1(x, y, \varepsilon) \tag{2.1}$$

in the canonically conjugate variables x, y, where $x = (x_1, \dots, x_m) \bmod 2\pi$ belongs to the m-dimensional torus $\mathbb{T}^m$, and $y = (y_1, \dots, y_m)$ lies in an open domain of $\mathbb{R}^m$. Usually it is enough to have a finite smoothness in ε.

The frequencies $\nu = \nu(y^0) = \partial H_0/\partial y(y^0)$ on the invariant torus

$$N_{y^0} = \{(x, y) : y = y^0\}$$

of the unperturbed system are said to be *Diophantine*[2] if there exist positive constants c, γ such that

$$|\langle k, \nu\rangle| \geq \frac{1}{c\|k\|^\gamma} \quad \text{for any nonzero vector } k \in \mathbb{Z}^m. \tag{2.2}$$

[1] A dynamical system is called ergodic with respect to an invariant probability measure on the phase space if the measure of any invariant set equals zero or one.

[2] The Diophantine property is discussed in Appendix 9.1.

D. Treschev, O. Zubelevich, *Introduction to the Perturbation Theory of Hamiltonian Systems*, Springer Monographs in Mathematics, DOI 10.1007/978-3-642-03028-4_2, © Springer-Verlag Berlin Heidelberg 2010

The meaning of conditions (2.2) is that the "small divisors" $\langle k, \nu\rangle$ are not too small. The norm $\|\cdot\|$ does not play an important role here. Usually it is taken as $\|v\| = \max_j |v_j|$. The torus N_{y^0} with Diophantine frequencies $\nu(y^0)$ is called Diophantine.

Theorem 2.1. *Suppose that the unperturbed system is nondegenerate at the point* y^0:

$$\det \frac{\partial^2 H_0}{\partial y^2}(y^0) \neq 0$$

and the torus N_{y^0} *is Diophantine. Then* N_{y^0} *survives the perturbation. It is just slightly deformed and as before carries quasiperiodic motions with the frequencies* ν.

Theorem 2.1 was formulated by Kolmogorov [68]. Kolmogorov is also the author of the idea that a success in the struggle with small divisors, appearing in series which determe the perturbed tori, can be achieved with the help of a rapidly converging method similar to Newton's method. A complete proof of Theorem 2.1 is given by Arnold [6] for real-analytic H. Moser proved a theorem on the preservation of quasiperiodic motions for reversible systems [93] and showed that Theorem 2.1 remains true also in the case of sufficiently smooth dependence of the Hamiltonian on phase variables.[3]

In the remarks presented below, we formulate some generalizations of the Kolmogorov theorem.

1. The deformation of an individual torus after a perturbation has the order $O(\varepsilon)$. Moser [93] noticed that, for a Hamiltonian (2.1) which is analytic in ε, the invariant tori are analytic in the small parameter. Expansions of perturbed quasiperiodic solutions in the small parameter can be constructed explicitly. Proofs of the convergence of these series need special technics, [30, 46, 50]. As is well known, the usual majorant method fails because of the existence of small denominators. However, as Eliasson noted [46], the most dangerous terms in these series compensate each other. This fact is used to establish the convergence.

2. KAM-tori can be also obtained by the method of renormalization [65, 69]. The renormalization technic was applied in [81] in the problem of the destruction of KAM-curves in 2-dimensional area-preserving maps.

3. The Kolmogorov tori form a smooth family [77, 107, 127]. In the case of an analytic nondegenerate system, tori with frequencies which satisfy the Diophantine conditions (2.2) with a fixed γ survive for some $c \sim 1/\sqrt{\varepsilon}$. Let $\Omega(c)$ denote this set of frequencies and let $\Omega = \Omega_\nu$ be the set of all unperturbed frequencies. Then there exists a diffeomorphism[4]

$$\Psi : \Omega_\nu \times \mathbb{T}^n_\vartheta \to \mathbb{R}^n_y \times \mathbb{T}^n_x,$$

[3] Takens [129] constructed an example of a one-parameter family of two-dimensional symplectic maps C^1-close to integrable which have no invariant curves close to the unperturbed ones. This means that for the KAM theory a sufficiently large smoothness is necessary. Smoothness of class C^{2m} is known to be sufficient [106].

[4] For simplicity we assume that the map $y \mapsto \nu(y)$ is a global diffeomorphism.

such that the image of the Cantorian set of Diophantine tori $\Omega(c) \times \mathbb{T}^n$ belongs to the set of the Kolmogorov tori. In the variables ν, ϑ for $\nu \in \Omega(c)$ the equations of motion can be written in the form

$$\dot{\nu} = 0, \quad \dot{\vartheta} = \nu.$$

For an analytic Hamiltonian (2.1) the diffeomorphism Ψ is analytic in ϑ and infinitely differentiable in ν. Nonintegrability of the perturbed system prevents analyticity of Ψ in all arguments.

4. The condition of nondegeneracy can be weakened considerably if we look not at an individual torus, but at the whole family of nonresonant tori. In particular, the following theorem was anounced by Rüssmann (the proof is given in [121, 122]).

Let the system be analytic and suppose that the image of the map $y \mapsto \nu(y)$ does not belong to any hyperplane in $\mathbb{R}^m_\nu$. Then for small ε the perturbed system has invariant tori. Moreover, in any compact domain of the phase space the measure of the set lying outside these tori tends to zero as $\varepsilon \to 0$.

5. The Diophantine condition can be also weakened. A continuous function Ω : $[1, \infty) \to \mathbb{R}$ is called approximating if

(1) Ω is non-decreasing and $\Omega(1) = 1$;
(2) $\int_1^\infty t^{-2} \log \Omega(t)\, dt < \infty$.

Rüssmann [115] has proved that in the Kolmogorov theorem the Diophantine condition for the vector ν can be replaced by the following one.

For some approximating function Ω

$$|\langle k, \nu\rangle| \geq 1/\Omega(|k|) \quad \text{for all nonzero } k \in \mathbb{Z} \tag{2.3}$$

In particular, the functions $\Omega(t) = t^\gamma$ are approximating for any $\gamma > 0$. In this case (2.3) coincides with the ordinary Diophantine conditions. If we take an approximating function in the form $\Omega(t) = \exp(t^\lambda - 1)$, $0 < \lambda < 1$, conditions (2.3) are much weaker than the usual ones.

A detailed discussion of various aspects of KAM theory is presented in [33].

Theorem 2.1 admits several different formulations. Below we present its nonautonomous, isoenergetic and discrete versions.

The nonautonomous version. Consider a nonautonomous Hamiltonian system with Hamiltonian

$$H(x, y, t, \varepsilon) = H_0(y) + \varepsilon H_1(x, y, t, \varepsilon) \tag{2.4}$$

in the canonical coordinates $x = (x_1, \ldots, x_m) \bmod 2\pi$, $y = (y_1, \ldots, y_m)$. The function (2.4) is assumed to be 2π-periodic in time t. Consider the unperturbed torus

$$\hat{N}_{y^0} = \{(x, y, t) : y = y^0\}.$$

In the nonautonomous case t should be considered as an additional angular variable and the corresponding frequency should be taken into account when we define the frequency vector $\hat{\nu}(y)$ on the torus $\hat{N}_{y^0}$. Hence

$$\hat{\nu}(y) = \begin{pmatrix} \nu \\ 1 \end{pmatrix}, \qquad \nu = \nu(y^0) = \frac{\partial H_0}{\partial y}(y^0).$$

Theorem 2.2. *Let the unperturbed system be nondegenerate at the point y^0 and let the frequency vector $\hat{\nu} = \hat{\nu}(y^0) \in \mathbb{R}^{m+1}$ be Diophantine. Then the invariant torus $\hat{N}_{y^0}$ of the unperturbed system survives the perturbation. It is just slightly deformed and as before carries quasiperiodic motions with the frequencies $\hat{\nu}$.*

The isoenergetic version. Again consider the system with Hamiltonian (2.1).

Theorem 2.3. *Suppose that the invariant torus N_{y^0} of the unperturbed system lies on the energy level $\{H_0 = h\}$, the unperturbed system is isoenergetically nondegenerate at y^0:*

$$\det \begin{pmatrix} \partial^2 H_0/\partial y^2(y^0) & \nu(y^0) \\ \nu^T(y^0) & 0 \end{pmatrix} \neq 0,$$

and the frequencies $\nu(y^0)$ are Diophantine. Then on the energy level $\{H = h\}$ of the perturbed system there is an invariant torus close to the original one. The frequencies on this torus are $\lambda\nu(y^0)$, where $\lambda = 1 + O(\varepsilon)$.

The discrete version. Now we formulate an analog of the Kolmogorov theorem for near-integrable symplectic maps. Consider the symplectic map

$$(x, y) \mapsto (X, Y) = T(x, y),$$
$$x = (x_1, \dots, x_m) \bmod 2\pi, \quad X = (X_1, \dots, X_m) \bmod 2\pi, \qquad y, Y \in \mathbb{R}^m,$$
$$X = x + \frac{\partial f(y)}{\partial y} + O(\varepsilon), \quad Y = y + O(\varepsilon).$$

The vector

$$\nu_* = \begin{pmatrix} \nu \\ 2\pi \end{pmatrix}, \qquad \nu = \frac{\partial f}{\partial y}(y^0),$$

is said to be the frequency vector on the invariant torus $\{y = y^0\}$ of the unperturbed system.

Theorem 2.4. *Suppose that the unperturbed (integrable) map is nondegenerate at the point y^0:*

$$\det \frac{\partial^2 f}{\partial y^2}(y^0) \neq 0$$

and the frequency vector $\nu_* \in \mathbb{R}^{m+1}$ is Diophantine. Then the perturbed system has an invariant torus close to the original one with the same frequencies.

2.2 A Reduction of Theorems 2.2–2.4 to the Standard Version

We do not prove Theorem 2.1 in this book. The reader can find the proof in, for example, [33, 93]. We will use the KAM-technique in Sect. 2.7 in a simpler problem.

Now we show that Theorems 2.2–2.4 follow from Theorem 2.1. The plan is as follows:

1. Theorem 2.1 $\Rightarrow$ Theorem 2.2.
2. Theorem 2.2 $\Rightarrow$ Theorem 2.3.
3. Theorem 2.2 $\Rightarrow$ Theorem 2.4.

1. Let us check the first implication. To this end we assume that the conditions of Theorem 2.2 hold. Consider the autonomous system with Hamiltonian $H(x, y, t, \varepsilon) + E$, where the function H satisfies (2.4), and the variable E is canonically conjugate to time t (autonomization, Sect. 1.1). The projections $(x, y, t, E) \mapsto (x, y, t)$ of solutions for this system coincide with solutions of the nonautonomous system with Hamiltonian H. The unperturbed Hamiltonian equals $H_0(y) + E$. The frequency vector corresponding to the torus $\{y = y^0, E = 0\}$ is $\hat{\nu}$.

The only obstacle for the application of Theorem 2.1 is a degeneracy of the unperturbed system. It is possible to remove the degeneracy by considering the system with Hamiltonian e^{H+E}. Indeed, trajectories of the systems with Hamiltonians $H + E$ and e^{H+E} are the same, since the passage from one Hamiltonian to another is equivalent to a change of time.[5]

Without loss of generality we can put $y^0 = 0$, $H_0(0) = 0$. Then the new unperturbed Hamiltonian is as follows:

$$
\begin{aligned}
e^{H_0+E} &= \exp\big(\langle \nu, y\rangle + \langle H_0'' y, y\rangle/2 + O_3(y) + E\big) \\
&= 1 + \langle \nu, y\rangle + E + \frac{1}{2}\langle H_0'' y, y\rangle + \frac{1}{2}(E + \langle \nu, y\rangle)^2 + O_3(y, E),
\end{aligned}
$$

where $H_0'' = \partial^2 H_0/\partial y^2(0)$. Hence, we have:

$$
\det \frac{\partial^2 e^{H_0+E}}{\partial (y, E)^2} = \det \begin{pmatrix} H_0'' + \nu\nu^T & \nu \\ \nu^T & 1 \end{pmatrix} = \det H_0'' \neq 0.
$$

The degeneracy is thus removed.

According to Theorem 2.1 the system with Hamiltonian e^{H+E} has an invariant torus with the frequencies $\hat{\nu}$. Therefore, the system with Hamiltonian $H + E$ has an invariant torus with frequencies $\tilde{\nu}$, proportional to the $\hat{\nu}$. Since the time frequency equals one, we have $\tilde{\nu} = \hat{\nu}$.

2. Assume that the conditions of Theorem 2.3 hold. Take the m-th component of the vector $\nu = \nu(y^0)$. According to the Diophantine conditions it does not vanish. Let us reduce the order of the system on the energy level $\{H = h\}$ (isoenergetic reduction). To this end we solve the equation $H(x, y, \varepsilon) = h$ with respect to y_m:

$$
y_m = -F(\tilde{x}, \tilde{y}, x_m, \varepsilon, h), \qquad \tilde{y} = (y_1, \dots, y_{m-1})^T, \quad \tilde{x} = (x_1, \dots, x_{m-1})^T.
$$

[5] Poincaré used this trick in the restricted three-body problem.

Since $\dot{x}_m = \partial H/\partial y_m \neq 0$, it is possible to perform the change of time $t \mapsto \tau = x_m$, $d/d\tau = (\cdot)'$. Solutions $(\tilde{x}(\tau), \tilde{y}(\tau))$ satisfy the equations

$$\tilde{x}' = \partial F/\partial \tilde{y}, \quad \tilde{y}' = -\partial F/\partial \tilde{x}.$$

Let us obtain the new Hamiltonian F explicitly. Again we can assume that $y^0 = 0$. The Taylor expansion of H has the form

$$H(\tilde{x}, x_m, \tilde{y}, y_m, \varepsilon) = H_0(0) + \langle \tilde{\nu}, \tilde{y} \rangle + \nu_m y_m + \langle \Pi y, y \rangle/2 + O_3(y) + O(\varepsilon),$$

where $\tilde{\nu} = (\nu_1, \ldots, \nu_{m-1})^T$ and $\Pi = \partial^2 H/\partial y^2(0)$. Recall that $h = H_0(0)$ according to the conditions of Theorem 2.3. Therefore,

$$y_m = -\langle \tilde{\nu}, \tilde{y} \rangle/\nu_m + O_2(\tilde{y}) + O(\varepsilon). \tag{2.5}$$

Let us put

$$\Pi = \begin{pmatrix} \tilde{\Pi} & p \\ p^T & \Pi_{mm} \end{pmatrix},$$

where $\tilde{\Pi}$ is an $(m-1) \times (m-1)$ matrix and $p \in \mathbb{R}^{m-1}$. Then the solution $y_m = y_m(\tilde{y}, \varepsilon)$ (2.5) can be made more precise:

$$y_m = -\frac{1}{\nu_m}\left(\langle \tilde{\nu}, \tilde{y} \rangle + \frac{1}{2}\langle \tilde{\Pi}\tilde{y}, \tilde{y} \rangle - \langle p, \tilde{y} \rangle \frac{\langle \tilde{\nu}, \tilde{y} \rangle}{\nu_m} + \frac{\Pi_{mm}}{2}\left(\frac{\langle \tilde{\nu}, \tilde{y} \rangle}{\nu_m} \right)^2 \right) + O_3(\tilde{y}) + O(\varepsilon).$$

(We have obtained this equation, looking for a solution of the equation $H = h$ in the form $y_m = -\langle \tilde{\nu}, \tilde{y} \rangle/\nu_m + \Phi(\tilde{y}) + O_3(\tilde{y}) + O(\varepsilon)$, where Φ is quadratic in $\tilde{y}$.) Hence,

$$\frac{\partial^2 F}{\partial \tilde{y}^2}(0) = \left(\frac{1}{\nu_m}\tilde{\Pi} - \frac{1}{\nu_m^2}(p\tilde{\nu}^T + \tilde{\nu}p^T) + \frac{\Pi_{mm}}{\nu_m^3}\tilde{\nu}\tilde{\nu}^T \right).$$

The proof of the following identity is a simple exercise in linear algebra:

$$\nu_m^{1+m} \det\left(\frac{1}{\nu_m}\tilde{\Pi} - \frac{1}{\nu_m^2}(p\tilde{\nu}^T + \tilde{\nu}p^T) + \frac{\Pi_{mm}}{\nu_m^3}\tilde{\nu}\tilde{\nu}^T \right) = -\det\begin{pmatrix} \Pi & \nu \\ \nu^T & 0 \end{pmatrix}.$$

Thus we have checked that the conditions of Theorem 2.2 hold for the unperturbed torus $\tilde{y} = 0$ of the system with Hamiltonian F. The invariant torus of the perturbed system with Hamiltonian F corresponds to the invariant torus we search for in the original system on the energy level $\{H = h\}$.

3. Theorem 2.4 can be reduced to the nonautonomous version of the Kolmogorov theorem with the help of an inclusion of the map into a flow. Indeed, the map T can be regarded as the Poincaré map in the system with Hamiltonian $H = f(y)/(2\pi) + O(\varepsilon)$. If T is smooth, H can be taken smooth; if T is analytic, H can be also taken analytic by Theorem 1.10 and Remarks 1.5–1.6. It remains to apply Theorem 2.2.

2.3 Lower-Dimensional Tori

In this section we consider autonomous Hamiltonian systems and study invariant tori whose dimension is less than the number of degrees of freedom: the so called *lower-dimensional tori*. By using methods of the previous section it is easy to construct isoenergetic, non-autonomous and discrete versions of results presented below.

Some results concerning lower-dimensional KAM theory can be found in [25, 47, 108]; finite-dimensional tori for near-integrable partial differential equations are studied in [64, 75] (see also references therein).

Let a $2m$-dimensional symplectic manifold (M, ω) be the phase space of an autonomous Hamiltonian system with Hamiltonian H. The torus $N \subset M$ is called a reducible lower-dimensional (invariant) torus of the system (M, ω, H) if $n < m$ and there exist local coordinates

$$\begin{aligned} &x = (x_1, \dots, x_n) \bmod 2\pi, \quad y = (y_1, \dots, y_n), \quad z = (z_1, \dots, z_{2l}), \\ &n + l = m, \end{aligned}$$

on M such that

$$\begin{aligned} \omega &= \sum_{j=1}^{n} dy_j \wedge dx_j + \sum_{q=1}^{l} dz_{l+q} \wedge dz_q, \\ N &= \{(x, y, z) : y = 0, z = 0\}, \\ H &= \langle \nu, y\rangle + \frac{1}{2}\langle Ay, y\rangle + \frac{1}{2}\langle Qz, z\rangle + O_3(y, z), \qquad A = A^T, \quad Q = Q^T. \end{aligned}$$

Hence x are coordinates on N while y and z are coordinates in transversal directions.

Consider the linear system $(\mathbb{R}^{2l}, \sum_{q=1}^{l} dz_{l+q} \wedge dz_q, \frac{1}{2}\langle Qz, z\rangle)$:

$$\dot{z} = JQz, \qquad J = \begin{pmatrix} 0_l & I_l \\ -I_l & 0_l \end{pmatrix}, \tag{2.6}$$

where I_l and 0_l are the unity and zero $l \times l$ matrices. Since (2.6) are Hamiltonian, the characteristic polynomial $f(\mu) = \det(JQ - \mu I_{2l})$ is even: $f(\mu) = f(-\mu)$. Let $\pm\mu_1, \dots, \pm\mu_l$ be the eigenvalues of JQ.

Definition 2.1. The reducible lower-dimensional torus N is said to be (partially) *hyperbolic* if the imaginary axis does not contain any μ_j, $j = 1, \dots, l$.

The problem of the survival of a lower-dimensional torus after a perturbation of the system belongs to KAM theory. In comparison with the case of m-dimensional tori in addition to $\langle k, \nu\rangle$ new "small divisors"

$$\text{(I)} \quad i\mu_j + \langle k, \nu\rangle \quad \text{and} \quad \text{(II)} \quad i(\mu_j - \mu_q) + \langle k, \nu\rangle$$

appear.

Expressions of type (I) can be small if there exists $\mu_j \in i\mathbb{R}$ (the torus is not hyperbolic). Small expressions of type (II) correspond to resonances between the frequences ν and $\operatorname{Im}\mu_j$. They do not only appear in the case when all μ_j are real and pairwise distinct.

Because of the presence of a new group of variables z and new small divisors, KAM theory for lower-dimensional tori is more complicated than the traditional one. In the general situation one has to assume that the system depends on some additional (exterior) parameters. Then it is possible to prove the existence of lower-dimensional tori which are close to the unperturbed ones for the majority of values of the parameters.

The situation becomes simpler in the case of hyperbolic tori. Here the divisors of type (I) are not small and the small divisors of type (II) can only spoil the complete reducibility which is not crucial in the hyperbolic case. Below in Sect. 2.4 we consider this case in more detail.

Examples. Lower-dimensional tori appear in the following situations.

A. Consider a neighborhood of an equilibrium of an autonomous Hamiltonian system with m degrees of freedom. Assume that the characteristic exponents of the equilibrium are pairwise distinct. Then the Hamiltonian of the system can be reduced to the following normal form (see, for example, [9]):

$$\begin{aligned} H &= \sum_{j=1}^{J} \frac{\nu_j}{2}(\alpha_j^2 + \beta_j^2) + \sum_{k=1}^{K} \lambda_k p_k q_k \\ &\quad + \sum_{l=1}^{L} \big(a_l(u_{2l-1}v_{2l-1} + u_{2l}v_{2l}) - b_l(u_{2l-1}v_{2l} - u_{2l}v_{2l-1})\big) \qquad (2.7) \\ &\quad + O_3(\alpha, \beta, p, q, u, v), \\ \omega &= d\alpha \wedge d\beta + dp \wedge dq + du \wedge dv. \end{aligned}$$

Here the equilibrium is situated at the origin:

$$\alpha = \beta = 0, \quad p = q = 0, \quad u = v = 0,$$

the quantities $\pm i\nu_j$, $\pm\lambda_k$, $\pm a_l \pm ib_l$ are characteristic exponents of the equilibrium, and $J + K + 2L = m$.

Neglecting the terms O_3, we obtain the Hamiltonian of the linearized system. This system has the invariant tori

$$\mathbb{T}_c^n = \{(\beta, \alpha, q, p, v, u) : \alpha_j^2 + \beta_j^2 = c_j \geq 0,\ p = q = 0,\ u = v = 0\}.$$

The dimension n of $\mathbb{T}_c^n$ is equal to the number of nonzero constants c_j. For definiteness suppose that $c_1, \dots, c_n \neq 0$ and $c_{n+1} = \dots = c_J = 0$.

It is possible to assume that the quantities λ_k and a_l are positive. Consider the variables $(x, y, \hat\beta, \hat\alpha, z_u, z_s)$ determined as follows:

$$\alpha_j = \sqrt{y_j/2}\cos x_j, \quad \beta_j = \sqrt{y_j/2}\sin x_j, \qquad j = 1, \dots, n,$$
$$\hat\beta = (\beta_{n+1}, \dots, \beta_J), \quad \hat\alpha = (\alpha_{n+1}, \dots, \alpha_J), \qquad z_s = (p, u), \quad z_u = (q, v).$$

Then the quadratic part of the Hamiltonian (2.7) takes the form

$$H = \sum_{j=1}^{J} \nu_j y_j + \sum_{j=n+1}^{J} \nu_j(\alpha_j^2 + \beta_j^2) + \langle z_s, \Omega z_u \rangle,$$

where the square $(K + 2L) \times (K + 2L)$ matrix Ω is as follows:

$$\Omega = \begin{pmatrix} \Omega_1 & 0 \\ 0 & \Omega_2 \end{pmatrix}, \qquad \Omega_1 = \operatorname{diag}(\lambda_1, \dots, \lambda_K),$$

$$\Omega_2 = \begin{pmatrix} a_1 & b_1 & \dots & 0 & 0 \\ -b_1 & a_1 & \dots & 0 & 0 \\ \vdots & \vdots & \ddots & \vdots & \vdots \\ 0 & 0 & \dots & a_L & b_L \\ 0 & 0 & \dots & -b_L & a_L \end{pmatrix}.$$

Eigenvalues of the matrix Ω have positive real parts. The torus $\mathbb{T}_c^n$ is hyperbolic iff $n = J$.

Note that, to prove that lower-dimensional tori close to $\mathbb{T}_c^n$ exist in the original (nonlinear) system, further preparatory work (including nonlinear normalization of the Hamiltonian) is needed.

B. Consider a Hamiltonian system which has a certain set of involutive first integrals. More precisely, we suppose that, in some canonical coordinates x, y, q, p,

$$x = (x_1, \dots, x_n) \bmod 2\pi, \quad y = (y_1, \dots, y_n),$$
$$q = (q_1, \dots, q_l), \quad p = (p_1, \dots, p_l), \qquad n + l = m,$$

the Hamiltonian of the system has the form $H = H(y, q, p)$. Let us fix y, say $y = 0$. Suppose that the point $(q, p) = (q^0, p^0)$ is critical for the function $H(0, q, p)$. It is possible to assume that $q^0 = p^0 = 0$. Then the expansion of the Hamiltonian H into the Taylor series is as follows:

$$H = H(0) + \langle \nu, y \rangle + \frac{1}{2}\left\langle \begin{pmatrix} q \\ p \end{pmatrix}, G\begin{pmatrix} q \\ p \end{pmatrix} \right\rangle + O_2(y) + O_3(q, p) + O(y)O(q, p).$$

The n-dimensional torus $N = \{y = 0, \ q = p = 0\}$ is invariant.

The terms $O(y)O(q, p)$ in this equation can be reduced to $O(y')O_2(q', p')$ by a canonical change $(x, y, q, p) \mapsto (x', y', q', p')$. Indeed, suppose that in the Hamiltonian H

$$O(y)O(q, p) = \langle p, G_1 y \rangle + \langle q, G_2 y \rangle + O(y)O_2(q, p) - O_2(y)O(q, p),$$

where G_1, G_2 are constant $l \times n$ matrices. Then as a result of the change with the generating function

$$S = \langle x, y' \rangle + \langle q, p' \rangle + \langle p', \Omega^{-1} G_1 y' \rangle + \langle q, -(\Omega^T)^{-1} G_2 y' \rangle,$$

we obtain:

$$H = \text{const} + \langle \nu, y' \rangle + \langle y', Ay' \rangle / 2 + \langle p', \Omega q' \rangle + O_3(y', q', p'),$$

where A is some constant $n \times n$ matrix.

C. As is well-known, any resonant Liouville torus of an integrable Hamiltonian system is foliated by nonresonant tori of a smaller dimension. It turns out that after a generic perturbation some (finite) number of these tori survive. To formulate the rigorous result, consider the Hamiltonian system with m degrees of freedom in the canonical coordinates $x \bmod 2\pi$, y with the real-analytic Hamiltonian

$$H(x, y, \varepsilon) = H_0(y) + \varepsilon H_1(x, y, \varepsilon). \tag{2.8}$$

The phase space of the unperturbed system is foliated by invariant m-dimensional tori $N^m_{y^0} = \{(x, y) : y = y^0\}$ with the frequencies $\nu(y) = \partial H_0 / \partial y$.

We put $\nu = \nu(y^0)$, $\nu = (\nu_1, \dots, \nu_m)^T$. Consider the subgroup g_ν of the commutative group $(\mathbb{Z}^m, +)$ such that for any vector $k \in g_\nu$ we have $\langle \nu, \tau \rangle = 0$. It is natural to call g_ν the resonant group. Let rank g_ν denote the number of generators of g_ν.

Proposition 2.1. *Suppose that* rank $g_\nu = l$, $n = m - l$. *Then any trajectory of the unperturbed system lying on the torus $N^m_{y^0}$, fills densely some n-dimensional subtorus of $N^m_{y^0}$. Moreover, $N^m_{y^0}$ is smoothly foliated by such nonresonant n-dimensional tori.*

Indeed, it is known from the theory of Abelian groups that there exists a set $k_1^*, \dots, k_n^*, k_1, \dots, k_l$ of vectors from $\mathbb{Z}^m$ such that the $(m \times m)$ matrix K_0, having these vectors as rows, is unimodular (i.e. $\det K_0 = 1$), and the vectors $k_1, \dots, k_l$ generate g_ν. Let K and K_* respectively be the $(m \times l)$ and $(m \times n)$ matrices such that the vectors $k_1, \dots, k_l$ and $k_1^*, \dots, k_n^*$ respectively are their rows. Obviously, rank $K = l$, rank $K_* = n$, $K^T \nu = 0$.

Let us perform the change of the angular variables: $q = K_0^T x$. Since the matrix K_0^T is unimodular, the variables $q \bmod 2\pi$ are coordinates on $N^m_{y^0}$. The equation $\dot{x} = \nu$ on the torus $\mathbb{T}^m_{y^0}$ in the new coordinates has the form $\dot{q} = K_0^T \nu$. The last l components of the vector $K_0^T \nu$ vanish, and the first n components form the vector

$$\nu^* = K_*^T \nu.$$

The resonant group $g_{K_0^T \nu}$ contains the subgroup $\mathbb{Z}_0^l = \{j \in \mathbb{Z}^m : j_1 = \cdots = j_n = 0\}$. Since the groups g_ν and $g_{K_0^T \nu}$ are isomorphic to one another, we have:

rank $g_{K_0^T \nu} = l$ and $g_{K_0^T \nu} = \mathbb{Z}_0^l$. Therefore, the frequencies $\nu^* = (\nu_1^*, \ldots, \nu_n^*)$ are rationally independent.

Any trajectory $\gamma \in N_{y^0}^m$ of the unperturbed system fills densely a torus

$$\{q \bmod 2\pi : (q_{n+1}, \ldots, q_m) = \text{const}\}, \tag{2.9}$$

and any point $q \in N_{y^0}^m$ belongs to one of the tori (2.9).

In the coordinates x, y the torus (2.9), containing the point x^0, has the form

$$N_{x^0, y^0}^n = \{(x, y) : K^T(x - x^0) = 0 \bmod 2\pi, \, y = y^0\}. \tag{2.10}$$

Below we present sufficient conditions under which N_{x^0,y^0}^n survives a perturbation of the system. Let us introduce the operation $\langle \cdot \rangle_{g_\nu}$ of the averaging corresponding to the subgroup g_ν. For any continuous function $f : \mathbb{T}^m \to \mathbb{R}$ we define

$$\langle f \rangle_{g_\nu}(x) = \lim_{T \to \infty} \frac{1}{T} \int_0^T f(x + \nu t)\, dt.$$

Proposition 2.2. *Let the Fourier expansion of the function f have the form*

$$f(x) = \sum_{j \in \mathbb{Z}^m} f_j e^{i\langle j, x \rangle}. \tag{2.11}$$

Then

$$\langle f \rangle_{g_\nu}(x) = \sum_{\mu \in \mathbb{Z}^l} f_{K\mu} e^{i\langle K\mu, x \rangle} = \sum_{j \in g_\nu} f_j e^{i\langle j, x \rangle}. \tag{2.12}$$

Proof. Putting $\varphi(q) = f((K_0^T)^{-1} q)$, we have:

$$\frac{1}{T} \int_0^T f(x + \nu t)\, dt = \frac{1}{T} \int_0^T \varphi(K_0^T x + K_0^T \nu t)\, dt.$$

Since $K_0^T \nu = \binom{\nu^*}{0}$ and the frequency vector $\nu^* \in \mathbb{R}^n$ is nonresonant, by the Weil theorem on the equality of the time and space averages [9], we get:

$$\begin{aligned} &\langle f \rangle_{g_\nu}(x) = \frac{1}{(2\pi)^n} \int_{\mathbb{T}^n} \varphi\left(K_0^T x + \begin{pmatrix} q' \\ 0 \end{pmatrix} \right) dq', \\ &q' \in \mathbb{T}^n, \qquad \begin{pmatrix} q' \\ 0 \end{pmatrix} \in \mathbb{T}^m. \end{aligned} \tag{2.13}$$

Since the Fourier expansion of φ has the form

$$\varphi(q) = \sum_{j \in \mathbb{Z}^m} f_j e^{i\langle j, (K_0^T)^{-1} q \rangle} = \sum_{\mu \in \mathbb{Z}^m} f_{K_0 \mu} e^{i\langle \mu, q \rangle},$$

the integral (2.13) equals

$$\sum_{\mu \in \mathbb{Z}_0^l} f_{K_0 \mu} e^{i\langle \mu, K_0^T x \rangle} = \sum_{\mu \in \mathbb{Z}^l} f_{K\mu} e^{i\langle K\mu, x \rangle}.$$

The first equation (2.12) is proved. The second equation (2.12) follows immediately from definition of the matrix K. □

Below we assume that the vector ν^* is Diophantine. The Diophantine condition for ν^* can be expressed in terms of the original frequency vector ν. Namely, we call the vector ν g_ν-Diophantine if there exist positive constants c, γ such that

$$|\langle k, \nu \rangle| \geq \frac{1}{c|k|^\gamma} \quad \text{for any } k \in \mathbb{Z}^m \setminus g_\nu.$$

This notion is discussed in Sect. 9.1.

Now we present a theorem on the existence of invariant tori close to $N^n_{x^0, y^0}$ for small $\varepsilon > 0$ in the hyperbolic case. The general case is more complicated, see for example, [28].

Theorem 2.5 ([132]). *Suppose that the following conditions hold.*

1. *The unperturbed system is nondegenerate, i.e.,*

$$\det \Pi \neq 0, \qquad \Pi = \frac{\partial^2 H_0}{\partial y^2}(y^0).$$

2. *The frequency vector ν is g_ν-Diophantine and* rank $g_\nu = l$.
3. *The point x^0 is critical for the function $h(x) = \langle H_1(y^0, x, 0)\rangle_{g_\nu}$ and the Hesse matrix $W = \partial^2 h/\partial x^2(x^0)$ is such that the matrix $W\Pi$ has l eigenvalues*[6] *off the semiaxis $\mathbb{R}_+ = \{\lambda \in \mathbb{R} : \lambda \geq 0\}$.*

Then for small $\varepsilon \geq 0$ there exist an analytic in ε for $\varepsilon > 0$, smooth in $\sqrt{\varepsilon}$ for $\varepsilon \geq 0$, family of n-dimensional hyperbolic tori $N^n_{y^0}(\varepsilon)$ of the system filled by quasi-periodic solutions with the frequencies ν^*, where $N^n_{y^0}(0) = N^n_{x^0, y^0}$.

Invariant asymptotic manifolds (see below) $\Gamma^{s,u}(\varepsilon)$ of the hyperbolic torus $N^n_{y^0}(\varepsilon)$ are close to one another: their local pieces contained in a neighborhood of the torus $N^m_{y^0}(\varepsilon)$ can be transformed to each other by a deformation of order $\sqrt{\varepsilon}$.

Bolotin proved [20, 21] that, if the matrix Π is positive definite, the manifolds $\Gamma^{s,u}(\varepsilon)$ intersect outside $N^n_{y^0}(\varepsilon)$ along several doubly asymptotic (homoclinic) trajectories. Note that in the vicinity of the torus $N^n_{y^0}(\varepsilon)$ the system is exponentially (in ε) close to an integrable one. In particular, the rate of the splitting of the manifolds $\Gamma^{s,u}(\varepsilon)$ is exponentially small.

---

[6] The matrix W is degenerate: rank $W \leq l$. Hence at least n eigenvalues of $W\Pi$ vanish.

2.4 Hyperbolic Tori

The notion of a hyperbolic torus is widely used in the literature (see for example [28, 55, 104, 132, 148]). The concept of hyperbolicity first appeared in the general theory of dynamical systems, where a hyperbolic set is defined in terms of the so called stable and unstable foliations. However, traditionally in the KAM theory hyperbolic tori are defined with the help of coordinates. Here, following [24], we will show that under certain natural conditions the definition of a hyperbolic torus conventional for the general theory of dynamical systems is quite suitable for needs of the KAM theory.

Consider the system (M, ω, H). We use the notation v_H for the corresponding Hamiltonian vector field on M and g^t for the phase flow of the system. Let $N \subset M$ be an invariant n-dimensional torus with frequency vector $\nu \in \mathbb{R}^n$. Thus N is the image of a smooth embedding

$$\Phi : \mathbb{T}^n \to M \tag{2.14}$$

and the equation

$$\partial_\nu \Phi = v_H \circ \Phi, \qquad \partial_\nu = \langle \nu, \partial/\partial x \rangle,$$

is satisfied. Equivalently, $\Phi(x+\nu t) = g^t(\Phi(x))$. Consider the linearized phase flow

$$Dg^t : T_N M \to T_N M, \qquad v \in T_x M \mapsto Dg^t(x)\, v \in T_{g^t(x)} M.$$

It is defined by a linear differential equation which is called the variational equation for the invariant torus N. The following definition was suggested in [23]. A slightly different invariant definition is used in [21].

Definition 2.2. The torus N is called hyperbolic[7] if $n < m$ and there exist two smooth l-dimensional ($l = m - n$) subbundles $E^{s,u}$ of the bundle $T_N M$ such that

1. $E^{s,u}$ are invariant for the linearized phase flow:

$$Dg^t(w) E^{s,u}_w = E^{s,u}_{g^t(w)}, \qquad w \in N, \quad t \in \mathbb{R}.$$

2. The linearized flow is contracting on E^s and expanding on E^u, i.e., for some positive constants C and λ, we have

$$\|Dg^t(w)|_{E^s_w}\| \le Ce^{-\lambda t}, \qquad w \in N, \quad t \ge 0,$$
$$\|Dg^{-t}(w)|_{E^u_w}\| \le Ce^{-\lambda t}, \qquad w \in N, \quad t \ge 0.$$

Here $\|\cdot\|$ is any reasonable norm for the operators $Dg^t\big|_{E^{s,u}} : E^{s,u}_w \to E^{s,u}_{g^t(w)}$. Passing to a finite covering of the torus N, without loss of generality we can assume

[7] In the general theory of dynamical systems such a torus would be called partially normally hyperbolic, because $E^u_x \oplus E^s_x \oplus T_x N \ne T_x M$. Here we follow the traditions of Hamiltonian dynamics and KAM theory.

that the bundles $E^{s,u}$ are oriented. The bundle E^s is called stable (or contracting), and E^u unstable (or expanding).

Definition 2.3. If the vector ν is nonresonant, the corresponding hyperbolic torus is called nonresonant. If ν is Diophantine, the torus is called Diophantine.

Definition 2.4. A hyperbolic torus N is called *nondegenerate* if all bounded solutions of the variational equation are tangent to N. Thus $\|Dg^t(x)v\| \le C$ for all t implies that $v \in T_x N$.

Let $L(\mathbb{R}^l)$ be the set of linear operators $\mathbb{R}^l \to \mathbb{R}^l$. For a smooth function $\Lambda : \mathbb{T}^n \to L(\mathbb{R}^l)$ consider a skew product system on $\mathbb{T}^n \times \mathbb{R}^l$:

$$\dot{x} = \nu, \quad \dot{v} = \Lambda(x)v, \qquad x \in \mathbb{T}^n, \quad v \in \mathbb{R}^l. \tag{2.15}$$

Its phase flow has the form

$$(x, v) \mapsto (x(t), v(t)) = (x + \nu t, \phi_t(x)v), \tag{2.16}$$

where $\phi_t : \mathbb{T}^n \to L(\mathbb{R}^l)$, $\phi_0(x) = I$, is the fundamental matrix. Hence,

$$\phi_{s+t}(x) = \phi_s(x + \nu t)\phi_t(x) \quad \text{for all } s, t \in \mathbb{R} \text{ and } x \in \mathbb{T}^n. \tag{2.17}$$

Definition 2.5. The function $\Lambda : \mathbb{T}^n \to L(\mathbb{R}^l)$ is said to be ν-positive definite if there exist positive constants C, λ such that $\|\phi_{-t}(x)\| \le Ce^{-\lambda t}$ for all $x \in \mathbb{T}^n$ and $t \ge 0$. The function Λ is said to be ν-negative definite if $\|\phi_t(x)\| \le Ce^{-\lambda t}$, $t \ge 0$.

In other words, Λ is ν-positive or ν-negative definite if the Lyapunov exponents of the skew product system (2.15) are positive or negative respectively.

Remark 2.1. The duality of the systems

$$\begin{cases} \dot{v} = \Lambda(x)v, \\ \dot{x} = \nu, \end{cases} \quad \text{and} \quad \begin{cases} \dot{u} = -\Lambda^T(x)u, \\ \dot{x} = \nu \end{cases}$$

implies that Λ is ν-positive definite if and only if $-\Lambda^T$ is ν-negative definite.

The simplest example of a ν-positive definite function is $\Lambda = \text{const}$, where the eigenvalues of Λ have positive real parts. An equivalent condition is: $\Lambda + \Lambda^*$ is positive definite with respect to some Euclidean metric on $\mathbb{R}^l$, where the operator Λ^* is conjugate to Λ with respect to the same metric. Here the property of ν-positive definiteness does not depend on ν. For $l = 1$ equations (2.15) can be solved explicitly. Then, for ν nonresonant, ν-positive definiteness of Λ is equivalent to the condition $\int_{\mathbb{T}^n} \Lambda(x)\, dx > 0$.

As a model example, consider a system on $P = \mathbb{T}^n_x \times \mathbb{R}^n_y \times \mathbb{R}^l_{z_u} \times \mathbb{R}^l_{z_s}$ with Hamiltonian

$$H(x, y, z_u, z_s) = \langle \nu, y\rangle + \langle Ay, y\rangle/2 + \langle z_s, \Omega(x)z_u\rangle + O_3(y, z) \tag{2.18}$$

and the symplectic structure

$$\omega = dy \wedge dx + dz_s \wedge dz_u,$$

where the symmetric $(n \times n)$ matrix A is constant, and the matrix function Ω is ν-positive definite.

Proposition 2.3. *The invariant torus* $N = \{(x, y, z_u, z_s) : y = 0, z_{s,u} = 0\}$ *is hyperbolic.*

Indeed, the Hamilton equations have the form

$$\begin{cases} \dot{x} = \nu + Ay + O_2(y, z), \\ \dot{y} = O_2(y, z), \\ \dot{z}_u = \Omega(x)z_u + O_2(y, z), \\ \dot{z}_s = -\Omega^T(x)z_s + O_2(y, z). \end{cases} \tag{2.19}$$

The variational equations on

$$T_N P = \mathbb{T}^n_x \times \mathbb{R}^n_\xi \times \mathbb{R}^n_\eta \times \mathbb{R}^l_{\zeta_u} \times \mathbb{R}^l_{\zeta_s}$$

take the form

$$\begin{cases} \dot{x} = \nu, \\ \dot{\xi} = A\eta, \\ \dot{\eta} = 0, \\ \dot{\zeta}_u = \Omega(x)\zeta_u, \\ \dot{\zeta}_s = -\Omega^T(x)\zeta_s. \end{cases} \tag{2.20}$$

The subbundles $E^u = \mathbb{T}^n \times \mathbb{R}^l_{\zeta_u}$ and $E^s = \mathbf{T}^n \times \mathbb{R}^l_{\zeta_s}$ are obviously invariant. They satisfy Definition 2.2 of a hyperbolic torus due to the condition of ν-positive definiteness of Ω.

Definition 2.6. Symplectic coordinates (x, y, z_u, z_s) in a finite covering of a neighborhood of a hyperbolic torus N are said to be canonical for N if H satisfies (2.18), up to a constant. If there exist smooth canonical coordinates for a hyperbolic torus, the torus is said to be *weakly reducible*. A hyperbolic torus is said to be *reducible* if it is weakly reducible with a constant matrix Ω.

A simple exercise is to check that this definition of a reducible hyperbolic torus is equivalent to Definition 2.1.

Proposition 2.4. *A weakly reducible hyperbolic torus is nondegenerate if and only if* $\det A \neq 0$.

Indeed, for any bounded solution of the variational equations (2.20), $\zeta_{s,u} = 0$ and $A\eta = 0$. All these solutions are tangent to N if and only if the matrix A is nondegenerate.

In [55, 132, 148] a hyperbolic torus was defined in terms of canonical coordinates, where the matrix $\Lambda(x) + \Lambda^T(x)$ is positive definite for any $x \in \mathbb{T}^n$. Up to a change of variables, this is equivalent to the weaker condition that, for some positive definite matrix G, the matrix $G\Lambda(x) + \Lambda^T(x)G$ is positive definite for any $x \in \mathbb{T}^n$. The condition of ν-positive definiteness is weaker.

Proposition 2.5. *The function $\Lambda : \mathbb{T}^n \to L(\mathbb{R}^l)$ is ν-positive definite if and only if for some smooth family of positive definite matrices $G : \mathbb{T}^n \to L(\mathbb{R}^l)$ the matrix*

$$\partial_\nu G + G\Lambda + \Lambda^T G \quad \textit{is positive definite for all } x \in \mathbb{T}^n. \tag{2.21}$$

The metric defined by G is called the Lyapunov metric.

Corollary 2.1. *A small perturbation of a ν-positive definite function is again ν-positive definite.*

Proof (of Proposition 2.5). This proposition is a version of a result of Lyapunov (a more general statement is presented in [116, 117]). Suppose that Λ is ν-positive definite. Take any family $K : \mathbb{T}^n \to L(\mathbb{R}^l)$ of symmetric positive definite matrices and put

$$G(x) = \int_{-\infty}^{0} \phi_s^T(x) K(x + \nu s)\phi_s(x)\, ds,$$

where ϕ_s is the fundamental matrix of system (2.15). The integral converges exponentially due to the property of ν-positive definiteness. By (2.17)

$$\begin{aligned}
&\big\langle G(x+\nu t)\phi_t(x)v, \phi_t(x)v\big\rangle \\
&= \int_{-\infty}^{0} \big\langle K(x+\nu(s+t))\phi_s(x+\nu t)\phi_t(x)v, \phi_s(x+\nu t)\phi_t(x)v\big\rangle\, ds \\
&= \int_{-\infty}^{0} \big\langle K(x+\nu(s+t))\phi_{s+t}(x)v, \phi_{s+t}(x)v\big\rangle\, ds \\
&= \int_{-\infty}^{t} \big\langle K(x+\nu\tau)\phi_\tau(x)v, \phi_\tau(x)v\big\rangle\, d\tau.
\end{aligned}$$

Therefore

$$\frac{d}{dt}\bigg|_{t=0} \big\langle G(x+\nu t)\phi_t(x)v, \phi_t(x)v\big\rangle = \langle K(x)v, v\rangle > 0 \quad \text{for any } v \neq 0.$$

This implies (2.21). On the other hand, if condition (2.21) holds, then

$$\frac{d}{dt}\bigg|_{t=0} \big\langle G(x+\nu t)\phi_t(x)v, \phi_t(x)v\big\rangle \geq \lambda\langle v, v\rangle \quad \text{for any } v \neq 0,$$

where λ is a positive constant. This implies the required ν-positive definiteness. □

Recall that a torus N is said to be isotropic if $\omega|_N = 0$.

Theorem 2.6. *Any Diophantine isotropic hyperbolic torus is weakly reducible. If $n = m - 1$ or $n = 1$, it is reducible. In the analytic case the canonical coordinates are analytic.*

Of course, this trivially holds also for $n = 0$ and $n = m$. Herman [59] proved that, if the symplectic form ω is exact, then any nonresonant invariant torus is isotropic. Theorem 2.6 gives a relation between the dynamical definition of a hyperbolic torus and the definition conventional for the KAM theory. We present a proof of the theorem in Sect. 2.5.

The variational system (2.20) has two invariant isotropic subbundles

$$E^u = \mathbb{T}^n_x \times \mathbb{R}^l_{\zeta_u} = \{\eta = 0, \zeta_s = 0\},$$
$$E^s = \mathbb{T}^n_x \times \mathbb{R}^l_{\zeta_s} = \{\eta = 0, \zeta_u = 0\}.$$

Solutions which belong to the first (respectively, to the second) one, approach the hyperbolic torus exponentially when $t \to -\infty$ (respectively, $t \to +\infty$). Manifolds with analogous properties exist in system (2.19) as well. Graff proved [55] that in an analytic system with Hamiltonian (2.18), where $\Omega + \Omega^T$ is positive definite, there are stable and unstable invariant Lagrangian manifolds $\Gamma^{s,u}$ containing N such that $T_x\Gamma^{s,u} = T_x N \oplus E^{s,u}_x$ for any $x \in N$. They can be represented as $\Gamma^{s,u} = f_{s,u}(E^{s,u})$, where $f_{s,u} : E^{s,u} \to \mathbb{T}^n_x \times \mathbb{R}^n_y \times \mathbb{R}^l_{z_u} \times \mathbb{R}^l_{z_s}$ are analytic maps of the form

$$f_u(\theta, \zeta) = (\theta + O_2(\zeta), O_2(\zeta), \zeta, O_2(\zeta)),$$
$$f_s(\theta, \zeta) = (\theta + O_2(\zeta), O_2(\zeta), O_2(\zeta), \zeta).$$

The Hamiltonian system restricted to $\Gamma^{s,u}$ has the form

$$\dot\vartheta = \nu, \quad \dot\zeta = \Lambda^u(\vartheta, \zeta)\zeta \qquad \text{on } \Gamma^u,$$
$$\dot\vartheta = \nu, \quad \dot\zeta = -\Lambda^s(\vartheta, \zeta)\zeta \qquad \text{on } \Gamma^s,$$

where the matrices $\Lambda^u + (\Lambda^u)^T$ and $\Lambda^s + (\Lambda^s)^T$ are positive definite for small $|\zeta|$.

Conjecture 2.1. Manifolds $\Gamma^{s,u}$ exist and are unique for a weakly reducible Diophantine torus. For an analytic system, they are analytic.

In this case, the matrices $\Lambda^{s,u}$ are ν-positive definite for small $|\zeta|$.

In the C^k category with large k, the existence (not uniqueness) of the stable and unstable manifolds is easy to prove by using Theorem 2.6 and standard hyperbolic technique. Let H be smooth. Fix a large integer k.

Theorem 2.7 ([24]). *Let N be a C^k-smooth isotropic Diophantine hyperbolic torus. Then N has m-dimensional Lagrangian stable and unstable manifolds $\Gamma^{s,u}$ of class C^k.*

For analytic H, this theorem does not imply that the manifolds $\Gamma^{s,u}$ are analytic.

Graff [55] (see also [148]) proved that reducible nondegenerate Diophantine hyperbolic tori of a real-analytic system survive a small perturbation. They just slightly deform and remain hyperbolic and real-analytic.

Conjecture 2.2. The Graff theorem holds for weakly reducible Diophantine tori.

We believe that the proof of both conjectures is straightforward and standard; however, the detailed argument is not short. Again, in the C^k category with large k, Conjecture 2.2 is an easy theorem.

Theorem 2.8 ([24]). *Let k be sufficiently large. Then an isotropic nondegenerate Diophantine hyperbolic torus survives a small C^k perturbation of the Hamiltonian.*

In the analytic case Theorem 2.8 doesn't imply that the perturbed torus is analytic. To prove this, one needs KAM theory.

Eliasson [48] obtained the following normal form for an analytic system near a hyperbolic torus for $n = m - 1$.

Theorem 2.9. *Let N be a reducible nondegenerate $(m-1)$-dimensional Diophantine hyperbolic torus. Then there exist analytic canonical coordinates (x, y, z_u, z_s) in a neighborhood of N such that*

$$H = \langle \nu, y \rangle + \lambda z_s z_u + O_2(y, z_s z_u).$$

According to Theorem 2.6, the assumption of the reducibility for the torus N always holds if the form ω is exact.

The proof of Theorem 2.9 is of KAM nature: it is based on a Newton-type iterative procedure. The Eliasson coordinates are convenient for studying the dynamics in a neighborhood of N.

2.5 Hyperbolic Tori: Weak Reducibility

In this section we prove Theorem 2.6. The proof is based on several auxiliary propositions.

Proposition 2.6. *Suppose that the form ω is exact. Let N be a nonresonant hyperbolic torus. Then*

1. *The torus N is isotropic, i.e., $\omega|_N = 0$.*
2. *For any $w \in N$ the subspaces $T_w N \oplus E^{s,u}_w$ are Lagrangian.*
3. *The form ω defines a nondegenerate bilinear form on $E^u_w \times E^s_w$: if $v_u \in E^u_w$ and $\omega(v_u, v_s) = 0$ for all $v_s \in E^s_w$, then $v_u = 0$.*

Proof. The first statement follows from the Herman theorem [59]. The argument is as follows. The form $\omega_0 = \Phi^*\omega$ (see (2.14)) on $\mathbb{T}^n$ is preserved by the differential equation $\dot{x} = \nu$. Let us put

$$\omega_0 = \sum_{j<k} \alpha_{jk}(x)\, dx_j \wedge dx_k.$$

The vector field $\langle \nu, \partial/\partial x \rangle$ is a restriction of the Hamiltonian vector field v_H to N. Therefore the derivative of ω_0 along $\langle \nu, \partial/\partial x \rangle$ vanishes. Moreover, since ν is non-resonant, the coefficients α_{jk} are constant. Since ω is exact, ω_0 is also exact. Therefore, $\alpha_{jk} = 0$.

To prove the second statement, take any two vectors $v, v' \in T_w N \oplus E^u_w$. Then

$$\omega(v, v') = \omega(Dg^t v, Dg^t v') \to 0, \qquad t \to -\infty.$$

Indeed, for any vector $v \in T_w N \oplus E^u_w$ let $u \in T_w N$ be the component of v tangent to the torus. Then

$$\|Dg^t v - Dg^t u\| \to 0, \qquad \|Dg^t u\| \le C.$$

The last statement follows from the first two and the nondegeneracy of ω. Indeed, suppose that for some $v_u \in E^u_w$ we have $\omega(v_u, v_s) = 0$ for any $v_s \in E^s_w$. From statement 2 we have: $\omega(v_u, v) = 0$ for any $v \in T_w N$. Hence, the space spanned by $T_w N \oplus E^s_w$ and v_u is isotropic. If $v_u \neq 0$, the dimension of this space equals $m + 1$. This contradicts the nondegeneracy of ω. □

Proposition 2.7. *In a finite covering $\widetilde{U}$ of a neighborhood U of N in M there exist coordinates $(x, y, z_u, z_s) \in \mathbb{T}^n \times \mathbb{R}^n \times \mathbb{R}^l_u \times \mathbb{R}^l_s$ such that:*

1. $\Phi(x) = (x, 0, 0, 0)$ *for any* $x \in \mathbb{T}^n$,
2. $\omega = dy \wedge dx + dz_u \wedge dz_s$,
3. $E^u = \mathbb{T}^n \times \mathbb{R}^l_u$, $E^s = \mathbb{T}^n \times \mathbb{R}^l_s$.

Here to avoid misunderstandings $\mathbb{R}^l_s$ and $\mathbb{R}^l_u$ denote two copies of conjugate spaces $\mathbb{R}^l$. Thus $\mathbb{R}^l_u = (\mathbb{R}^l_s)^*$.

Proof. By definition, $N = \Phi(\mathbb{T}^n)$. Let x be the standard coordinate in $\mathbb{T}^n$. After passing to a finite covering, the bundles E^s and E^u become trivial. Since the bundle E^s is trivial, for any $w = \Phi(x)$ we have an isomorphism $E^s_w = \mathbb{R}^l_{z_s}$ smoothly depending on $x \in \mathbb{T}^n$. We fix this isomorphism. By Proposition 2.6, $V = E^u_w \oplus E^s_w$ is a symplectic space. Hence $E^u_w = (E^s_w)^*$ with the isomorphism given by the symplectic form ω. Thus $V = \mathbb{R}^l_u \oplus \mathbb{R}^l_s$ with the standard symplectic structure $dz_s \wedge dz_u$. Let

$$V^\perp = \{v \in T_w M \mid \omega(v, u) = 0 \text{ for any } u \in V\}$$

be the symplectic complement of V. Then $V^\perp$ is a symplectic space of dimension $2n$, and it contains the Lagrangian subspace $T_w N = \mathbb{R}^n$. Hence there exists a symplectic isomorphism $V^\perp = T_w N \oplus T^*_w N = \mathbb{R}^n \oplus (\mathbb{R}^n)^*$ with the standard symplectic structure $dy \wedge dx$. Thus, for any $w = \Phi(x) \in N$, we have a symplectic isomorphism

$$T_w M = V^\perp \oplus V = \mathbb{R}^n_x \oplus \mathbb{R}^{n*}_y \oplus (\mathbb{R}^l_u)_{z_u} \oplus (\mathbb{R}^l_s)_{z_s}$$

smoothly depending on $x \in \mathbb{T}^n$. The bilinear form $\omega|_{T_w M}$ equals $dy \wedge dx + dz_s \wedge dz_u$.

Using the exponential map $T_w M \to M$, we obtain a smooth map

$$f : \mathbb{T}^n_x \times \mathbb{R}^n_y \times (\mathbb{R}^l_u)_{z_u} \times (\mathbb{R}^l_s)_{z_s} \to M$$

defining the coordinates x, y, z_u, z_s in a tubular neighborhood of the torus N such that

$$f^*\omega = dy \wedge dx + dz_u \wedge dz_s + O(y, z).$$

Put

$$\Omega = dy \wedge dx + dz_u \wedge dz_s.$$

The symplectic structures $f^*\omega$ and Ω coincide on the torus $\mathbb{T}^n$ and are homotopic in a neighborhood of $\mathbb{T}^n$ in the class of symplectic structures. For example, the homotopy can be chosen as $af^*\omega + (1-a)\Omega, 0 \le a \le 1$. According to Theorem 1.2, there exists a diffeomorphism g in a neighborhood of the torus, such that g and Dg equal the identity on $\mathbb{T}^n$ and $g^*(f^*\omega) = \Omega$.

The map $f \circ g$ defines the coordinates we need. □

Proposition 2.8. *Suppose that for some vector $\nu \in \mathbb{R}^n$ the function $\Lambda : \mathbb{T}^n \to L(\mathbb{R}^l)$ is ν-positive definite or ν-negative definite. Then for any smooth function $v : \mathbb{T}^n \to \mathbb{R}^l$ there exists a unique smooth solution $u : \mathbb{T}^n \to \mathbb{R}^l$ of the equation*

$$\partial_\nu u(x) - \Lambda(x)u(x) = v(x). \tag{2.22}$$

For $\Lambda \equiv 0$ equation (2.22) has a unique solution provided that the mean values of u and v over the torus $\mathbb{T}^n$ vanish and the vector ν is Diophantine.

If v and Λ are real-analytic, the function u is also real-analytic.

Proof. Consider the case of a ν-negative definite function Λ. Let $\phi_t(x)$ be the fundamental matrix of system (2.15). Since Λ is ν-negative definite, the norm of $\phi_t(x)$ satisfies the estimate

$$\|\phi_t^{-1}(x)\| \le Ce^{-\lambda t}, \qquad t > 0, \tag{2.23}$$

for some positive constant λ. We will show that

$$u(x) = \int_{-\infty}^{0} \phi_s^{-1}(x)v(x + \nu s)\, ds. \tag{2.24}$$

Indeed, equation (2.22) is equivalent to the following one:

$$\dot{u}(x + \nu t) - \Lambda(x + \nu t)u(x + \nu t) = v(x + \nu t).$$

Searching for a solution in the form $u(x + \nu t) = \phi_t(x)\theta(x, t)$, we get

$$\theta(x, t) = \int_{-\infty}^{t} \phi_\tau^{-1}(x)v(x + \nu\tau)\, d\tau.$$

Putting $t = 0$, we obtain (2.24). The integral converges exponentially due to (2.23).

The solution (2.24) is unique since all nontrivial solutions of the homogeneous equation $\partial_\nu u(x) - \Lambda(x)u(x) = 0$ are unbounded on $\mathbb{T}^n$.

If υ and Λ are real-analytic, $\phi_t(x)$ is also real-analytic in both x and t. The integral (2.24) converges exponentially in a complex neighborhood of $\mathbb{T}^n$. Hence, $u(x)$ is also real-analytic.

The case of a ν-positive definite function Λ is analogous. For $\Lambda \equiv 0$ and ν Diophantine, equation (2.22) can be easily solved by using Fourier expansions. □

Proposition 2.9. *The coordinates* (x, y, z_u, z_s) *from Proposition 2.7 can be transformed to canonical ones for the torus* N.

Proof. Since $y = z_s = z_u = 0$ is an invariant torus with frequency ν, in the coordinates (x, y, z_u, z_s) the Hamiltonian $H = \langle \nu, y\rangle + O_2(y, z)$ has the form

$$
\begin{aligned}
H = {} & \langle \nu, y\rangle + \langle A(x)y, y\rangle/2 + \langle z_u, B_u(x)y\rangle + \langle z_s, B_s(x)y\rangle \\
& + \langle z_s, C_s(x)z_s\rangle/2 + \langle z_s, C(x)z_u\rangle + \langle z_u, C_u(x)z_u\rangle/2 \\
& + O_3(y, z_{s,u}),
\end{aligned}
$$

where the $n \times n$ matrix A, the $l \times n$ matrices $B_{s,u}$, and the $l \times l$ matrices C, $C_{s,u}$ are functions of $x \in \mathbb{T}^n$. The matrices A and $C_{s,u}$ are symmetric.

The corresponding differential equations are as follows

$$
\begin{cases}
\dot{x} = \nu + Ay + B_u^T z_u + B_s^T z_s + O_2(y, z), \\
\dot{y} = O_2(y, z), \\
\dot{z}_u = C_s z_s + C z_u + B_s y + O_2(y, z), \\
\dot{z}_s = -C^T z_s - C_u z_u + B_u y + O_2(y, z).
\end{cases}
$$

By Proposition 2.7 the manifolds $\{y = 0, z_s = 0\}$ and $\{y = 0, z_u = 0\}$ are invariant in linear approximation. This means that $C_s \equiv C_u \equiv 0$. According to the definition of a hyperbolic torus the function C is ν-positive definite.

To eliminate the terms containing B_s and B_u, we perform the canonical change $(x, y, z_u, z_s) \mapsto (\hat{x}, \hat{y}, \hat{z}_u, \hat{z}_s)$ with the generating function

$$
S(x, z_u, \hat{y}, \hat{z}_s) = \langle x, \hat{y}\rangle + \langle z_u, \hat{z}_s\rangle + \langle z_u, b_u(x)\hat{y}\rangle + \langle z_s, b_s(x)\hat{y}\rangle,
$$

where the matrices $b_{s,u}(x)$ are solutions of the equations

$$
\partial_\nu b_{s,u}(x) \pm C(x) b_{s,u}(x) + B_{s,u}(x) = 0.
$$

In the new variables

$$
H = \langle \nu, \hat{y}\rangle + \langle \hat{A}(\hat{x})\hat{y}, \hat{y}\rangle/2 + \langle \hat{z}_s, C(\hat{x})\hat{z}_u\rangle + O_2(\hat{y}, \hat{z}_{s,u})
$$

with some matrix function $\hat{A}$.

The change $(\hat{x}, \hat{y}, \hat{z}_u, \hat{z}_s) \mapsto (x, y, z_u, z_s)$ with the generating function

$$
S(\hat{x}, \hat{z}_u, y, z_s) = \langle \hat{x}, y\rangle + \langle \hat{z}_s, z_u\rangle + \langle a(\hat{x})y, y\rangle,
$$

where the matrix a is a solution of the equation

$$\partial_\nu a(x) + \hat{A}(x) = \overline{A}, \qquad \overline{A} = \frac{1}{(2\pi)^n} \int_{\mathbb{T}^n} \hat{A}(x)\, dx,$$

defines coordinates canonical for N. Indeed,

$$H = \langle \nu, y\rangle + \langle \overline{A}y, y\rangle/2 + \langle z_s, C(x)z_u\rangle + O_2(y, z). \quad \square$$

Now suppose that $l = 1$. Then positive definiteness of C is equivalent to the condition $\int_{\mathbb{T}^n} C(x)\, dx = \overline{C} > 0$. Let us perform the change $(x, y, z_u, z_s) \mapsto (X, Y, Z_u, Z_s)$ with generating function $S = \langle x, Y\rangle + u(x)z_u Z_s$, where

$$\partial_\nu u(x) + u(x)(C(x) - \overline{C}) = 0.$$

This equation has a positive solution on the torus $\mathbb{T}^n$. In the new variables

$$H = \langle \nu, Y\rangle + \langle \overline{A}Y, Y\rangle/2 + \overline{C} Z_u Z_s + O_2(Y, Z).$$

This implies reducibility of N for $l = 1$. The reducibility in the case $n = 1$ follows from the Floquet theorem.

2.6 Applications of the KAM Theory

1. The Kolmogorov theorem implies the existence of a large set Q_ε of quasiperiodic motions in near-integrable systems. Since the measure of this set is positive, the systems cannot be ergodic neither in the whole phase space nor on a non-singular energy level. Let $\mathcal{N}_\varepsilon$ be the subset of the phase space complementary to Q_ε. It is interesting to establish an asymptotic estimate for the measure of the set $\mathcal{N}_\varepsilon$ for $\varepsilon \to 0$.

Let D be an open domain in the phase space such that its closure $\overline{D}$ is compact and any point of the set $\overline{D}$ lies on an invariant torus of the unperturbed system. Suppose that the unperturbed system is non-degenerate in D. Then according to [94, 107] the measure of the set $D \cap \mathcal{N}_\varepsilon$ is $O(\sqrt{\varepsilon})$.

If we do not include into $\mathcal{N}_\varepsilon$ only points lying on tori existing according to the ordinary Kolmogorov theorem then, in general, the measure of $D \cap \mathcal{N}_\varepsilon$ has order not less than $O(\sqrt{\varepsilon})$. This estimate follows from the fact that the perturbation generates "holes" of area $\sim\sqrt{\varepsilon}$ in the vicinity of resonant unperturbed tori (see Figs. 2.1–2.2, where about 180 trajectories of the Standard Chirikov Map (1.5), Chap. 1, are presented for $\varepsilon = 2\pi \cdot 0.1$ and for $\varepsilon = 2\pi \cdot 0.14$. As usual, the coordinate y is vertical and x is horizontal).

However, it is easy to see that these "holes" in turn are filled with invariant tori rather densely. Taking this phenomenon into account, one can be quite sure that in the case of 2 degrees of freedom the measure of the set $D \cap \mathcal{N}_\varepsilon$ is exponentially small in ε, but as far as we know this statement has not been proved yet. In the case

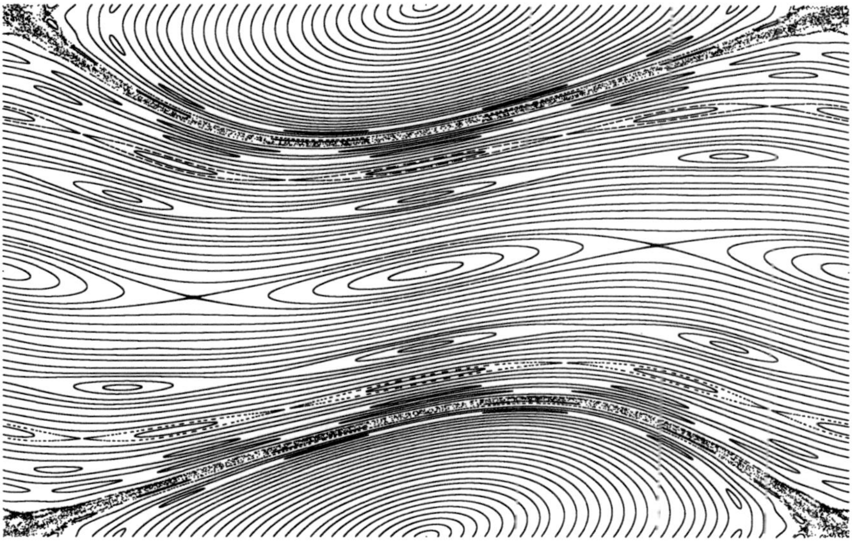

Fig. 2.1 Phase portrait of the standard map for $\varepsilon = 2\pi \cdot 0.1$.

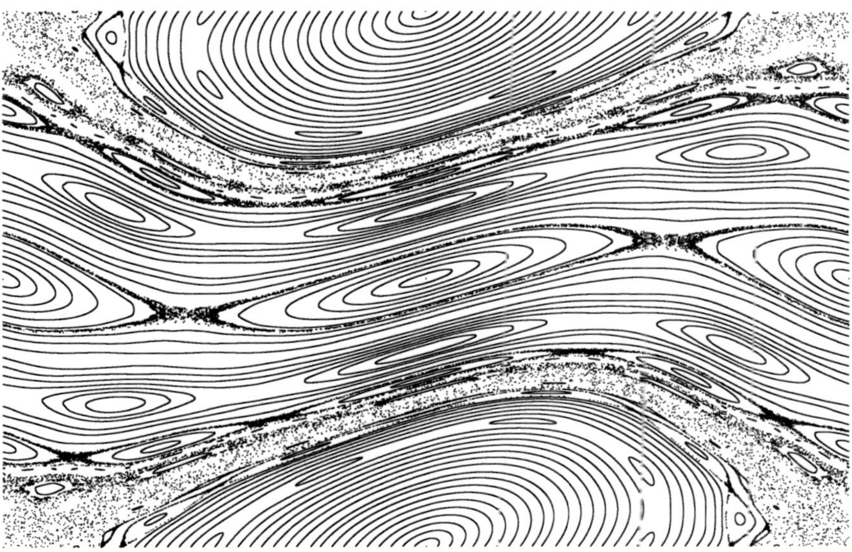

Fig. 2.2 Phase portrait of the standard map for $\varepsilon = 2\pi \cdot 0.14$.

of 3 and more degrees of freedom the estimate of the measure must be polynomial in ε because of the influence of multiple resonances.

As is well-known, in near-integrable systems chaos develops essentially in the vicinity of separatrices of the unperturbed system. In particular, if the domain D has a nonempty intersection with asymptotic manifolds of the unperturbed system, the measure of the set $D \cap \mathcal{N}_\varepsilon$ in general is expected to have the order $\varepsilon|\log\varepsilon|$ (see Chap. 5).

2. In 1994 the first author learned from V.I. Arnold about the following problem. Consider a continuous family of invariant curves of an integrable two-dimensional exact symplectic map, and a resonant curve from this family. Take a perturbed map, which is also exact symplectic. Generically the resonant curve is destroyed, and in its neighborhood a domain of chaotic motions—the so called stochastic layer—appears. The boundary of the stochastic layer contains a pair of nonresonant curves which appear as a result of a deformation of curves from the initial family. What is the difference of frequencies on these boundary curves if the perturbation has the order $0 < \varepsilon \ll 1$?

The answer turns out to be not trivial, although it can be obtained as a combination of standard facts (see Sect. 9.3. If the system is analytic, the difference of the frequencies is of order ε. This result can not be obtained directly from the analysis of the Taylor expansion of the map in the perturbing parameter. It is well known that the boundary invariant curves we discuss go essentially on the distance $\sim\sqrt{\varepsilon}$ from each other, but in some places this distance is exponentially small. Formally speaking, the answer is obtained as a result of the following calculation: $\sqrt{\varepsilon}/\log(e^{-c/\sqrt{\varepsilon}}) \sim \varepsilon$. If the system has a finite order of smoothness, the frequency difference is of order $\sqrt{\varepsilon}/\log\varepsilon$, because the argument of the logarithm above should be replaced by ε^N, where N is the order of smoothness.

3. The KAM theory gave a tool to prove the Lyapunov stability for typical elliptic periodic solutions in autonomous Hamiltonian systems with two degrees of freedom.[8] Indeed, let us reduce the order of such a system on the energy level M_h in a neighborhood of an elliptic periodic trajectory γ. Passing on to the (linear) normal form, we obtain the nonautonomous system with Hamiltonian

$$H(x, y, t) = \mu(x^2 + y^2)/2 + O_3(x, y), \tag{2.25}$$

where x and y are real canonically conjugate variables. The function (2.25) is 2π-periodic in time t, and the constant μ is positive and distinct from $2\pi k$ for any integer k. The trajectory γ has the form $\{(x, y, t \bmod 2\pi) : x = y = 0\}$.

In a neighborhood of the curve γ the term O_3 in the Hamiltonian H can be regarded as a small perturbation. Degeneracy of the unperturbed integrable linear system with Hamiltonian $\mu(x^2 + y^2)/2$ can be removed by the normalization in H of the third and fourth order terms. More precisely, suppose that the following conditions hold:

$$\mu \neq 2\pi n/3, \quad \mu \neq \pi k/2, \qquad n, k \in \mathbb{Z}. \tag{2.26}$$

Then by the Birkhoff transformation Hamiltonian (2.25) can be reduced to the form

$$H(x, y, t) = \mu(x^2 + y^2)/2 + \mu_*(x^2 + y^2)^2 + O_5(x, y). \tag{2.27}$$

[8] Recall that a periodic solution of a Hamiltonian system is called elliptic if all its multipliers are not real and lie on the unit circle.

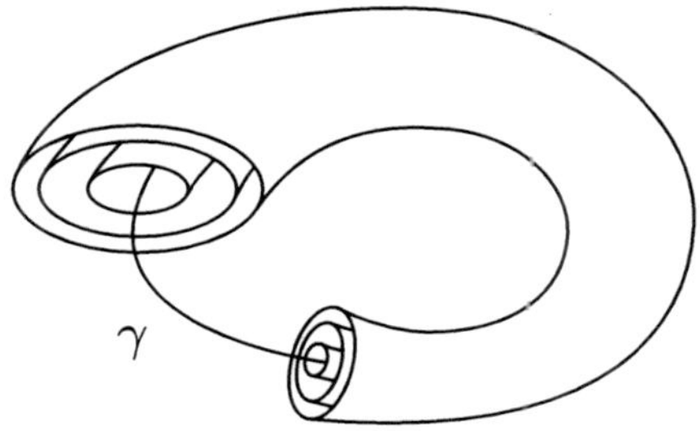

Fig. 2.3 Invariant tori surrounding the curve γ on M_h.

Here μ_* is a constant and the new canonical variables are again denoted by x, y. Now we can consider the system with the Hamiltonian $\mu(x^2+y^2)/2+\mu_*(x^2+y^2)^2$ as the unperturbed integrable system. In the case $\mu_* \neq 0$ we have nondegeneracy for small values of $(x^2 + y^2)$.

Note that the existence of a large number of invariant tori in the system with Hamiltonian (2.27) does not follow directly from theorems formulated in previous sections. However, by using the usual methods of the KAM theory, it is possible to prove that for arbitrary small $r > 0$ there are two-dimensional invariant tori of the form

$$\mathbb{T}_r^2 = \{(x, y, t \bmod 2\pi) : x^2 + y^2 + O_5(x, y) = r^2\}.$$

Now return to the original system with two degrees of freedom. The tori $\mathbb{T}_r^2$, like the periodic solution γ, lie on the energy level M_h. Each torus divides the three-dimensional manifold M_h into two invariant sets: the interior of the solid torus (containing, in particular, the curve γ) and its exterior (see Fig. 2.3). Since for $r \to 0$ the tori $\mathbb{T}_r^2$ come arbitrary close to $\gamma = \gamma_h$, the periodic solution γ is orbitally Lyapunov stable on the energy level M_h.

Since the solution γ is nondegenerate, on neighboring energy levels $M_{h'}$ the picture is analogous: the periodic solutions $\gamma_{h'}$ close to γ are surrounded by invariant tori. This implies the orbital stability of the solution γ for the full system.

Note that, if either the nonresonant conditions (2.26) do not hold or $\mu_* = 0$, the solution γ can be unstable (see [11, 86]).

An analogous idea can be used to prove the Lyapunov stability of elliptic equilibrium positions in Hamiltonian systems with two degrees of freedom[9] [7].

4. Consider the problem of the evolution of the action variables in near-integrable Hamiltonian systems, known also as the problem of Arnold diffusion. Discussion of the Arnold diffusion as a dynamical phenomenon was initiated by the famous paper [8]. However, by the present time the term has no precise meaning. It expresses the general idea that, in Hamiltonian systems with more than 2 degrees of freedom trajectories have no generic obstacles to travel in the phase space except (in the autonomous case) the energy conservation.

[9] In this case the problem of the stability is nontrivial only when the equilibrium is not a strict extremum of H, in other words if the Hamiltonian is not a Lyapunov function.

According to [60] the Arnold conjecture about the transitivity[10] of generic multidimensional Hamiltonian systems on compact connected components of generic energy levels is false. However, in these examples Hamiltonians are not presented as a sum of kinetic and potential energies and can not be regarded as perturbations of Liouville integrable ones. Hence, the situation is still unclear in the "physical" case.

Usually the Arnold diffusion is discussed for real-analytic near-integrable systems.[11] Here one should distinguish various settings:

(1) the unperturbed Hamiltonian depends only on the action variables,
(2) the unperturbed system contains a family of hyperbolic tori,[12]
(3) the problem usually becomes much simpler if the system contains two or more (may be, dependent) small parameters.

In [31] cases (1) and (2) are called *a priori stable* and *a priori unstable* respectively. A system of the third type can be regarded as either a priori stable or a priori unstable depending on what system is called unperturbed.

The existence of the Arnold diffusion in the strong sense, i.e., transitivity on an energy level, has not been proved in any example of type (1)–(3). However, there are some results [8, 14, 85] where an evolution of a slow (action) variable has been established. All these results concern systems of type (3).

The mechanism of the Arnold diffusion contains both local and global aspects. The local aspect means that a neighborhood of a hyperbolic torus and of some (not small) pieces of the corresponding asymptotic surfaces should be considered. The splitting of these surfaces as a rule generates chaotic dynamics in the neighborhood. The global aspect means that one needs to join different neighborhoods into a transition chain which takes a diffusion trajectory far from the initial state.

Trying to realize this (in fact, Arnold's) program, one meets problems of a different kind. On the local stage the problem of splitting the asymptotic surfaces appears. For systems of type (2) the splitting is given by the standard Poincaré–Melnikov method (see Sect. 3.3). In systems of the first type hyperbolic tori appear only after the perturbation. The splitting is exponentially small with respect to the perturbation rate, and the standard technics fail. To get round this difficulty, an additional small parameter is frequently introduced which moves the system to class (3). Usually after this the set up becomes less natural. There are several attempts to estimate how large the additional small parameter can be in the splitting problem, where the splitting can be analyzed [35, 85]. It was found that if the additional small parameter does not exceed a positive power of the main one, the naive application of the Poincaré–Melnikov theory usually gives correct asymptotics of the splitting. The problem of asymptotic surface splitting in the multi-frequency case looks hopeless if the naive Poincaré–Melnikov integral does not give the correct asymptotics of

[10] A dynamical system is called transitive if it has a dense trajectory.

[11] The smooth case is simpler, see [41].

[12] A typical example of such an (unperturbed) system is a direct product of a system of type (1) with a pendulum. Hyperbolic tori are generated by the hyperbolic equilibrium position of the pendulum.

the splitting. Here we mean that, apparently, in this case the asymptotics can not be obtained in a form suitable for further application.

Another (simpler) local problem is to find a trajectory passing through a neighborhood of asymptotic surfaces. Usually this is almost equivalent to constructing a hyperbolic invariant set in this neighborhood.

Concerning the global aspect, the problem of gaps appears. It is connected with the well-known fact that hyperbolic tori generically form a Cantor set on an energy level. So, if between two "neighboring" tori there is a large gap, it is difficult to pass from a neighborhood of one torus to a neighborhood of the other. In other words, it is difficult to establish a heteroclinic connection. Some ideas and results on overcoming the large gap problem are presented in [37].

There are several attempts to avoid the cumbersome KAM technique in the problem of diffusion. A variational approach was proposed in [29], where the genericity of the diffusion in a priori unstable systems with two and a half degrees of freedom was established.

In [139–141] (see also [103]) the approach based on the method of the separatrix map was used. The main result of these papers is that the diffusion in a priori unstable systems with two and a half degrees of freedom is generic, and moreover there are orbits along which the average velocity of the slow variable drift is of order $\varepsilon/\log|\varepsilon|$. It can be proved that "faster orbits" do not exist.

There are still no results on genericity of the diffusion in systems with more than one slow variable even in the a priori unstable case. Here the result should be as follows.

For a generic system let γ be an arbitrary smooth curve in the space of slow variables. Then there exists an orbit whose projection to this space goes in a small tubular neighborhood of γ with average velocity along γ of order $\varepsilon/\log|\varepsilon|$.

For a priori stable systems there are estimates from above for the maximal velocity of the diffusion. In real-analytic systems the estimate is exponentially small with respect to the rate of the perturbation [80, 96, 109].

5. J. Mather suggested a general variational method for constructing connecting orbits for invariant sets of positive definite time periodic Hamiltonian systems [89]. He conjectured that the following statement is true.

Let M be a compact Riemannian manifold with Riemannian metric $\|\cdot\|$. Consider a classical Hamiltonian system with smooth time-periodic Hamiltonian

$$H(q,p,t) = \frac{1}{2}\|p\|^2 + V(q,t) \tag{2.28}$$

on $T^*M\times\mathbb{T}$. Then, for a generic potential V, there exists a trajectory with $\|p(t)\| \to \infty$ as $t \to \infty$.

In an unpublished manuscript [89] Mather obtained the following result.

Let M be a two-dimensional torus. Then there exists a minimizing closed geodesic σ in an arbitrary simple homotopy class Γ of closed curves in M. If the minimizing geodesic in Γ is unique, it follows from the results of Morse [91] that σ

possesses a homoclinic[13] trajectory γ. Suppose that σ is nondegenerate, and the homoclinic trajectory γ is transversal.[14] There exist $a, b \in \mathbb{R}$ such that

$$\begin{aligned} &\operatorname{dist}(\gamma(t), \sigma(t+a)) \to 0 \quad \text{as } t \to -\infty, \quad \text{and} \\ &\operatorname{dist}(\gamma(t), \sigma(t+b)) \to 0 \quad \text{as } t \to \infty. \end{aligned}$$

The numbers a, b are defined mod l, where l is the period of the geodesic σ. Without loss of generality suppose that the average of V over σ vanishes: $\int_0^l V(\sigma(\tau), t)\, d\tau = 0$ for all t. Then the Poincaré function

$$I(t) = \lim_{T\to\infty} \left(\int_{-T}^{T} V(\gamma(\tau), t)\, d\tau - \int_{-T+a}^{T+b} V(\sigma(\tau), t)\, d\tau \right), \qquad t \in \mathbb{T}, \quad (2.29)$$

is well-defined, i.e., it doesn't depend on the choice of a, b. Mather proved that, if I is nonconstant, then there exists an orbit such that $\|p(t)\| \to \infty$ as $t \to \infty$. Note that for a generic metric and a generic potential all these conditions are satisfied.

The proof is of the variational nature and is based on a combination of the Peierls barrier method of [88] and the classical Poincaré–Melnikov method.

In [23] a multidimensional version of this result is presented. Hamiltonians of a more general form compared with (2.28) are studied, and the periodic orbit is replaced by a hyperbolic torus. Instead of the variational methods, a more traditional approach, based on KAM theory and the classical Poincaré–Melnikov method, is used. Analogous ideas were applied to the Mather problem in [36]. A proof, partially using the original Mather's variational ideas, is presented in [63]. In fact, there is nothing magic in these results: a small parameter naturally appears as a ratio of potential and kinetic energy.

There is an obvious upper estimate for the diffusion velocity. It is based on the equation $dH/dt = \partial H/\partial t \sim 1$. By using a multidimensional version of the separatrix map, Piftankin [102] (see also [103]) constructed orbits in the Mather problem in which the average velocity of the energy growth is of order 1.

2.7 Perturbations of a Quasi-Periodic Flow on a Torus

In this and in the next sections we present some technical details of the traditional KAM method. The problem to be considered on the reduction of a flow on a torus is taken from [5]. Choosing this model problem we intended to eliminate technicalities from the procedure of quadratic convergence. It is this procedure that forms a core of the traditional KAM method. We hope that a reader who understands the proof of Theorem 2.10 will not have big problems in an analysis of other KAM-type theorems performed by the method of quadratic convergence.

---

[13] I.e., doubly asymptotic to σ.

[14] The last assumption means that stable and unstable asymptotic manifolds of σ intersect along γ transversely.

Consider the ordinary differential equation

$$\dot{x} = \nu + \varepsilon(f(x, \varepsilon) + \lambda), \tag{2.30}$$

where $x = (x_1, \ldots, x_m) \bmod 2\pi$ is a point of the m-dimensional torus $\mathbb{T}^m$, the function f is periodic in x, the vectors $\nu, \lambda \in \mathbb{R}^m$ are constant, and ε is as usual a small parameter. The problem is to reduce system (2.30) to the form

$$\dot{\xi} = \nu \tag{2.31}$$

by the change of the variables

$$x = \xi + \varepsilon Q(\xi, \varepsilon) \tag{2.32}$$

where Q is periodic in ξ.

First, let us clear up the role of the parameters λ. It is easy to see that without these parameters the problem, as a rule, has no solution. Indeed, let us put $\lambda \equiv 0$ and $f = \text{const} \neq 0$. Then for small ε systems (2.30) and (2.31) can not be transformed to each other because of the difference in the topological structure of their solutions.[15] Therefore, to have a positive solution for the problem we have to seek a change of variables (2.32) together with the function $\lambda(\varepsilon)$, for which this change exists.[16]

Recall that the frequencies ν are called Diophantine if

$$|\langle k, \nu\rangle| \geq \alpha \|k\|^{-\gamma} \quad \text{for all } 0 \neq k \in \mathbb{Z}^m. \tag{2.33}$$

Theorem 2.10. *Suppose that the frequency vector* ν *is Diophantine and the function* f *is analytic in* x *and* ε. *Then there exist analytic functions* $\lambda(\varepsilon)$, $Q(\xi, \varepsilon)$ *such that for small* ε *the system* (2.30)$|_{\lambda=\lambda(\varepsilon)}$ *is transformed to* (2.31) *by the change* (2.32).

The change of variables (2.32) is constructed below as a composition of an infinite number of changes. After each change the system approaches closer to (2.31). It is possible to organize the procedure so that at each step the perturbation (an analog of the function f obtained at a given step of the procedure) has a rate which is, roughly speaking, the square of the previous perturbation rate. This provides the convergence of the composition.

The quadratic convergence is usually associated with the Newton method for solving the equation $F(y) = 0$, where $F : D \to \mathbb{R}^m$ is a smooth function, $D \subset \mathbb{R}^m$. Recall the construction of the Newton method. Let y_* be the solution we search for, where the Jacobi matrix $\partial F/\partial y$ is supposed to be nondegenerate at the point y_*. Suppose also that the solution y_* is known approximately, i.e., for a given $y = y_0$ the approximate equality $y_0 \approx y_*$ holds. Then consider the sequence

---

[15] For example, if the vectors ν and f are not parallel, for arbitrary small ε the groups of resonances for the frequency vectors ν and $\nu + \varepsilon f$ have different ranks.

[16] Usually in KAM-type theorems, analogs of the parameters λ can be found among inner parameters of the system. For example, in the ordinary Kolmogorov theorem, λ appear when one uses the possibility to shift the action variables $y \mapsto y + \varepsilon\lambda$.

$$y_0, y_1, \dots \tag{2.34}$$

such that for any nonnegative $j \in \mathbb{Z}$

$$y_{j+1} = y_j - \left(\frac{\partial F}{\partial y}(y_j)\right)^{-1} F(y_j).$$

Simple estimates show that the sequence (2.34) converges to y_* and moreover, for some constant C, the following inequalities hold:

$$|y_{j+1} - y_*| < C|y_j - y_*|^2, \qquad j = 0, 1, \dots.$$

The convergence speed for this procedure is quadratic. It is considerably faster than (an exponential) convergence speed of the contraction method.

Now we show how it is possible to provide fast convergence in the problem of the reduction of a flow on a torus. As we have mentioned above, the change of variables (2.32) $x \mapsto \xi$ is constructed as a limit of the composition

$$x \equiv x_0 \mapsto x_1 \mapsto x_2 \mapsto \cdots, \qquad \xi = x_\infty.$$

Below it is convenient to present the function $\lambda(\varepsilon)$ as the sum

$$\lambda = \sum_{j=0}^{\infty} \lambda_j, \qquad \lambda_j = \lambda_j(\varepsilon).$$

We will regard λ_j as the piece of the function $\lambda(\varepsilon)$, used for the j-th change of variables. We put

$$f = f_0, \quad \hat{\lambda}_n = \sum_{j=n}^{\infty} \lambda_j, \qquad n = 0, 1, \dots.$$

Suppose that after n changes system (2.30) takes the form

$$\dot{y} = \nu + \varepsilon\hat{\lambda}_n + \varepsilon f_n(y, \varepsilon, \hat{\lambda}_n), \tag{2.35}$$

where $y \equiv x_n$ are new variables. The $(n+1)$-st change is constructed as follows:

$$y = \eta + \varepsilon q_n(\eta, \varepsilon, \hat{\lambda}_n). \tag{2.36}$$

The function q_n will be defined below. The change (2.36) transforms (2.35) in the following way:

$$\dot{\eta} + \varepsilon \sum_{j=1}^{m} (q_n)_{\eta_j} \dot{\eta}_j = \nu + \varepsilon\hat{\lambda}_n + \varepsilon f_n(\eta + \varepsilon q_n, \varepsilon, \hat{\lambda}_n). \tag{2.37}$$

Here the symbols $(q_n)_{\eta_j}$ denote the partial derivatives $\partial q_n/\partial \eta_j$. We want the equation (2.37) to be of the form $\dot{\eta} = \nu$. Hence, we should take q_n and $\hat{\lambda}_n$ satisfying the equation

$$\partial_\nu q_n = \hat{\lambda}_n + f_n(\eta + \varepsilon q_n, \varepsilon, \hat{\lambda}_n), \tag{2.38}$$

where the differential operator ∂_ν is as follows:

$$\partial_\nu = \langle \nu, \partial/\partial\eta \rangle = \sum_{j=1}^{m} \nu_j \partial/\partial \eta_j.$$

Equation (2.38) is nonlinear and can not be solved at once.[17] Hence, consider instead of it the simpler one:

$$\partial_\nu q_n = \lambda_n + f_n(\eta, \varepsilon, \lambda_n). \tag{2.39}$$

This equation is called *homologic*. Before we solve it and pass to rigorous estimates, we show on an informal level that the sequence of the variable changes we construct really converges quadratically.

Indeed, since $\hat{\lambda}_n = \hat{\lambda}_{n+1} + \lambda_n$ then as a result of the change (2.36) and (2.39), the system (2.37) takes the form

$$\begin{aligned} \dot{\eta} &= (I + \varepsilon (q_n)_\eta)^{-1} (\nu + \varepsilon \hat{\lambda}_n + \varepsilon f_n(\eta + \varepsilon q_n, \varepsilon, \hat{\lambda}_n)) \\ &= \nu + \varepsilon \hat{\lambda}_{n+1} + \varepsilon f_{n+1}(\eta, \varepsilon, \hat{\lambda}_{n+1}), \end{aligned}$$

where I is the identity $m \times m$ matrix and

$$\begin{aligned} f_{n+1}(\eta, \varepsilon, \hat{\lambda}_{n+1}) &= (I + \varepsilon (q_n)_\eta)^{-1} \\ &\quad \times (-\varepsilon (q_n)_\eta \hat{\lambda}_{n+1} + f_n(\eta + \varepsilon q_n, \varepsilon, \hat{\lambda}_n) - f_n(\eta, \varepsilon, \lambda_n)). \end{aligned} \tag{2.40}$$

Let us estimate (roughly) the function f_{n+1}. Since q_n and λ_n are solutions of the linear equation (2.39), we can expect that q_n, λ_n, and f_n have the same order of smallness: $q_n \sim f_n$, $\lambda_n \sim f_n$. Since the functions f_n are bounded (and moreover are small beginning from $n = 1$) then $(I + \varepsilon (q_n)_\eta)^{-1} \approx I$. Therefore $\hat{\lambda}_{n+1}$ is of the order $|\lambda_{n+1}| < |\lambda_n|$ and $(q_n)_\eta \hat{\lambda}_{n+1} \sim f_n^2$. We also have:

$$f_n(\eta + \varepsilon q_n, \varepsilon, \hat{\lambda}_n) - f_n(\eta, \varepsilon, \lambda_n) \sim \varepsilon (f_n)_\eta q_n + (f_n)_\lambda \hat{\lambda}_{n+1} \sim f_n^2.$$

Thus, $f_{n+1} \sim f_n^2$, which means that we have quadratic convergence.

2.8 Proof of the Theorem on the Reduction of a Flow

In this section we establish the convergence of the procedure which reduces the system (2.30) to the form (2.31). To this end we show that the sequences $f_n(\eta, \varepsilon, 0)$,

[17] Solving equation (2.38) is equivalent to the original problem of reduction of the flow on a torus.

q_n and $\hat{\lambda}_n$ rapidly converge to zero. Actually the convergence of these sequences is not quadratic as was promised above. However, it is very close to quadratic in the sense that in each of these sequences the norm of a term does not exceed a constant multiplied by the norm of the previous term to a power κ, where the quantity $\kappa \in (1, 2)$ is arbitrary. Below we take $\kappa = 3/2$.

The proof uses essentially the analyticity of the original system. To make estimates we need some notation. Let $D_\rho \subset \mathbb{C}^m$ be the following domain:

$$D_\rho = \{\eta \in \mathbb{C}^m : |\mathrm{Im}\, \eta_j| \leq \rho, \ j = 1, \ldots, m\}.$$

Let $|\cdot|$ be the modulus of a scalar. In the case of a vector argument $v = (v_1, \ldots, v_m)$ we put $|v| = \sum_j |v_j|$. If the argument of $|\cdot|$ is a matrix $A = (a_{jk})$, we assume that $|A| = \max_{j,k} |a_{jk}|$. These conventions are not really important. They are accepted only for convenience. Note that for any vector $v \in \mathbb{R}^m$ and any $(m \times m)$ matrix A

$$|Av| \leq m|A||v|.$$

For any scalar, vector, or matrix function $\varphi(\eta)$ analytic and 2π-periodic in D_ρ, we put

$$\|\varphi\|_\rho = \max_{D_\rho} |\varphi|.$$

Below we will have to estimate the norm $\|\cdot\|_\rho$ of the derivative of an analytic function in terms of the norm of the function itself. The main tool for this is the following lemma.

Lemma 2.1 (The Cauchy estimate). *Let the function $g(z)$ be analytic in the ball $B_s \subset \mathbb{C}$ of the radius $s > 0$:*

$$g : B_s \to \mathbb{C}, \qquad B_s = \{z \in \mathbb{C} : |z| \leq s\}$$

and let $|g| \leq G$ in B_s. Then for any positive $u < s$ and any $l \in \mathbb{N}$ the derivative $g^{(l)} = d^l g/dz^l$ satisfies the estimate

$$|g^{(l)}(z)| \leq l!G/u^l \quad \textit{for any } z \in B_{s-u}.$$

Corollary 2.2. *For any $z_0 \in B_{s-u}$ consider the Taylor expansion of the function g: $g(z) = \sum_{l=0}^{\infty} g_l (z - z_0)^l$. Then the coefficients g_l satisfy the inequality $|g_l| < G/u^l$.*

The proof of the lemma is based on the Cauchy integral formula. For any $z \in B_{s-u}$

$$g^{(l)}(z) = \frac{l!}{2\pi i} \int_\gamma \frac{g(\zeta)\, d\zeta}{(\zeta - z)^{l+1}},$$

where $g^{(l)} = d^l g/dz^l$ and $\gamma = \{\zeta \in \mathbb{C} : |\zeta - z| = u\}$. We have the estimate

$$|g^{(l)}(z)| \leq \frac{l!}{2\pi} \int_\gamma \frac{G\, d\zeta}{u^{l+1}} \leq l! \frac{G}{u^l}.$$

Note that Lemma 2.1 and its proof remain true also in the case of vector- and matrix-valued functions g.

The following assertion allows us to estimate Fourier coefficients of an analytic function in terms of the norm of the function in the domain D_ρ.

Lemma 2.2. *Let the function $\varphi : D_\rho \to \mathbb{C}$ be analytic, 2π-periodic, and let $\|\varphi\|_\rho = M$. Suppose that its Fourier expansion has the form $\varphi(\eta) = \sum_{k\in\mathbb{Z}^m} \varphi_k e^{i\langle k,\eta\rangle}$. Then*

$$|\varphi_k| \le e^{-|k|\rho} M, \qquad |k| = \sum_{j=1}^{m} |k_j|.$$

Proof. Let us put $\sigma = (\sigma_1, \ldots, \sigma_m)^T$, $\sigma_j = \operatorname{sign} k_j$. We have the equations

$$\varphi_k = \frac{1}{(2\pi)^m} \int_{\mathbb{T}^m} e^{-i\langle k,x\rangle} \varphi(x) dx = \frac{1}{(2\pi)^m} \int_{\mathbb{T}^m} e^{-i\langle k,x-i\sigma\rho\rangle} \varphi(x - i\sigma\rho)\, dx.$$

Since $|\varphi|$ does not exceed M and $\langle k, \sigma\rangle = |k|$, we get $|\varphi_k| \le e^{-|k|\rho} M$. ◻

Now we turn to an analysis of the sequence of changes (2.36). First, note that the functions f_n are linear in $\hat\lambda_n$. This fact can be easily proved by induction. Indeed, the function $f_0 = f$ does not depend on $\hat\lambda_0 = \lambda$. Suppose that f_n depends on $\hat\lambda_n$ linearly. Equation (2.39) does not contain $\hat\lambda_{n+1}$. Hence, the function q_n does not depend on $\hat\lambda_{n+1}$. According to the induction assumption, equation (2.40) generates the function f_{n+1} which depends on $\hat\lambda_{n+1}$ linearly. Note that linearity of the functions f_n with respect to $\hat\lambda_n$ is not very important. It only makes estimates slightly simpler.

Some properties of solutions of the homologic equation are presented in the following lemma.

Lemma 2.3. *Suppose that $f_n(\eta, \varepsilon, \lambda) = f'(\eta, \varepsilon) + f''(\eta, \varepsilon)\lambda$, where $\|f'\|_\rho = \delta'$, $\|f''\|_\rho = \delta'' < a$, and a is a small positive constant.*[18] *Then there exists a solution of* (2.39) *satisfying the estimates*

$$|\lambda_n| \le 2m\delta', \qquad \|q_n\|_{\rho-\sigma} \le \frac{c\delta'}{\alpha\sigma^{\gamma+m}},$$

where the constant c depends only on the dimension m.

Proof. Let us drop the subscript n in q_n and λ_n for brevity. We put

$$f' = \sum_{k\in\mathbb{Z}^m} f'_{(k)} e^{i\langle k,\eta\rangle}, \quad f'' = \sum_{k\in\mathbb{Z}^m} f''_{(k)} e^{i\langle k,\eta\rangle},$$
$$q = \sum_{k\in\mathbb{Z}^m} q_{(k)} e^{i\langle k,\eta\rangle}, \quad q_{(0)} = 0.$$

[18] Not exceeding $1/(2m^2)$ and such that for any $(m \times m)$ matrix A with norm $|A| \le a$ the inequality $|(I + A)^{-1}| < 2$ holds.

Then by using (2.39) we get $\lambda - (I + f''_{(0)})^{-1} f'_{(0)} = 0$. Since the constant a is small, we can assume that $|(I + f''_{(0)})^{-1}| < 2$. Therefore,

$$|\lambda| = |(I + f''_{(0)})^{-1} f'_{(0)}| \leq 2m\delta'.$$

The Fourier coefficients $q_{(k)}$ can be obtained as follows:

$$q_{(k)} = \frac{f'_{(k)} + f''_{(k)}\lambda}{i\langle \nu, k\rangle}.$$

Since the constant a is small, we have $2m^2\delta'' \leq 1$. By using Lemma 2.2 and the Diophantine conditions (2.33), we obtain the estimate

$$|q_{(k)}| \leq e^{-|k|\rho} \frac{\delta' + 2m\delta''\delta'}{|\langle \nu, k\rangle|} \leq e^{-|k|\rho} \frac{2\delta'|k|^{\gamma}}{\alpha}.$$

Suppose that $\eta \in D_{\rho-\sigma}$. Then

$$|q(\eta, \varepsilon)| \leq \left| \sum_{k\neq 0} \frac{2\delta'|k|^{\gamma}}{\alpha} e^{-|k|\rho} e^{i\langle k, \eta\rangle} \right| \leq \sum_{k\neq 0} \frac{2\delta'|k|^{\gamma}}{\alpha} e^{-\sigma|k|}.$$

The last sum can be estimated with the help of an integral: it does not exceed

$$2^m \int_{\mathbb{R}^m} \frac{2\delta'}{\alpha} |k|^{\gamma} e^{-\sigma|k|}\, dk \leq \frac{c\delta'}{\alpha\sigma^{\gamma+m}}.$$

for some $c = c(m)$. □

The passage from the n-th to $n+1$-th step is described by the following lemma.

Lemma 2.4 (Inductive Lemma). *Suppose that*

$$f_n(x, \varepsilon, \hat{\lambda}_n) = f'_n(x, \varepsilon) + f''_n(x, \varepsilon)\hat{\lambda}_n, \qquad \|f'_n\|_{\rho} = \delta', \quad \|f''_n\|_{\rho} = \delta'' < a.$$

Assume also that for some small constant $b < 1$, depending only on m,

$$\frac{\varepsilon c\delta'}{\alpha\sigma^{\gamma+m+1}} < b. \tag{2.41}$$

Then $f_{n+1}(\eta, \varepsilon, \hat{\lambda}_{n+1}) = f'_{n+1}(\eta, \varepsilon) + f''_{n+1}(\eta, \varepsilon)\hat{\lambda}_{n+1}$ satisfies the estimates

$$\|f'_{n+1}\|_{\rho-2\sigma} \leq \frac{2m^2c}{\alpha} \frac{\varepsilon(\delta')^2}{\sigma^{\gamma+m+1}}, \quad \|f''_{n+1}\|_{\rho-2\sigma} \leq \delta'' + \frac{4m^2c}{\alpha} \frac{\varepsilon\delta'}{\sigma^{\gamma+m+1}}.$$

Proof. The functions f'_{n+1}, f''_{n+1} can be obtained from (2.40):

$$\begin{aligned} f'_{n+1} &= (I + \varepsilon q_{\eta})^{-1}(f'_n(\eta + \varepsilon q, \varepsilon) - f'_n(\eta, \varepsilon)), \\ f''_{n+1} &= (I + \varepsilon q_{\eta})^{-1}(-\varepsilon q_{\eta} + f''_n(\eta + \varepsilon q, \varepsilon)). \end{aligned} \tag{2.42}$$

Here again for brevity we write q instead of q_n. In particular,

$$f''_{n+1}(\eta, \varepsilon) - f''_n(\eta + \varepsilon q, \varepsilon) = -(I + \varepsilon q_\eta)^{-1} \cdot \varepsilon q_\eta \cdot (I + f''_n(\eta + \varepsilon q, \varepsilon)). \quad (2.43)$$

According to Lemma 2.3 and inequality (2.41) we have

$$\|\varepsilon q\|_{\rho-\sigma} \leq \frac{\varepsilon c \delta'}{\alpha \sigma^{\gamma+m}} < b\sigma < \sigma.$$

Therefore, $\eta + \varepsilon q \in D_\rho$ for $\eta \in D_{\rho-\sigma}$.

By using the Cauchy estimate we obtain the inequality

$$\|\varepsilon q_\eta\|_{\rho-2\sigma} \leq \frac{\varepsilon c \delta'}{\alpha \sigma^{\gamma+m}} \frac{1}{\sigma} < b.$$

Hence, for sufficiently small b

$$\|(I + \varepsilon q_\eta)^{-1} - I\|_{\rho-2\sigma} \leq 1,$$

$$\|f'_n(\eta + \varepsilon q, \varepsilon) - f'_n(\eta, \varepsilon)\|_{\rho-2\sigma} \leq \|(f'_n)_\eta\|_{\rho-\sigma} \|\varepsilon q\|_{\rho-2\sigma} \leq m \frac{\delta'}{\sigma} \frac{\varepsilon c \delta'}{\alpha \sigma^{\gamma+m}}.$$

These inequalities and equations (2.42) imply the estimate

$$|f'_{n+1}|_{\rho-2\sigma} \leq 2m^2 \frac{\varepsilon c (\delta')^2}{\alpha \sigma^{\gamma+m+1}}.$$

By using (2.43) we estimate the function f''_{n+1}:

$$\|f''_{n+1}\|_{\rho-2\sigma} - \|f''_n\|_{\rho-2\sigma} \leq 2m^2 \frac{\varepsilon c \delta'}{\alpha \sigma^{\gamma+m+1}} (1 + a).$$

It remains to use the inequality $a < 1$. □

We see that the sequences δ'_n, δ''_n satisfy the inequalities

$$\delta'_{n+1} \leq \frac{c'}{\alpha} \frac{\varepsilon (\delta'_n)^2}{\sigma_n^{\gamma+m+1}}, \quad (2.44)$$

$$\delta''_{n+1} \leq \delta''_n + \frac{c''}{\alpha} \frac{\varepsilon \delta'_n}{\sigma_n^{\gamma+m+1}}, \quad (2.45)$$

where the constants c' and c'' depend only on m,

$$\delta'_0 = \|f\|_{\rho_0}, \quad \delta''_0 = 0, \quad \rho_{n+1} = \rho_n - \sigma_n.$$

Lemma 2.5. *Suppose that*

$$\sigma_n = \frac{\pi^2 \rho_0}{12(n+1)^2}.$$

Then for sufficiently small ε

$$\rho_n > \rho_0/2, \tag{2.46}$$

$$\delta'_n \le \delta'_0 \varepsilon^{(3/2)^n - 1}, \tag{2.47}$$

$$\delta''_{n+1} - \delta''_n \le 2^{-1-n} a. \tag{2.48}$$

Lemma 2.5 implies Theorem 2.10, since the estimates (2.46)–(2.48) guarantee the convergence of the infinite sequence of coordinate changes we construct. For $n \to \infty$ we have:

$$\delta'_n \to 0, \quad \delta''_n \to \delta''_\infty \le a, \quad \rho_n \to \rho_\infty \ge \rho_0/2.$$

In particular, the change $\xi = x_\infty \mapsto x_0$ is defined and real-analytic in the domain $\{|\mathrm{Im}\,\xi| \le \rho_0/2\}$.

Proof (of Lemma 2.5). Inequality (2.46) follows from definition of σ_n and from the equation $\rho_{n+1} = \rho_n - \sigma_n$.

To verify the estimates (2.47)–(2.48), we use induction in n. For $n = 0$ the inequalities obviously hold. Suppose that they are valid for $n = l$. By using (2.44) and the induction assumption, we get

$$\delta'_{l+1} \le \frac{c'}{\alpha} \varepsilon (\delta'_0)^2 \varepsilon^{2(3/2)^l - 2} \left(\frac{\pi^2 \rho_0}{12(l+1)^2} \right)^{-\gamma - m - 1}.$$

For small ε this expression does not exceed $\delta'_0 \varepsilon^{(3/2)^{l+1} - 1}$.

By using (2.45), we obtain

$$\delta''_{l+1} - \delta''_l \le \frac{c''}{\alpha} \delta'_0 \varepsilon^{(3/2)^l} \left(\frac{\pi^2 \rho_0}{12(l+1)^2} \right)^{-\gamma - m - 1}.$$

The last expression does not exceed $2^{-1-l} a$ for small ε. □

Chapter 3
Splitting of Asymptotic Manifolds

The behavior of manifolds which are asymptotic to equilibriums, to periodic solutions, or in general to hyperbolic tori determines many features of chaos in dynamical systems. In the present chapter we present the Poincaré–Melnikov theory of splitting of asymptotic manifolds (separatrices) in Hamiltonian systems with one and a half degrees of freedom and in two-dimensional symplectic maps. Then we discuss the multidimensional version of this theory.

3.1 Normal Coordinates

Consider the system with Hamiltonian $H(x, y, t)$, where $(x, y) \in D \subset \mathbb{R}^2$ are canonically conjugated variables and the function H is τ_0-periodic in time: $H(x, y, t) = H(x, y, t + \tau_0)$. At the time moment τ_0, let the solution with initial conditions $(x, y, 0)$ have the form $(T(x, y), \tau_0)$. Then T is called the first return map, the time-τ_0 map, or the Poincaré map. It is symplectic, i.e., it preserves the area form $dy \wedge dx$. Below we denote $z = (x, y) \in D$.

Let $\sigma(t) = (x(t), y(t))$ be a τ_0-periodic solution. Then $T(\sigma(0)) = \sigma(0)$. Stability properties of the fixed point $\sigma = \sigma(0)$ in the linear approximation are determined by the monodromy matrix $M = \partial T/\partial z|_{z=\sigma}$, because the linear part of the map T near σ is the multiplication by M. Eigenvalues of the monodromy matrix are called the multipliers of the solution $\sigma(t)$. Since T preserves the area, $\det M = 1$. Therefore, the characteristic polynomial for M has the form $\mu^2 - \mu \operatorname{tr} M + 1$.

There are four possibilities.

- $|\operatorname{tr} M| > 2$: the multipliers are real and distinct. The periodic solution is called *hyperbolic*.
- $|\operatorname{tr} M| < 2$: the multipliers are distinct and lie on the unit circle. The solution is called *elliptic*.
- $\operatorname{tr} M = -2$: the multipliers equal -1. The solution is called *parabolic*.
- $\operatorname{tr} M = 2$: the multipliers equal 1. The solution is called *degenerate*.

D. Treschev, O. Zubelevich, *Introduction to the Perturbation Theory of Hamiltonian Systems*, Springer Monographs in Mathematics,
DOI 10.1007/978-3-642-03028-4_3, © Springer-Verlag Berlin Heidelberg 2010

Proposition 3.1. *A small perturbation of the system (in the class of systems with Hamiltonians which are τ_0-periodic in t) does not destroy a nondegenerate solution. The solution is just slightly deformed.*

Formally, this statement is similar to a KAM-type theorem. Its proof is however much simpler because of the absence of small divisors. Indeed, suppose that the system depends on a parameter c and the unperturbed system is determined by the equation $c = 0$. The Poincaré map also depends on the parameter: $T = T_c(z)$. The initial condition $z = z_c$ for a τ_0-periodic solution should satisfy the equation $z - T_c(z) = 0$. This equation has the solution $(z, c) = (\sigma, 0)$. The Jacobi matrix

$$\left.\frac{\partial(z - T_c(z))}{\partial z}\right|_{(z,c)=(\sigma,0)} = I - M$$

is nondegenerate due to the nondegeneracy of the periodic solution $\sigma(t)$. Therefore, according to the implicit function theorem, the function z_c is defined for small $|c|$.

Below in this chapter we consider only hyperbolic periodic solutions. We assume that the corresponding multipliers are positive.[1] In the vicinity of such a solution there exist the so-called normal coordinates.

Theorem 3.1. *Let σ be a hyperbolic fixed point of a smooth symplectic map T. Then there are smooth symplectic coordinates (q, p) and a function $\mathcal{M}(qp)$ in a neighborhood of the point σ such that $\sigma = (0, 0)$ and T has the form*

$$(q, p) \mapsto (q\mathcal{M}(qp), p/\mathcal{M}(qp)), \tag{3.1}$$

where $\mu = \mathcal{M}(0) > 1$ and μ^{-1} are the multipliers of σ.

Consider a Hamiltonian system with one and a half degrees of freedom, the phase space $D \subset \mathbb{R}^2$ and the Hamiltonian $H(z, t)$, $z \in D$, $t \in \mathbb{T} = \mathbb{R}/(\tau_0\mathbb{Z})$. The following theorem is an analogue of Theorem 3.1.

Theorem 3.2. *Let $(\sigma(t), t)$ be a hyperbolic periodic solution of the Hamiltonian system with Hamiltonian H, and let $\sigma(0) = \sigma(\tau_0)$. Then in a neighborhood of the curve $(\sigma(t), t)$ on the extended phase space $D \times \mathbb{T}_t$ there are smooth symplectic coordinates (q, p, t) such that*

(1) *the time t coincides with the original time;*
(2) *the periodic solution is $(\sigma(t), t) = \{q = 0, p = 0, t\}$;*
(3) *the Hamiltonian depends only on the product qp: $\mathcal{H} = \mathcal{H}(pq)$.*

Remarks.

(1) The coordinates defined in Theorems 3.1–3.2 are said to be normal. The existence of a transformation to normal coordinates was established by Birkhoff on a formal level [16]. The convergence of the transformation was proved by

[1] The case of negative multipliers can be reduced to the previous one by considering $z(t)$ as a $2\tau_0$-periodic solution which implies the replacement of the multipliers by their squares.

Moser in the real-analytic case [92]. The case of finite smoothness was treated in [13].

(2) The smoothness of the normal coordinates and of the functions $\mathcal{M}$ depends linearly on the smoothness of T. It follows from [13] that, if T is of class $C^{\mathbf{r}}$, then the map defining the change of coordinates can be taken to be C^s-smooth, where $s \leq (\mathbf{r} - 3)/2$.

(3) The curves $\Gamma^s = \{(q, p) : q = 0\}$ and $\Gamma^u = \{(q, p) : p = 0\}$ are invariant for the system (3.1). Every solution located on Γ^s (respectively, on Γ^u) tends exponentially to σ as $n \to +\infty$ (respectively, as $n \to -\infty$). The asymptotic curves Γ^s and Γ^u are called the *separatrices*.

(4) In normal coordinates the Hamiltonian equations have the form

$$\dot{q} = \mathcal{H}'(qp)\, q, \quad \dot{p} = -\mathcal{H}'(qp)\, p, \qquad \mathcal{H}'(\rho) = d\mathcal{H}(\rho)/d\rho. \tag{3.2}$$

In particular, the function qp is a first integral and $e^{\pm\tau_0\lambda}$ $(\lambda = \mathcal{H}'(0))$ are multipliers of the periodic solution γ. We can assume that $\lambda > 0$. Indeed, $\lambda \neq 0$ due to the nondegeneracy of γ, and in the case $\lambda < 0$ we can perform the change $q \mapsto p$, $p \mapsto -q$, $\lambda \mapsto -\lambda$.
The surfaces $\Gamma^s = \{(q, p, t \bmod \tau_0) : q = 0\}$ and $\Gamma^u = \{(q, p, t \bmod \tau_0) : p = 0\}$ are invariant for the system (3.2). Any solution lying on Γ^s (respectively, on Γ^u) tends exponentially to γ when $t \to +\infty$ (respectively, when $t \to -\infty$). The asymptotic surfaces $\Gamma^{s,u}$ are called the *separatrices*.

(5) The normal coordinates (q, p) are not uniquely defined. For example, the coordinates $(q', p') = (-p, q)$ are also normal. Moreover, for any smooth function $r(\rho)$ $(r(0) \neq 0)$ the coordinates $q' = q/r(qp)$, $p' = pr(qp)$ are normal.

(6) If the map T depends (smoothly or analytically) on the parameter ε, then while the fixed point $\sigma = \sigma_\varepsilon$ remains hyperbolic, normal coordinates also can be regarded as depending (respectively, smoothly or analytically) on ε.

(7) Theorems 3.1 and 3.2 are equivalent to each other. Indeed, Theorem 3.2 can be reduced to Theorem 3.1 by using the Poincaré map, while Theorem 3.1 can be reduced to Theorem 3.2 by using inclusion of the map T into a Hamiltonian flow.

3.2 The Poincaré–Melnikov Method: Traditional Aspect

Consider a nonautonomous system with smooth Hamiltonian $H = H(x, y, t, \varepsilon)$, where the pair of canonically conjugate coordinates (x, y) determines a point of the two-dimensional domain D, and H is 2π-periodic in t. The domain D can be regarded as a subset of a plane. If the variable x is angular: $x = x \bmod 2\pi$, then D is a subset of the cylinder $\mathbb{R} \times \mathbb{T}$.

We assume that the unperturbed $(\varepsilon = 0)$ system is integrable. The integrability is understood in the sense of the existence of a smooth locally nonconstant first integral. For simplicity we assume that the unperturbed Hamiltonian does not depend explicitly on time. Then we can take as an integral the function H_0.

Let $\sigma = (x_0, y_0)$ be a hyperbolic equilibrium of the unperturbed system. The matrix determining the linearization of the system at the point σ has nonzero real eigenvalues $\pm\lambda$, $\lambda > 0$. In the extended phase space $D \times \mathbb{T} = \{x, y, t\}$, instead of the equilibrium σ, we have the 2π-periodic solution $\gamma_0 = \{x, y, t\} = \{x_0, y_0, t \bmod 2\pi\}$. Its multipliers are $e^{\pm 2\pi\lambda}$.

The hyperbolic periodic solution γ_0 generates the two-dimensional asymptotic surfaces (the separatrices)

$$\Gamma_0^{s,u} \subset \{(x, y) : H_0(x, y) = H_0(x_0, y_0)\} \times \mathbb{T}_t.$$

If the energy level $\{H_0 = H_0(x_0, y_0)\}$ is compact, the stable and unstable separatrices, as a rule, coincide: $\Gamma_0^s = \Gamma_0^u$.

According to Proposition 3.1, for small $\varepsilon \neq 0$ the system has a hyperbolic periodic solution γ_ε close to γ_0. However, the corresponding asymptotic surfaces $\Gamma_\varepsilon^{s,u}$, in general, no longer coincide because of nonintegrability of the perturbed system.

In this section we introduce the main tool for studying separatrix splitting in near-integrable systems, the Poincaré–Melnikov integral [90, 105]. Suppose that the set $(\Gamma_0^u \cap \Gamma_0^s) \setminus \gamma_0$ is nonempty. Let Γ_0 be one of its connected components. Then Γ_0 is a two-dimensional surface diffeomorphic to a cylinder. The surface Γ_0 is foliated by homoclinic[2] solutions of the form

$$(\hat{x}(t + \tau), \hat{y}(t + \tau), t \bmod 2\pi),$$

where τ is a parameter, specifying the solutions, and $(\hat{x}(t), \hat{y}(t))$ is a solution of the autonomous system with Hamiltonian H_0, $(\hat{x}(t), \hat{y}(t)) \to \sigma$ as $t \to \pm\infty$.

The perturbed separatrices[3] are foliated by solutions of the form

$$\begin{gathered} \left(\hat{x}_\varepsilon^{s,u}(t + \tau, \tau), \hat{y}_\varepsilon^{s,u}(t + \tau, \tau), t \bmod 2\pi\right), \\ \hat{x}_0^{s,u}(t, \tau) = \hat{x}(t), \quad \hat{y}_0^{s,u}(t, \tau) = \hat{y}(t), \end{gathered} \tag{3.3}$$

where t is again the time on a solution and τ is a parameter.

Solutions (3.3), corresponding to the index "u" (respectively, "s"), are asymptotic to γ_ε for $t \to -\infty$ (respectively, for $t \to +\infty$).

Consider the following differential operators on Γ_0:

$$\partial_{s,u} = \left(\frac{\partial}{\partial\varepsilon}\Big|_{\varepsilon=0} \hat{x}_\varepsilon^{s,u}\right)\frac{\partial}{\partial x} + \left(\frac{\partial}{\partial\varepsilon}\Big|_{\varepsilon=0} \hat{y}_\varepsilon^{s,u}\right)\frac{\partial}{\partial y}.$$

They associate with any smooth function in a neighborhood of Γ_0 a smooth function on Γ_0. The operators $\partial_{s,u}$ correspond to the differentiation along the vector fields that to the first approximation in ε, transform the surface Γ_0 to $\Gamma_\varepsilon^{s,u}$.

Putting $H = H_0(x, y) + \varepsilon H_1(x, y, t) + O(\varepsilon^2)$, we define the function

[2] I.e., doubly asymptotic to γ_0.

[3] More precisely, their branches appearing as a result of the perturbation of the surface Γ_0.

$$I(\tau) = \int_{-\infty}^{\infty} \{H_0, H_1\}(\hat{x}(t+\tau), \hat{y}(t+\tau), t)\, dt, \tag{3.4}$$

called the Poincaré–Melnikov integral. The integral converges rapidly due to the hyperbolicity of the periodic solution γ_0. The function I is 2π-periodic in τ. This can be easily checked with the help of the change of the variable $t = \tilde{t} - \tau$ in the integral.

Theorem 3.3. *For any real τ and ϑ*

$$(\partial_s H_0 - \partial_u H_0)(\hat{x}(\vartheta + \tau), \hat{y}(\vartheta + \tau), \vartheta) = I(\tau). \tag{3.5}$$

The left-hand side of (3.5) can be regarded as ε^{-1} multiplied by the first approximation of the distance between the perturbed separatrices in a neighborhood of the point $(\hat{x}(\vartheta + \tau), \hat{y}(\vartheta + \tau), \vartheta \bmod 2\pi)$. The distance is measured by the function H_0. Figure 3.1 presents a part of the section of the extended phase space $D \times \mathbb{T} = \{x, y, t\}$ by the plane $t = \vartheta = \text{const}$. The bold curves correspond to the separatrices Γ_0 and $\Gamma_\varepsilon^{s,u}$. The thin horizontal lines are level lines of the function H_0. The coordinates of the presented points are as follows:

$$A = (\hat{x}(\vartheta + \tau), \hat{y}(\vartheta + \tau)), \quad A_\varepsilon^{s,u} = (\hat{x}_\varepsilon^{s,u}(\vartheta + \tau, \tau), \hat{y}_\varepsilon^{s,u}(\vartheta + \tau, \tau)).$$

Hence, we have

$$\frac{1}{\varepsilon}(H_0(A_\varepsilon^s) - H_0(A_\varepsilon^u)) = (\partial_s H_0 - \partial_u H_0)(A, \vartheta) + O(\varepsilon) = I(\tau) + O(\varepsilon).$$

Here the first equation follows from the definition of the differential operators $\partial_{s,u}$, and the second one follows from (3.5).

Corollary 3.1. *If $I(\tau) \not\equiv 0$, the separatrices $\Gamma_\varepsilon^{s,u}$ split in the first order in ε.*

Proof (of Theorem 3.3). Let d/dt be the derivative along solutions of the perturbed system. Then

$$\frac{d}{dt} H_0(x, y, t, \varepsilon) = \{H, H_0\}(x, y, t, \varepsilon). \tag{3.6}$$

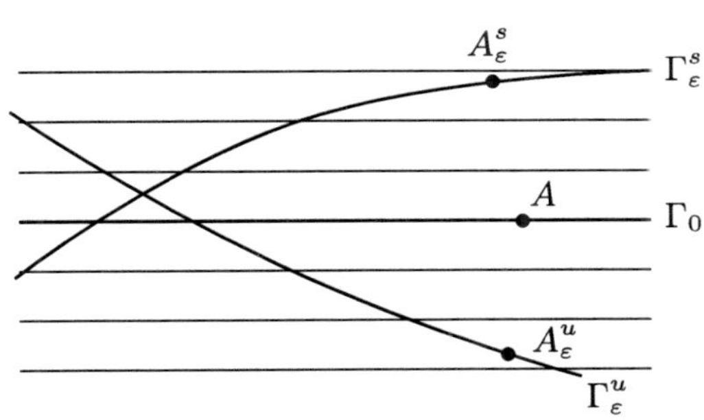

Fig. 3.1 Perturbed separatrices and level lines of the energy.

Let us take $(\hat{x}^u_\varepsilon(t+\tau,\tau), \hat{y}^u_\varepsilon(t+\tau,\tau), t)$ as arguments of the functions (3.6) and differentiate equation (3.6) with respect to ε at the point $\varepsilon = 0$:

$$\frac{d}{dt}\frac{d}{d\varepsilon}\bigg|_{\varepsilon=0} H_0(\hat{x}^u_\varepsilon(t+\tau,\tau), \hat{y}^u_\varepsilon(t+\tau,\tau), t) = \{H_1, H_0\}(\hat{x}(t+\tau), \hat{y}(t+\tau), t). \quad (3.7)$$

The left-hand side of (3.7) reads

$$\frac{d}{dt}(\partial_u H_0)(\hat{x}(t+\tau), \hat{y}(t+\tau), t).$$

Integrating (3.7) with respect to time, we get

$$(\partial_u H_0)(\hat{x}(\vartheta+\tau), \hat{y}(\vartheta+\tau), \vartheta) = -\int_{-\infty}^{\vartheta} \{H_0, H_1\}(\hat{x}(t+\tau), \hat{y}(t+\tau), t)\, dt. \quad (3.8)$$

Here we have used the equation

$$\lim_{t\to-\infty} \partial_u H_0(\hat{x}(t+\tau), \hat{y}(t+\tau), t) = 0,$$

which holds for any τ because

$$\lim_{t\to-\infty} (\hat{x}(t+\tau), \hat{y}(t+\tau)) = (x_0, y_0) \quad \text{and} \quad dH_0(x_0, y_0) = 0.$$

Analogously we have

$$(\partial_s H_0)(\hat{x}(\vartheta+\tau), \hat{y}(\vartheta+\tau), \vartheta) = \int_{\vartheta}^{+\infty} \{H_0, H_1\}(\hat{x}(t+\tau), \hat{y}(t+\tau), t)\, dt. \quad (3.9)$$

Combining equations (3.8) and (3.9), we obtain (3.5). □

Simple zeros of the function I correspond to homoclinic solutions of the perturbed system. More precisely, the following proposition is valid.

Proposition 3.2. *Suppose that $I(\tau_0) = 0$ and $\frac{dI}{d\tau}(\tau_0) \neq 0$. Then for small $\varepsilon \neq 0$ the system has a solution of the form*

$$\hat{\gamma}_{\varepsilon,\tau_0}(t) = (\hat{x}(t+\tau_0), \hat{y}(t+\tau_0), t \bmod 2\pi) + O(\varepsilon),$$

which is doubly asymptotic to γ_ε. Moreover, the surfaces Γ^u_ε and Γ^s_ε intersect along $\hat{\gamma}_{\varepsilon,\tau_0}$ transversely, and the angle between them at a point of the curve $\hat{\gamma}_{\varepsilon,\tau_0}$ is of order[4] ε.

[4] To define an angle between the separatrices, one should fix a metric. In the phase space $D \times \mathbb{T}^1$ there is no natural metric. Therefore, an angle between the separatrices is not a good measure of the splitting. Note also that the quantity (*an angle between* Γ^u_ε *and* Γ^s_ε)$/\varepsilon$ is not uniformly bounded on the whole curve $\hat{\gamma}_{\varepsilon,\tau_0}$ for $\varepsilon \to 0$. On the other hand, the statement "an angle is of order ε" does not depend on choice of the metric.

Proof. To construct $\hat{\gamma}_{\varepsilon,\tau_0}$, we will find the constants $\tau^{s,u} = \tau^{s,u}(\varepsilon)$ such that solutions of the perturbed system

$$(\hat{x}_\varepsilon^{s,u}(t+\tau^{s,u},\tau^{s,u}),\ \hat{y}_\varepsilon^{s,u}(t+\tau^{s,u},\tau^{s,u}),\ t \bmod 2\pi)$$

coincide. Putting $t = 0$, we obtain the vector equation

$$(\hat{x}_\varepsilon^u(\tau^u,\tau^u),\ \hat{y}_\varepsilon^u(\tau^u,\tau^u)) = (\hat{x}_\varepsilon^s(\tau^s,\tau^s),\ \hat{y}_\varepsilon^s(\tau^s,\tau^s)). \tag{3.10}$$

Consider these vectors in a convenient local coordinate system (ρ, h) on D in a neighborhood of the separatrix Γ_0.

We take the function H_0 as the coordinate h, and take ρ canonically conjugate to h, i.e., $dh \wedge d\rho = dy \wedge dx$. This condition does not determine ρ uniquely. We specify the coordinate ρ in the following way. Let $\mathscr{C} \in D$ be a curve passing through the point $A_0 = (\hat{x}(\vartheta), \hat{y}(\vartheta))$ and transversal to level lines of the function H_0. Let $B \in D$ be any point lying sufficiently close to A_0. Consider the solution $z_B(t)$ of the system with Hamiltonian H_0 having initial condition at the point B. A certain interval of the solution $z_B(t)$ lying in a small neighborhood of the point A_0 intersects the curve $\mathscr{C}$. We denote this intersection point by B' and put

$$\rho(B) = \vartheta + \text{the time of the motion along } z_B(t) \text{ from } B' \text{ to } B.$$

In particular,

$$\rho|_{\mathscr{C}} = \vartheta, \qquad \rho(\hat{x}(t), \hat{y}(t)) = t. \tag{3.11}$$

Then (3.10) is equivalent to the pair of scalar equations:

$$\begin{aligned}
&\Phi_1(\tau^u,\tau^s,\varepsilon) = 0, \quad \Phi_2(\tau^u,\tau^s,\varepsilon) = 0.\\
&\Phi_1 = \rho(\hat{x}_\varepsilon^s(\tau^s,\tau^s),\ \hat{y}_\varepsilon^s(\tau^s,\tau^s)) - \rho(\hat{x}_\varepsilon^u(\tau^u,\tau^u),\ \hat{y}_\varepsilon^u(\tau^u,\tau^u)),\\
&\Phi_2 = \frac{1}{\varepsilon}\big(h(\hat{x}_\varepsilon^s(\tau^s,\tau^s),\ \hat{y}_\varepsilon^s(\tau^s,\tau^s)) - h(\hat{x}_\varepsilon^u(\tau^u,\tau^u),\ \hat{y}_\varepsilon^u(\tau^u,\tau^u))\big).
\end{aligned} \tag{3.12}$$

The function Φ_2, extended by continuity to the point $\varepsilon = 0$, is smooth in ε.

According to the definition of the coordinate ρ and to equations (3.3), (3.11), we have

$$\Phi_1 = \rho(\hat{x}(\tau^s), \hat{y}(\tau^s)) - \rho(\hat{x}(\tau^u), \hat{y}(\tau^u)) + O(\varepsilon) = \tau^s - \tau^u + O(\varepsilon).$$

The function Φ_2 can be presented as follows:

$$\begin{aligned}
\Phi_2 &= \frac{1}{\varepsilon}\big(H_0(\hat{x}_\varepsilon^s(\tau^s,\tau^s),\ \hat{y}_\varepsilon^s(\tau^s,\tau^s)) - H_0(\hat{x}(\tau^s), \hat{y}(\tau^s))\big)\\
&\quad - \frac{1}{\varepsilon}\big(H_0(\hat{x}_\varepsilon^u(\tau^u,\tau^u),\ \hat{y}_\varepsilon^u(\tau^u,\tau^u)) - H_0(\hat{x}(\tau^u), \hat{y}(\tau^u))\big)\\
&= \partial_s H_0(\hat{x}(\tau^s), \hat{y}(\tau^s), 0) - \partial_u H_0(\hat{x}(\tau^u), \hat{y}(\tau^u), 0) + O(\varepsilon)\\
&= I(\tau^u) + O(\tau^s - \tau^u) + O(\varepsilon).
\end{aligned}$$

Hence, system (3.12) takes the form

$$\tau^s - \tau^u + O(\varepsilon) = 0, \qquad I(\tau^u) + O(\tau^s - \tau^u) + O(\varepsilon) = 0. \tag{3.13}$$

For $\varepsilon = 0$ system (3.13) has the solution $\tau^u = \tau^s = \tau_0$. Moreover,

$$\left. \frac{\partial(\Phi_1, \Phi_2)}{\partial(\tau^u, \tau^s)} \right|_{(\tau^u, \tau^s, \varepsilon) = (\tau_0, \tau_0, 0)} = -\frac{dI}{d\tau}(\tau_0) \neq 0.$$

Therefore, the existence of the homoclinic solution $\hat{\gamma}_{\varepsilon,\tau_0}$ follows from the implicit function theorem.

In any independent of ε metric on D we obtain that the separatrix splitting (for example, an angle between them at a point of the curve $\hat{\gamma}_{\varepsilon,\tau_0}$) is of order ε. □

It is convenient to measure the splitting of the asymptotic surfaces $\Gamma_\varepsilon^{s,u}$ by some symplectic invariant which does not require for its definition additional structures on the phase space. Such invariants can be introduced in different ways (see for example [53, 135]).[5] One of the simplest possibilities is as follows. Consider a section Σ_ϑ of the extended phase space $D \times \mathbb{T}$ by the plane

$$\Pi_\vartheta = \{(x, y, t) : t = \vartheta \bmod 2\pi\}.$$

Let $\mathscr{D}_\vartheta(\zeta) \subset \Sigma_\vartheta$ be the domain bounded by segments of the curves $\Gamma_{\varepsilon,\vartheta}^{s,u} = \Gamma_\varepsilon^{s,u} \cap \Pi_\vartheta$, from a homoclinic point $\zeta \in \Sigma_\vartheta$ to the neighboring homoclinic point ζ' (see Fig. 3.2). Here we assume that the direction of motion along the curve $\Gamma_{\varepsilon,\vartheta}^{s,u}$ from ζ to ζ' is positive in the sense of the natural (dynamical) orientation of the curves.

Let $\mathscr{A}(\zeta)$ be the oriented area of $\mathscr{D}_\vartheta(\zeta)$:

$$\mathscr{A}(\zeta) = \int_{\mathscr{D}_\vartheta(\zeta)} dx \wedge dy = -\int_{\mathscr{D}_\vartheta(\zeta)} \omega.$$

The quantity $\mathscr{A}(\zeta)$ is positive if the orientation of the boundary of the domain $\mathscr{D}_\vartheta(\zeta)$ in the positive direction along the curve $\Gamma_{\varepsilon,\vartheta}^u$ and in the negative direction along the curve $\Gamma_{\varepsilon,\vartheta}^s$ is consistent with the orientation of the domain D. Otherwise, $\mathscr{A}(\zeta) < 0$.

Proposition 3.3. *The area $\mathscr{A}(\zeta)$ does not change if ϑ and ζ are continuously shifted along a homoclinic solution.*

Proof. Let the points $\zeta_1 \in \Sigma_{\theta_1}$ and $\zeta_2 \in \Sigma_{\vartheta_2}$ lie on the same homoclinic solution. Then the corresponding points $\zeta_1' \in \Sigma_{\vartheta_1}$ and $\zeta_2' \in \Sigma_{\vartheta_2}$ also lie on the same homoclinic solution. Hence, the domain $\mathscr{D}_1 = \mathscr{D}_{\vartheta_1}(\zeta_1)$ can be obtained from $\mathscr{D}_2 = \mathscr{D}_{\vartheta_2}(\zeta_2)$ by the shift $g^{\vartheta_2 - \vartheta_1} : D \times \mathbb{T} \to D \times \mathbb{T}$ along solutions of the system by the time $t = \vartheta_2 - \vartheta_1$.

The symplectic structure $\omega = dy \wedge dx$ can be continued from the domain D to the closed 2-form $\tilde{\omega} = dy \wedge dx$ on the whole extended phase space.[6] The map

[5] In the second paper the multidimensional case is considered.

[6] $\tilde{\omega} = \mathrm{pr}^*\omega$, where $\mathrm{pr} : D \times \mathbb{T} \to D$ is the projection to the first multiplier.

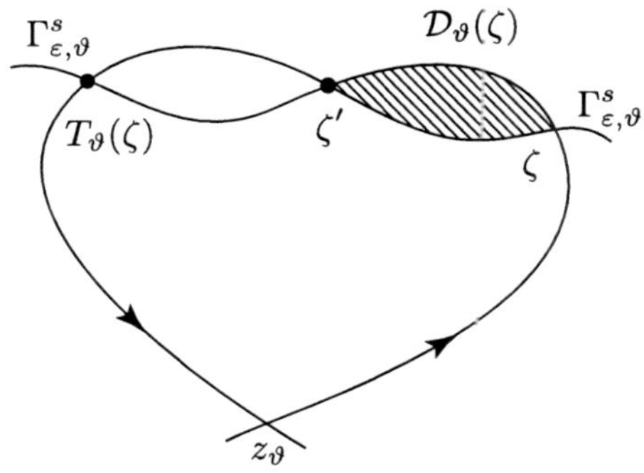

Fig. 3.2 Homoclinic points ζ, ζ' and $T_\vartheta(\zeta)$.

$g^{\vartheta_2-\vartheta_1}$ preserves $\tilde{\omega}$. Therefore,

$$\int_{\mathscr{D}_1}\tilde{\omega} = \int_{\mathscr{D}_1}(g^{\vartheta_2-\vartheta_1})^*\tilde{\omega} = \int_{g^{\vartheta_2-\vartheta_1}(\mathscr{D}_1)}\tilde{\omega} = \int_{\mathscr{D}_2}\tilde{\omega}. \quad \square$$

Let $T_\vartheta : \Sigma_\vartheta \to \Sigma_\vartheta$ be the Poincaré map.

Proposition 3.4. *Suppose that the symplectic structure is an exact form. Consider the intersection of the segments of the curves $\Gamma^s_{\varepsilon,\vartheta}$ and $\Gamma^u_{\varepsilon,\vartheta}$ between ζ and $T_\vartheta(\zeta)$. Let this intersection consist of the homoclinic points $\zeta = \zeta_0, \zeta_1, \zeta_2, \ldots, \zeta_k = T_\vartheta(\zeta)$. Suppose that $k < \infty$. Then $\sum_{j=0}^{k-1}\mathscr{A}(\zeta_j) = 0$.*

Corollary 3.2. $k > 1$.

Proof. By Proposition 1.1, the map T_ϑ is exact symplectic. This means that the primitive ν of the symplectic structure $\omega_\vartheta = \tilde{\omega}|_{\Sigma_\vartheta}$ is transformed as follows: $T^*_\vartheta\nu = \nu + dS$, $S : \Sigma_\vartheta \to \mathbb{R}$.

The point $z_\vartheta = (\gamma_\varepsilon(\vartheta), \vartheta) \in \Sigma_\vartheta$ is fixed for T_ϑ. Consider the closed curve Λ_0 that comprises the piece of the unstable separatrix $\Gamma^u_{\varepsilon,\vartheta}$ from the point z_ϑ to ζ and the piece of the curve $\Gamma^s_{\varepsilon,\vartheta}$ from ζ to z_ϑ (see Fig. 3.2). The image of the curve Λ_0 under the map T_ϑ is the curve Λ that comprises the piece of the unstable separatrix $\Gamma^u_{\varepsilon,\vartheta}$ from the point z_ϑ to $T_\vartheta(\zeta)$ and the piece of the curve $\Gamma^s_{\varepsilon,\vartheta}$ from $T_\vartheta(\zeta)$ to z_ϑ. Therefore,

$$\int_{\Lambda_0}\nu = \int_{\Lambda_0}(\nu + dS) = \int_{\Lambda_0}T^*_\vartheta\nu = \int_{T_\vartheta(\Lambda_0)}\nu = \int_{\Lambda}\nu.$$

Hence, $\int_{\Lambda_0}\nu - \int_\Lambda \nu = 0$. According to the Stokes formula, this difference is equal (possibly up to the sign) to the sum $\sum_{j=0}^{k-1}\mathscr{A}(\zeta_j)$. $\square$

If the angles between the curves $\Gamma^{s,u}_{\varepsilon,\vartheta}$ at the points ζ and ζ' on Σ_ϑ are of order ε then according to Proposition 3.2 the coordinates of the points ζ and ζ' on Σ_ϑ can be presented as follows:

$$\zeta = (\hat{x}(\vartheta + \tau), \hat{y}(\vartheta + \tau)) + O(\varepsilon), \quad \zeta' = (\hat{x}(\vartheta + \tau'), \hat{y}(\vartheta + \tau')) + O(\varepsilon), \tag{3.14}$$

where $\tau < \tau'$ are two neighboring nondegenerate solutions of the equation $I = 0$. Here the nondegeneracy is understood in the sense that the derivative $dI/d\tau$ does not vanish at these points.

Proposition 3.5. *For small values of the parameter* ε

$$\mathscr{A}(\zeta) = \varepsilon \int_{\tau}^{\tau'} I(t)\, dt + O(\varepsilon^2).$$

Proof. We use notation introduced in the proof of Theorem 3.3. The curves $\Gamma_{\varepsilon,\vartheta}^{s,u}$ have the form

$$(x, y) = (\hat{x}_{\varepsilon}^{s,u}(\vartheta + \tau, \tau), \hat{y}_{\varepsilon}^{s,u}(\vartheta + \tau, \tau)), \qquad \tau \in \mathbb{R}, \tag{3.15}$$

(see (3.3)). We define $\tau_{s,u}$, $\tau'_{s,u}$ as follows:

$$\begin{aligned} \zeta &= (\hat{x}_{\varepsilon}^{s,u}(\vartheta + \tau_{s,u}, \tau_{s,u}), \hat{y}_{\varepsilon}^{s,u}(\vartheta + \tau_{s,u}, \tau_{s,u})), \\ \zeta' &= (\hat{x}_{\varepsilon}^{s,u}(\vartheta + \tau'_{s,u}, \tau'_{s,u}), \hat{y}_{\varepsilon}^{s,u}(\vartheta + \tau'_{s,u}, \tau'_{s,u})). \end{aligned}$$

In the coordinates (ρ, h), the curves $\Gamma_{\varepsilon,\vartheta}^{s,u}$ look as follows:

$$\big(\vartheta + \tau + O(\varepsilon), H_0(\hat{x}(\vartheta+\tau), \hat{y}(\vartheta+\tau)) + \varepsilon\partial_{s,u} H_0(\hat{x}(\vartheta+\tau), \hat{y}(\vartheta+\tau), \vartheta)\big) + O(\varepsilon^2).$$

According to (3.14) $\tau_{s,u} = \tau + O(\varepsilon)$ and $\tau'_{s,u} = \tau' + O(\varepsilon)$. Hence,

$$\begin{aligned} \mathscr{A}(\zeta) &= \varepsilon \int_{\tau}^{\tau'} (\partial_s - \partial_u) H_0(\hat{x}(\vartheta + \lambda), \hat{y}(\vartheta + \lambda), \vartheta)\, d\lambda + O(\varepsilon^2) \\ &= \varepsilon \int_{\tau}^{\tau'} I(t)\, dt + O(\varepsilon^2). \quad \square \end{aligned}$$

3.3 The Poincaré–Melnikov Method: Multidimensional Aspect

Most of the examples where the splitting of asymptotic manifolds was studied deal with Hamiltonian systems with one and a half or two degrees of freedom and with 2-dimensional symplectic maps. However, people working in Hamiltonian dynamics always understood that analogous phenomena take place in the multidimensional situation. Here one has to study manifolds asymptotic to hyperbolic tori. The dimensions of such tori can be any integer between zero and the number of degrees of freedom (including zero). First attempts to present a multidimensional analog of the Poincaré–Melnikov integral (3.4) [62, 146] were not quite satisfactory because in general an integral along a trajectory homoclinic to a hyperbolic torus does not converge. Hence some tricks or non-generic assumptions were used. In [133] the convergence is achieved by adding to the integral a certain quasi-periodic function. Below in this section we present the corresponding constructions from [133] and [34].

Let $F_1, \dots, F_m$ be first integrals in involution of a Liouville integrable Hamiltonian system (M, ω, H_0), let $M_0 = \{F_1 = \dots = F_m = 0\} \subset M$ be their common zero level, and let $N \subset M_0$ be an n-dimensional non-degenerate Diophantine hyperbolic torus (see Definitions 2.2–2.4). Below we use canonical coordinates x, y, z_u, z_s, see Definition 2.6. In canonical coordinates it is easy to see that the integrals $F_1, \dots, F_m$ are dependent at points of N; hence M_0 is a critical level.

The torus N lies in the intersection of the invariant Lagrangian asymptotic manifolds $\Gamma^s, \Gamma^u \subset M_0$ such that any trajectory on $\Gamma^u(\Gamma^s)$ tends to N exponentially as $t \to -\infty$ $(t \to +\infty)$ (Sect. 2.4). Let $\Gamma \subset \Gamma^u \cap \Gamma^s$ be a connected Lagrangian manifold, doubly asymptotic to N. Existence of Γ means that the asymptotic manifolds Γ^u and Γ^s are doubled. We assume that $F_1, \dots, F_m$ are independent at points of $\Gamma \backslash N$. This condition is equivalent to the linear independence of the Hamiltonian vector fields v_{F_s}, $1 \le s \le m$, on $\Gamma \backslash N$. All these vector fields are tangent to Γ.

Let $H = H_0 + \varepsilon H_1 + O(\varepsilon^2)$ be the perturbed Hamiltonian. Then (see Theorem 2.8) for small ε there exists a hyperbolic torus N_ε of the perturbed system near N such that N_ε lies in the intersection of two asymptotic Lagrangian manifolds $\Gamma_\varepsilon^{s,u}$.

For $\varepsilon \neq 0$ the set $\Gamma_\varepsilon^s \cap \Gamma_\varepsilon^u$ in general does not contain any Lagrangian manifold. To formulate the corresponding result, for any homoclinic solution $\gamma(t) \subset \Gamma$ consider the quantity

$$I = \lim_{T \to +\infty} \left(- \int_{-T}^{T} (H_1(\gamma(t)) - h_1)\, dt + \chi(\gamma(-T)) - \chi(\gamma(T)) \right), \tag{3.16}$$

where

$$h_1 = \frac{1}{(2\pi)^n} \int_{\mathbb{T}^n} H_1(x, 0, 0, 0)\, dx, \tag{3.17}$$

and the function $\chi = \chi(x)$, expressed in the canonical coordinates for N, satisfies the equation

$$\partial_\nu \chi(x) + H_1(x, 0, 0, 0) = h_1, \qquad \partial_\nu = \left\langle \nu, \frac{\partial}{\partial x} \right\rangle. \tag{3.18}$$

If H_1 is sufficiently smooth, then there exists a solution of (3.18) unique up to an additive constant.

We see that I can be regarded as a function on Γ, invariant with respect to the unperturbed dynamics:

$$I : \Gamma \to \mathbb{R}, \quad v_{H_0} I = 0.$$

Derivatives along other directions on Γ are given by the formulas

$$v_{F_{j_1}} \cdots v_{F_{j_k}} I = \lim_{T\to+\infty}\Big(-\int_{-T}^{T}\{F_{j_1}\cdots\{F_{j_k},H_1\}\cdots\}(\gamma(t))\,dt + \{F_{j_1}\cdots\{F_{j_k},\chi\}\cdots\}(\gamma(-T)) - \{F_{j_1}\cdots\{F_{j_k},\chi\}\cdots\}(\gamma(T))\Big), \tag{3.19}$$

where $\{\,,\}$ denotes the Poisson bracket associated with the symplectic structure ω. The vector fields $v_{F_j}, \dots, v_{F_{j_k}}$ commute because the functions $F_1, \dots, F_k$ are in involutions.

The function I is called *the Poincaré–Melnikov potential.*

Theorem 3.4.

(1) *Limit* (3.16) *exists.*
(2) *If any of the quantities* (3.19) *does not vanish, then for small ε the manifolds $\Gamma^{s,u}$ split, i.e., their intersection does not contain a doubly asymptotic Lagrangian manifold.*
(3) *If for some $\zeta^0 \in \gamma \subset \Gamma$ we have $v_{F_1}I(\zeta^0) = \cdots = v_{F_m}I(\zeta^0) = 0$ and the rank of the matrix $(v_{F_j}v_{F_r}\, I(\zeta^0))$ equals $m-1$, then for small ε the manifolds Γ^u_ε and Γ^s_ε are transversal at the energy level along a homoclinic trajectory γ_ε, where $\gamma_\varepsilon \to \gamma$ as $\varepsilon \to 0$.*

Remarks.

1. In [18] there is an analog of Theorem 3.4 for $n = 1$.
2. In [18, 44, 132] some results are presented on non-integrability, generated by the splitting of multidimensional asymptotic manifolds.
3. By using autonomization it is easy to obtain a non-autonomous version of Theorem 3.4.

Proof (of Theorem 3.4). First recall that by Theorem 1.1

in a neighborhood $\mathcal{U}$ of any point on a Lagrangian submanifold L of a symplectic manifold (M,ω) there exist coordinates p, q such that $\omega = dp \wedge dq$ and the set $L \cap \mathcal{U}$ is given by the equations $p = 0$.

Let p, q be such coordinates, associated with $L = \Gamma$. The perturbed asymptotic manifolds $\Gamma^{s,u}_\varepsilon$ are Lagrangian. Therefore they can be presented as graphs

$$p = \varepsilon f^{s,u}(q,\varepsilon), \quad df \wedge dq = \varepsilon^{-1}\omega|_{\Gamma^{s,u}_\varepsilon} = 0.$$

Since $\mathcal{U}$ can be assumed to be simply connected, by the Poincaré lemma we have $f^{s,u} = \partial S^{s,u}(q,\varepsilon)/\partial q$.

Lemma 3.1. *The functions $S^{s,u}_0 = S^{s,u}|_{\varepsilon=0} : \Gamma \mapsto \mathbb{R}$ are well defined, i.e., they do not depend on the choice of the local coordinates p, q.*

We postpone the proof of Lemma 3.1 to the end of this section.

The manifolds $\Gamma_\varepsilon^{s,u}$ lie on the same energy level, so that the functions $S^{s,u}(q,\varepsilon)$ satisfy the Hamilton-Jacobi equation

$$H_\varepsilon\left(\varepsilon\frac{\partial S^{s,u}}{\partial q}, q, \varepsilon\right) = h(\varepsilon). \tag{3.20}$$

In the first approximation in ε we have

$$\frac{\partial H_0}{\partial p}(0,q)\,\frac{\partial S_0^{s,u}}{\partial q}(q) + H_1(0,q) = h_1, \qquad h_1 = \frac{d}{d\varepsilon}\bigg|_{\varepsilon=0} h. \tag{3.21}$$

We get: $\frac{d}{dt}S_0^{s,u}(\gamma(t)) + H_1(\gamma(t)) = h_1$, where $\frac{d}{dt}$ is the derivative along the unperturbed system. Hence,

$$S_0^{s,u}(\gamma(t_2)) - S_0^{s,u}(\gamma(t_1)) = -\int_{t_1}^{t_2}(H_1(\gamma(t)) - h_1)\,dt. \tag{3.22}$$

Lemma 3.2. *In the canonical coordinates (x, y, z) for N, the constant h_1 in* (3.21) *satisfies* (3.17) *and*

$$S_0^{s,u} - \chi = \text{const} + O(\varepsilon, y, z).$$

The proof of Lemma 3.2 also appears at the end of this section.

Necessary conditions for coincidence of the manifolds Γ_ε^u and Γ_ε^s are

$$\frac{\partial^l}{\partial q_{j_1}\cdots\partial q_{j_l}}(S_0^s - S_0^u) = 0. \tag{3.23}$$

The Hamiltonian fields v_{F_j}, $1 \le j \le m$, are independent on Γ. Therefore equations (3.23) are equivalent to the following ones:

$$v_{F_{k_1}}\cdots v_{F_{k_l}}(S_0^s - S_0^u) \equiv \{F_{k_1}\cdots\{F_{k_l}, S_0^s - S_0^u\}\cdots\} = 0, \qquad 1 \le k_j \le m. \tag{3.24}$$

From (3.22) and Lemma 3.2 it follows that

$$\begin{aligned} -S_0^u(\gamma(0)) &= -\int_0^T H_1(\gamma(t))\,dt - S_0^u(\gamma(T)) \\ &= \lim_{T\to+\infty}\left(-\int_0^T H_1(\gamma(t))\,dt - \chi(\gamma(T))\right). \end{aligned}$$

Analogously

$$S_0^s(\gamma(0)) = \lim_{T\to+\infty}\left(-\int_{-T}^0 H_1(\gamma(t))\,dt - \chi(\gamma(-T))\right).$$

Thus,

$$(S_0^s - S_0^u)(\gamma(0)) = I, \tag{3.25}$$

the limit I exists and conditions (3.24) take the form $v_{F_{k_1}} \cdots v_{F_{k_l}} I = 0$. Statements 1 and 2 of Theorem 3.4 are proved.

The proof of statement 3 is based on the implicit function theorem. Let us find the solution q^0 of the equations

$$\frac{\partial S^s}{\partial q_j}(q^0, \varepsilon) = \frac{\partial S^u}{\partial q_j}(q^0, \varepsilon), \qquad 1 \le j \le m. \tag{3.26}$$

The functions S^s and S^u satisfy the same Hamilton-Jacobi equation (3.20). Hence, one of the equations (3.26) depends on the others.

Equations (3.26) are equivalent to the following ones:

$$v_{F_j}(S^s - S^u)(q^0, \varepsilon) = 0, \qquad 1 \le j \le m. \tag{3.27}$$

From (3.25) it follows that for $\varepsilon = 0$ equations (3.27) have the form $v_{F_1} I = \cdots = v_{F_m} I = 0$ at the point $(0, q^0)$. Existence of a solution of equations (3.27) for small ε depends on the properties of the Jacobian matrix

$$J = \left(\frac{\partial}{\partial q_k} v_{F_l}(S_0^s - S_0^u) \right)(q^0).$$

Equations (3.27) are dependent. Therefore J is degenerate. On the other hand, the condition rank $J = m - 1$ guarantees the existence of a one-parameter family of solutions to (3.27). This family is the curve γ_ε. Since rank $J = \operatorname{rank}(v_{F_j} v_{F_l} I)$, the proof of Theorem 3.4 is complete. □

If $F_1 = H_0$, then for any homoclinic solution γ we have $v_{F_1} I = 0$, and hence $v_{F_1} v_{F_l} I = v_{F_l} v_{F_1} I = 0, \ 1 \le l \le m$.

Corollary 3.3. *If* $F_1 = H_0$, *then the condition* rank $(I_{jr}) = m - 1; \ j, r = 1, \ldots, m$, *is equivalent to the condition* $\det(I_{jr}) \ne 0; \ j, r = 2, \ldots, m$.

Finally we prove Lemmas 3.1 and 3.2.

Proof (of Lemma 3.1). Let $\mathscr{U}$ be a connected, simply connected domain in M. Let p, q and $\hat p, \hat q$ be two coordinate systems in $\mathscr{U}$ such that

$$\omega|_{\mathscr{U}} = dp \wedge dq = d\hat p \wedge d\hat q \quad \text{and} \quad \Gamma \cap \mathscr{U} = \{(p, q) : p = 0\} = \{(\hat p, \hat q) : \hat p = 0\}.$$

Suppose that the Lagrangian manifold Γ_ε is given by the equations

$$\hat p = \varepsilon \partial \hat S_0(\hat q)/\partial \hat q + O(\varepsilon^2) \quad \text{or} \quad p = \varepsilon \partial S_0(q)/\partial q + O(\varepsilon^2).$$

Let us show that

$$\hat S_0(\hat q) = S_0(q) + \text{const}, \tag{3.28}$$

where the points $(\hat p = 0, \hat q)$ and $(p = 0, q)$ coincide.

Indeed, the coordinates (p, q) and $(\hat p, \hat q)$ are related by the equations

$$p = \frac{\partial W}{\partial q}, \quad \hat{q} = \frac{\partial W}{\partial \hat{p}}, \qquad W = W(\hat{p}, q).$$

Since the equations $p = 0$ and $\hat{p} = 0$ are equivalent, we have $\partial W(0, q)/\partial q = 0$. Hence

$$W(\hat{p}, q) = \text{const} + \sum_{1}^{m} a_l(q)\, \hat{p}_l + O_2(\hat{p}).$$

The manifold Γ_ε is given by the equations

$$\begin{aligned} p &= \frac{\partial W}{\partial q}\left(\varepsilon \frac{\partial \hat{S}_0(\hat{q})}{\partial \hat{q}}, q\right) + O(\varepsilon) = \varepsilon \sum_{l=1}^{m} \frac{\partial \hat{S}_0}{\partial \hat{q}_l} \frac{\partial a_l}{\partial q} + O(\varepsilon^2) \\ &= \varepsilon \sum_{l=1}^{m} \frac{\partial \hat{S}_0}{\partial \hat{q}_l} \frac{\partial \hat{q}_l}{\partial q}\bigg|_{\hat{p}=0} + O(\varepsilon^2) = \varepsilon \frac{\partial S_0}{\partial q} + O(\varepsilon^2). \end{aligned}$$

Since Γ is connected, the constant in (3.28) is a global constant on Γ. □

Proof (of Lemma 3.2). In a neighborhood $\mathscr{U}$ of the torus N the manifolds $\Gamma^{s,u} = \Gamma_0^{s,u}$ can be parameterized as follows:

$$\begin{aligned} \Gamma^u &: y = \hat{V}^u(x, z_u), \quad z_s = \hat{W}^u(x, z_u), \\ \Gamma^s &: y = \hat{V}^s(x, z_s), \quad z_u = \hat{W}^s(x, z_s). \end{aligned}$$

Since the manifolds $y = 0$, $z_s = 0$ and $y = 0$, $z_u = 0$ are tangent at N to Γ^u and Γ^s respectively (see Chap. 2, Sect. 2.4), the functions $\hat{V}^{s,u}$ and $\hat{W}^{s,u}$ are of order $O_2(z_{s,u})$. In the coordinates

$$p = (\bar{y}, \bar{z}_s), \quad q = (x, z_u), \quad \bar{y} = y - \hat{V}^u(x, z_u), \quad \bar{z}_s = z_s - \hat{W}^u(x, z_u),$$

we have $\Gamma^u = \{(p, q) : p = 0\}$ and $\omega = dp \wedge dq$. Indeed, the first equation is evident and the second one follows from the fact that Γ^u is Lagrangian. Thus, the manifold Γ_ε^u has the form

$$\Gamma_\varepsilon^u = \left\{(p, q) : p = \varepsilon \frac{\partial S_0^u}{\partial q}(q) + O(\varepsilon^2)\right\},$$

and in the original coordinates we obtain

$$\Gamma_\varepsilon^u = \left\{(x, y, z) : \ y = \varepsilon \frac{\partial S_0^u}{\partial x}\bigg|_N + O_2(\varepsilon, z_u), \ z_s = \varepsilon \frac{\partial S_0^u}{\partial z_u}\bigg|_N + O_2(\varepsilon, z_u)\right\}.$$

Similarly,

$$\Gamma_\varepsilon^s = \left\{(x, y, z) : \ y = \varepsilon \frac{\partial S_0^s}{\partial x}\bigg|_N + O_2(\varepsilon, z_s), \ z_u = \varepsilon \frac{\partial S_0^s}{\partial z_s}\bigg|_N + O_2(\varepsilon, z_s)\right\}.$$

Since $N_\varepsilon \subset \mathcal{U} \cap \Gamma_\varepsilon^u \cap \Gamma_\varepsilon^s$, we obtain

$$N_\varepsilon = \left\{ (x, y, z) : z_u = \varepsilon \frac{\partial S_0^s}{\partial z_s}\bigg|_N + O(\varepsilon^2),\ z_s = \varepsilon \frac{\partial S_0^u}{\partial z_u}\bigg|_N + O(\varepsilon^2), \right.$$
$$\left. y = \varepsilon \frac{\partial S_0^u}{\partial x}\bigg|_N + O(\varepsilon^2) = \varepsilon \frac{\partial S_0^s}{\partial x}\bigg|_N + O(\varepsilon^2) \right\}.$$

This implies that

$$\frac{\partial}{\partial x}(S_0^u - S_0^s)\bigg|_N = 0.$$

Recall that $H = \langle v, y\rangle + \langle Ay, y\rangle/2 + \langle z_s, \Omega(x) z_u\rangle + O_3(y, z)$. Thus the condition $H|_{N_\varepsilon} = \text{const.}$ in the first approximation in ε has the form

$$\partial_v S_0^{s,u}(x, 0) + H_1(x, 0, 0, 0) = h_1,$$

and so the equations for the functions $S_0^{s,u}|_N$ and χ (see (3.18)) coincide and h_1 satisfies (3.17). □

Chapter 4
The Separatrix Map

The separatrix map was invented to study dynamical systems near asymptotic manifolds. It was introduced by Zaslavsky and Filonenko [147] (see also [32, 49]) for near-integrable Hamiltonian systems with one-and-a-half degrees of freedom and independently by Shilnikov [123] in generic systems. The main difference between these two approaches is as follows. The Zaslavsky separatrix map determines the dynamics globally near the unperturbed separatrices, but needs the system to be near-integrable. The Shilnikov separatrix map does not need any closeness to integrability, but deals with the dynamics in a neighborhood of a homoclinic orbit.

In the present chapter we obtain explicit formulas for the Zaslavsky separatrix maps. These results are used in Chap. 5 for studying of the dynamics in the stochastic layer.

4.1 Definition and Formulas

4.1.1 Two-Dimensional Symplectic Map

Consider a symplectic map $\hat{T}$, defined on a two-dimensional domain $D \subset \mathbb{R}^2_z$. We assume that $\hat{T}$ satisfies the following conditions I–IV.

I. The map $\hat{T}$ has a hyperbolic fixed point σ.
II. $\hat{T}$ is integrable, i.e., there is a smooth locally non-constant function $F\colon D \to \mathbb{R}$ such that $F \circ \hat{T} = F$.

The dynamics of the unperturbed map $\hat{T}$ is simple. The domain D is fibred into invariant level curves of the integral F. The hyperbolic fixed point gives rise to four asymptotic branches (separatrices) $\Gamma^{u\pm}$ and $\Gamma^{s\pm}$, two unstable and two stable. The separatrices lie on the same level curve of the integral F.

III. $\Gamma^{s\pm} = \Gamma^{u\pm} = \Gamma^{\pm}$; thus $\Gamma^+ \cup \Gamma^- \cup \sigma$ is a figure-eight curve (Fig. 4.1, left part).

D. Treschev, O. Zubelevich, *Introduction to the Perturbation Theory of Hamiltonian Systems*, Springer Monographs in Mathematics,
DOI 10.1007/978-3-642-03028-4_4, © Springer-Verlag Berlin Heidelberg 2010

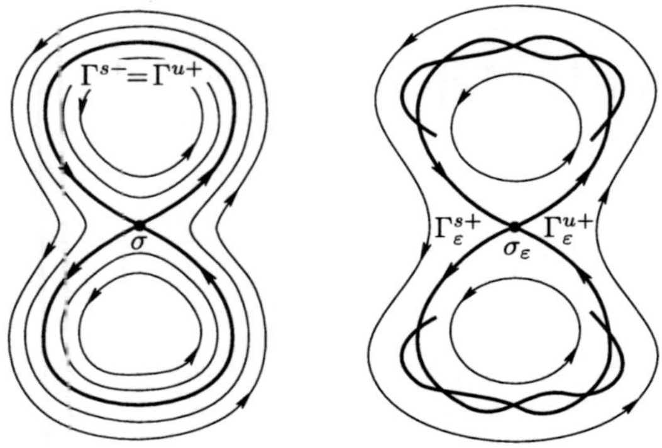

Fig. 4.1 The phase planes of the unperturbed and perturbed systems.

The curves Γ^+ and Γ^- are said to be the upper and lower loops of the figure-eight, respectively. Without loss of generality we may assume that $F(\sigma) = 0$. It is simple to show that $dF|_{z=\sigma} = 0$. We assume that σ is a non-degenerate critical point of F, i.e.,

IV. $\det \frac{\partial^2 F}{\partial z^2}(\sigma) \neq 0$.

Let $\mu > 1$ be the larger multiplier of the fixed point σ and let $\lambda = \log \mu$. Consider the perturbation $T_\varepsilon : D \to D$, $T_0 = \hat{T}$. For small ε, T_ε has a hyperbolic fixed point σ_ε, $\sigma_0 = \sigma$, smoothly depending on ε. Let $\Gamma_\varepsilon^{u\pm}$ and $\Gamma_\varepsilon^{s\pm}$ be the separatrix branches of σ_ε. These branches need not coincide for $\varepsilon \neq 0$ (Fig. 4.1, right part). When studying the dynamics in a neighborhood of the separatrices, it is convenient to use the separatrix map. Below we give the definition of this map.

Consider a neighborhood of the separatrices $U_c = \{z \in D : |F| < c\}$ with small $c > 0$. Let Λ^+ and Λ^- be curves going from one component of the boundary to another and transversal to the upper and lower separatrix loop respectively. Let $\Delta_\varepsilon^\pm$ be the subdomain of U_c between the curves $\Lambda^\pm$ and $T_\varepsilon(\Lambda^\pm)$ (Fig. 4.2). We write $\Delta_\varepsilon = \Delta_\varepsilon^+ \cup \Delta_\varepsilon^-$. For small c we have $T_\varepsilon(\Delta_\varepsilon) \cap \Delta_\varepsilon = \varnothing$. For any $z \in \Delta_\varepsilon$ we define[1]

$$n_r(z) = \min\{n \in \mathbb{N} : T_\varepsilon(z), \ldots, T_\varepsilon^{n-1}(z) \in U_c,\ T_\varepsilon^n(z) \in \Delta_\varepsilon\},$$
$$\Delta_\varepsilon^r = \{z \in \Delta_\varepsilon : n_r(z) \text{ is defined}\}.$$

The separatrix map $\mathscr{SM}_\varepsilon$ is defined as follows:

$$\mathscr{SM}_\varepsilon : \Delta_\varepsilon^r \to \Delta_\varepsilon, \qquad \mathscr{SM}_\varepsilon(z) = T_\varepsilon^{n_r}(z).$$

We write $\overline{\Delta_\varepsilon^\pm} = \Delta_\varepsilon^\pm \cup \Lambda^\pm \cup T_\varepsilon(\Lambda^\pm)$ and note that $\mathscr{SM}_\varepsilon$, extended by continuity to $\overline{\Delta_\varepsilon^\pm}$, satisfies $\mathscr{SM}_\varepsilon(z) = \mathscr{SM}_\varepsilon(T_\varepsilon(z))$, $z \in \Lambda^\pm$. Therefore, it is natural to as-

[1] The index r in the symbols n_r and Δ_ε^r stands for *return*.

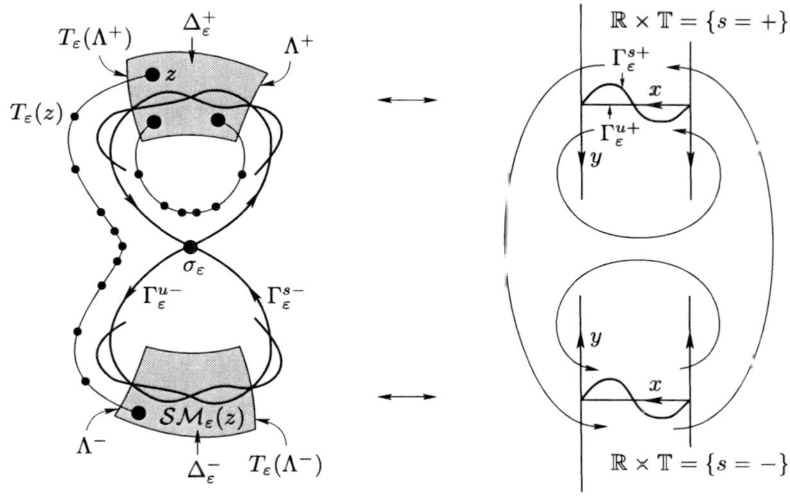

Fig. 4.2 Definition of the separatrix map. The *left part* of the figure is represented in the coordinates (z_1, z_2) and the *right part* in the coordinates (x, y, s).

sume that the separatrix map is defined on a subset of the cylinders $\overline{\Delta}_\varepsilon^\pm/\!\sim$, where the factorization by $\sim$ means the identification of the points $z \in \Lambda^\pm$ and $T_\varepsilon(z) \in T_\varepsilon(\Lambda^\pm)$.

The map $\mathcal{SM}_\varepsilon$ is convenient for studying the dynamics in a neighborhood of the separatrices, because long "uninteresting parts" of the trajectory passing near the point σ are automatically omitted. An important fact is that the separatrix maps $\mathcal{SM}_\varepsilon$ admit simple formulas in convenient coordinates for small values of ε.

Theorem 4.1. *Suppose that $\hat{T} = T_0$ satisfies conditions* I–IV. *Then on $\overline{\Delta}_\varepsilon^\pm/\!\sim$ there are coordinates $x \in \mathbb{T} = \mathbb{R}/\mathbb{Z}$, $y \in \mathbb{R}$, $s \in \{\pm\}$, depending smoothly on ε, such that*:

(1) $\overline{\Delta}_\varepsilon^+/\!\sim\; = \{s = +\}$, $\overline{\Delta}_\varepsilon^-/\!\sim\; = \{s = -\}$, *$x$ is the angular coordinate on the cylinders $\overline{\Delta}_\varepsilon^\pm/\!\sim$, and the geometric meaning of the variable εy is the (signed) distance to the unstable separatrix*;
(2) *the symplectic structure is $dz_2 \wedge dz_1 = \varepsilon\, dy \wedge dx$*;
(3) *there are smooth functions $\nu_\pm(x)$ and constants $\alpha_\pm > 0$ such that for any $(x^+, y^+, s^+) = \mathcal{SM}_\varepsilon(x, y, s)$ we have*

$$\begin{cases} x^+ = x + \frac{1+O(\varepsilon)}{\lambda}\left(\log \frac{\varepsilon}{\alpha_s^2\lambda} + \log|y^+|\right), \\ y^+ = y + \lambda\nu_s(x) + O(\varepsilon), \\ s^+ = s \cdot \operatorname{sign}(y^+), \end{cases} \tag{4.1}$$

where $\Delta_\varepsilon^r = \{(x, y) \in \Delta_\varepsilon : y^+ \neq 0\}$;

(4) $\{y = 0, s = \pm\} \subset \Gamma_\varepsilon^{u\pm}$ *and* $\{y^+ = 0, s = \pm\} \subset \Gamma_\varepsilon^{s\pm}$.

The separatrix map is not defined on the circles $y^+(x, y, \pm) = 0$ (the inverse map is not defined on the circles $y = 0$). Indeed, for instance, the points lying on the separatrices $\Gamma_\varepsilon^{s\pm}$ do not return to $\overline{\Delta}_\varepsilon^\pm$ but tend to σ_ε under the iterations of the map T_ε. Thus, the dynamics for the map $\mathscr{SM}_\varepsilon$ can be considered only for points on whose trajectories the variable y never vanishes (in other words, away from the separatrices). The complement of this set (the trace of the separatrices in $\overline{\Delta}_\varepsilon^\pm$) consists of countably many smooth curves and therefore has zero measure.

A surprising property of the separatrix map is that the dependence on $\log\varepsilon$ turns out to be periodic in the leading approximation. Indeed, neglecting the small terms of the form $O(\varepsilon)$, we obtain a map which does not change when ε is replaced by $\varepsilon\mu^n = \varepsilon e^{\lambda n}$ for any integer n.

For large values of the variable y, $\mathscr{SM}_\varepsilon$ is close to an integrable map. Indeed, putting $y = y_0(1 + v)$, where $|y_0|$ is a large parameter and the variable v is small, we obtain

$$\begin{aligned}
x^+ &= x + \frac{1 + O(\varepsilon)}{\lambda}\left(\log\frac{\varepsilon y_0}{\alpha_s^2\lambda} + \log|1 + v^+|\right),\\
v^+ &= v + O(|y_0|^{-1} + |\varepsilon|),\\
s^+ &= s \cdot \operatorname{sign}(y_0).
\end{aligned}$$

For $\varepsilon = y_0^{-1} = 0$ this map has the first integral v.

Corollary 4.1. *As a simple application we show that the perturbed system in general has no real-analytic first integral. Suppose that T_ε is real-analytic and has a real-analytic first integral. One can easily show (see Corollary 4.3 below) that every analytic first integral in normal coordinates is a function of y. Hence, every integrable map $\mathscr{SM}_\varepsilon$ preserves y. Thus, according to (4.1), the equations $\nu_\pm \equiv 0$ are necessary integrability conditions for the perturbed map T_ε (cf. [149]).*

4.1.2 Hamiltonian System with One and a Half Degrees of Freedom

Consider a Hamiltonian system with one and a half degrees of freedom in a neighborhood of the surfaces asymptotic to a hyperbolic periodic solution. We assume that $t \in \mathbb{R}/\mathbb{Z}$ and the solution is 1-periodic. In this case the separatrix map is defined naturally by passing to the period-1 map, but it can also be defined directly as the Poincaré map in the extended phase space. Namely, consider the Hamiltonian system

$$\dot z_1 = \frac{\partial H}{\partial z_2}, \quad \dot z_2 = -\frac{\partial H}{\partial z_1}, \qquad H(z, t, \varepsilon) = H_0(z) + \varepsilon H_1(z, t, \varepsilon), \quad z \in \mathbb{R}^2. \tag{4.2}$$

For $\varepsilon = 0$ we have a system with one degree of freedom and Hamiltonian H_0 (the *unperturbed* system). Suppose that:

(1) *the unperturbed system has a hyperbolic equilibrium state* σ, $dH_0|_{z=\sigma} = 0$;
(2) *the separatrices issuing from the point* σ *form a figure-eight (see Fig.* 4.1).

Let $\lambda > 0$ be the Lyapunov exponent of the unperturbed hyperbolic 1-periodic solution $(z(t), t) = (\sigma, t)$ and let $\Gamma^{\pm}(t)$ be the separatrix solutions of the unperturbed system. We call one of the loops of the figure-eight the upper loop and the other the lower loop. Let $\Lambda^+, \Lambda^- \subset \mathbb{R}^2_z$ be curves transversal to the upper and lower separatrix loops, respectively. Let $g^t_\varepsilon : \mathbb{R}^2_z \times \mathbb{T}_t \hookleftarrow$ be the phase flow of system (4.2). The separatrix map is the Poincaré map of the surface $\Delta = (\Lambda^+ \times \mathbb{T}_t) \cup (\Lambda^- \times \mathbb{T}_t)$ in the extended phase space $D \times \mathbb{T}_t$. More precisely, for any point $(z, t) \in \Delta$ we have

$$t_r(z) = \min\{t > 0 : g^t_\varepsilon(z) \in \Delta\}, \qquad \Delta^r_\varepsilon = \{(z, t) \in \Delta : t_r(z) < +\infty\}.$$

Then by definition,

$$\mathscr{SM}_\varepsilon : \Delta^r_\varepsilon \to \Delta, \qquad \mathscr{SM}_\varepsilon(z) = g^{t_r}_\varepsilon(z).$$

Without loss of generality we may assume that the separatrix loop is the zero level of the Hamiltonian H_0.

Theorem 4.2. *On the surface* Δ *there are coordinates* $(h, t, s) \in \mathbb{R} \times \mathbb{T} \times \{+, -\}$ *such that*:

(1) $\Lambda^+ \times \mathbb{T}_t = \{s = +\}$ *and* $\Lambda^- \times \mathbb{T}_t = \{s = -\}$;
(2) $H_0(z) = \varepsilon h + O_2(\varepsilon)$;
(3) t *is the time*;
(4) *the separatrix map has the form* $(h^+, t^+, s^+) = \mathscr{SM}_\varepsilon(h, t, s)$ *with*

$$\begin{cases} h^+ = h + I_s(t) + O(\varepsilon), \\ t^+ = t + \frac{1+O(\varepsilon)}{\lambda}\left(\log \frac{\varepsilon}{\alpha_s^2 \lambda} + \log|h^+|\right), \\ s^+ = s \cdot \operatorname{sign}(h^+), \end{cases} \tag{4.3}$$

where $\alpha_\pm > 0$ *are constants, the functions* $I_\pm$ *are periodic and can be expressed in terms of the Poincaré–Melnikov integral:*

$$I_s(t) = -\int_{-\infty}^{+\infty} \{H_0, H_1\}\left(\Gamma^s\left(t - \frac{\log \alpha_\pm}{\lambda} + \tau\right), \tau\right) d\tau.$$

Theorems 4.1 and 4.2 are in fact equivalent. Indeed, Theorem 4.2 can be obtained by applying Theorem 4.1 to the period-1 map $g^t_\varepsilon|_{t=1}$. Conversely, since one can embed any 2-dimensional symplectic map isotopic to the identity map into the flow of a Hamiltonian system with one and a half degrees of freedom, one can reduce Theorem 4.1 to Theorem 4.2.

4.2 Proof of Theorem 4.1

Note that equations similar to (4.1) and (4.3) are widely used in the physical literature, but without error terms. A rigorous upper estimate of these terms presents the main feature of our approach.

Gluing maps: definition. Let $(q, p) = (q_\varepsilon, p_\varepsilon)$ be the normal coordinates, defined in Theorem 3.1. In these coordinates T_ε has the form

$$\mathscr{T}_\varepsilon(q, p) = (q\mathscr{M}_\varepsilon(pq), p/\mathscr{M}_\varepsilon(pq)), \tag{4.4}$$

where $\sigma_\varepsilon = (0, 0)$ and $\mu_\varepsilon = \mathscr{M}_\varepsilon(0) > 1$.

Remark 4.1. In what follows we assume that the normal coordinates are chosen in such a way that $q > 0$ and $p > 0$ inside the upper loop of the "figure-eight" (see Fig. 4.3).

The normal coordinates, which were initially defined only in a small neighborhood of the hyperbolic point, can be extended along the separatrices by using the following inductive procedure. Let the coordinates of a point $z \in D$ be known and equal to (q, p). Then we define the coordinates of the points $T_\varepsilon(z)$ and $T_\varepsilon^{-1}(z)$ to be $\mathscr{T}_\varepsilon(q, p)$ and $\mathscr{T}_\varepsilon^{-1}(q, p)$, respectively. Far from the fixed point σ_ε we obtain two distinct extensions of the normal coordinates, namely, along the stable and unstable separatrices. The identification of coordinates thus obtained is called the gluing map. More formally, this construction is as follows.

Let $\mathscr{U}_\varepsilon$ be a neighborhood of σ_ε in which the normal coordinates (q, p) are initially defined. Let $\mathscr{N}_\varepsilon : U_\varepsilon \to \mathscr{U}_\varepsilon$ be the map introducing the normal coordinates, $U_\varepsilon \subset \mathbb{R}^2_{q,p}$. We fix positive integers n_u, n_s and define

$$U_\varepsilon^{n_s,n_u} = \bigcup_{-n_s \le n \le n_u} \mathscr{T}_\varepsilon^n(U_\varepsilon), \qquad \mathscr{U}_\varepsilon^{n_s,n_u} = \bigcup_{-n_s \le n \le n_u} T_\varepsilon^n(\mathscr{U}_\varepsilon). \tag{4.5}$$

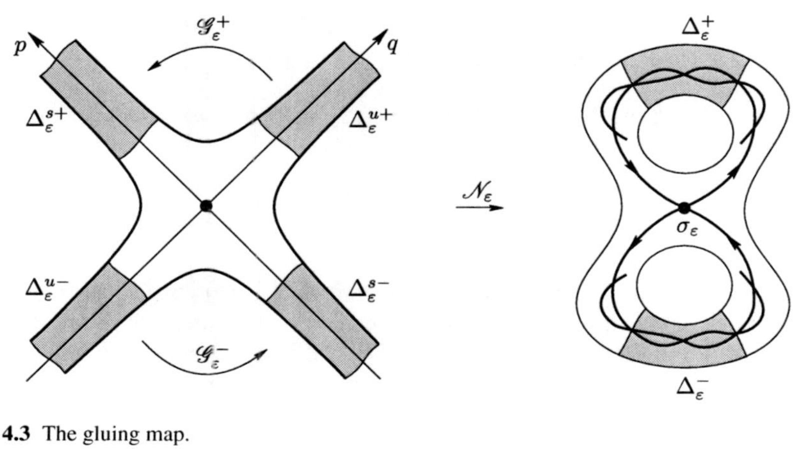

Fig. 4.3 The gluing map.

We extend the map $\mathcal{N}_\varepsilon$ to $U_\varepsilon^{n_s,n_u}$ in such a way that the following diagram is commutative:

$$\begin{array}{ccc} U_\varepsilon & \xrightarrow{\mathcal{T}_\varepsilon^n} & \mathcal{T}_\varepsilon^n(U_\varepsilon) \\ \mathcal{N}_\varepsilon \downarrow & & \downarrow \mathcal{N}_\varepsilon \\ \mathcal{U}_\varepsilon & \xrightarrow{T_\varepsilon^n} & T_\varepsilon^n(\mathcal{U}_\varepsilon) \end{array} \tag{4.6}$$

This extension is well defined because $\mathcal{N}_\varepsilon \circ \mathcal{T}_\varepsilon = T_\varepsilon \circ \mathcal{N}_\varepsilon$ on U_ε.

For large values of n_s and n_u the map $\mathcal{N}_\varepsilon$ is no longer injective. We choose n_s and n_u in such a way that $\mathcal{N}_\varepsilon$ determines a two-sheeted covering of the fundamental domains $\Delta_\varepsilon^\pm$.

Let $\Delta_\varepsilon^{u\pm}$ and $\Delta_\varepsilon^{s\pm}$ be the preimages of $\Delta_\varepsilon^\pm$ that are contained in a neighborhood of the unstable and stable separatrices respectively:

$$\Delta_\varepsilon^{u\pm} \subset \{(q,p) : c < \pm q < c^{-1},\ |p| \text{ is small}\},$$
$$\Delta_\varepsilon^{s\pm} \subset \{(q,p) :\ q \text{ is small},\ c < \pm p < c^{-1}\}.$$

The definition of the gluing map becomes

$$\mathcal{G}_\varepsilon^\pm : \Delta_\varepsilon^{u\pm} \to \Delta_\varepsilon^{s\pm}, \qquad \mathcal{G}_\varepsilon^\pm = (\mathcal{N}_\varepsilon|_{\Delta_\varepsilon^{s\pm}})^{-1} \circ \mathcal{N}_\varepsilon|_{\Delta_\varepsilon^{u\pm}}.$$

If (q_u, p_u) and (q_s, p_s) denote the normal coordinates of a point $z \in \Delta_\varepsilon^\mp$ obtained by the extensions along the unstable and stable separatrices respectively, i.e.,

$$\mathcal{N}_\varepsilon(q_u, p_u) = z, \qquad (q_u, p_u) \in \Delta_\varepsilon^{u\pm} \quad \text{and}$$
$$\mathcal{N}_\varepsilon(q_s, p_s) = z, \qquad (q_s, p_s) \in \Delta_\varepsilon^{s\pm},$$

then these coordinates are identified by the gluing map, $\mathcal{G}_\varepsilon^\pm(q_u, p_u) = (q_s, p_s)$.

Having explicit formulas for $\mathcal{G}_\varepsilon^\pm$, one can study the dynamics of the map T_ε in a neighborhood of the separatrices in the normal coordinates. Indeed, the normal coordinates define a smooth map of the cross-shaped domain shown to the left in Fig. 4.3 onto a neighborhood of the separatrices of the point σ_ε. Let the initial conditions be given in normal coordinates. To construct the trajectory of the system, we first apply the hyperbolic rotation $\mathcal{T}_\varepsilon$ several times. The iterations of the point begin moving away along the northeast or southwest branch of the cross and eventually fall into the domain of one of the gluing maps. After applying the map $\mathcal{G}_\varepsilon^+$ or $\mathcal{G}_\varepsilon^-$, we apply the map $\mathcal{T}_\varepsilon$ again, and so on. It is clear here that almost all information about the dynamics is contained in the gluing maps because the formulas for a hyperbolic rotation differ unessentially for different systems.

Gluing maps: formulas.

Proposition 4.1. *Let a C^2-smooth function F be a first integral of the hyperbolic rotation $\mathcal{T}(q,p) = (q\mathcal{M}(pq), p/\mathcal{M}(pq))$. Then $F(q,p) = F_0 + G(qp)$, $F_0 =$ const.*

Proof. Obviously, $F = F_0 = \text{const}$ on the separatrices $\{q = 0\}$ and $\{p = 0\}$. Therefore, $F = F_0 + O(qp)$. □

Corollary 4.2. *If F is a C^4-smooth first integral of $\mathcal{T}$, then*

$$F(q, p) = F_0 + F_1 qp + O_2(qp), \qquad F_0, F_1 = \text{const}. \tag{4.7}$$

Indeed, qp is a first integral of $\mathcal{T}$. Hence by Proposition 4.1 $F = F_0 + qp\,\Phi(q, p)$, where $\Phi \in C^2$ is a first integral. The equation $\Phi = \text{const.} + O(qp)$ implies (4.7). Inductive application of this argument proves the following corollary.

Corollary 4.3. *If both $\mathcal{T}$ and the first integral F are real-analytic, then $F(q, p) = \mathcal{F}(qp)$.*

The gluing maps corresponding to integrable systems have a rather simple form.

Proposition 4.2. *Suppose that $\hat{T} = T_0$ satisfies conditions* I–IV *with a C^4-smooth first integral F. Then the gluing maps $\mathcal{G}_0^\pm$ are of the form $(q_s, p_s) = \mathcal{G}_0^\pm(q_u, p_u)$ with*

$$\begin{cases} q_s = \dfrac{q_u^2 p_u}{\alpha_\pm^2} + O_2(q_u p_u), \\ p_s = \dfrac{\alpha_\pm^2}{q_u} + O(q_u p_u), \end{cases} \tag{4.8}$$

where $\alpha_\pm$ are positive constants.

Proof. Let $\mathcal{G}_0^+(q_u, p_u) = (q_s, p_s)$. Then

$$F(q_s, p_s) = F(q_u, p_u), \qquad dp_s \wedge dq_s = dp_u \wedge dq_u, \tag{4.9}$$

where the first integral F satisfies (4.7). Condition IV implies $F_1 \neq 0$. Hence,

$$q_s p_s = q_u p_u + O_2(q_u p_u). \tag{4.10}$$

Substituting the equation $q_s = (q_u p_u + O_2(q_u p_u))/p_s$ in (4.9), we obtain the following partial differential equation:

$$p_u \frac{\partial p_s}{\partial p_u} - q_u \frac{\partial p_s}{\partial q_u} = p_s(1 + O(q_u p_u)).$$

The solutions $p_s = p_s(q_u, p_u)$ smooth at $p_u = 0$ are $p_s = q_u^{-1}(a_0 + O(q_u p_u))$. The inequality $a_0 > 0$ follows from the convention in Remark 4.1 about the choice of normal coordinates. This implies the second equation in (4.8). The first equation in (4.8) now follows from (4.10). □

The gluing maps $\mathcal{G}_\varepsilon^\pm$ are obtained as perturbations of (4.8).

Proposition 4.3. *Suppose that $\hat{T} = T_0$ satisfies conditions I–IV with $F \in C^4$. Then $\mathcal{G}_\varepsilon^\pm$ are of the form $(q_s, p_s) = \mathcal{G}_\varepsilon^\pm(q_u, p_u)$ with*

$$\begin{cases} q_s = q_u \frac{q_u p_u + \varepsilon \nu_\pm (\log|q_u|/\log\mu)}{\alpha_\pm^2} + O_2(q_u p_u, \varepsilon), \\ p_s = \frac{\alpha_\pm^2}{q_u} + O(q_u p_u, \varepsilon), \end{cases} \tag{4.11}$$

where $\nu_\pm$ are periodic functions with period 1.

Proof. Indeed,

$$\mathscr{G}_\varepsilon^\pm(q_u, p_u) = \mathscr{G}_0^\pm(q_u, p_u) + \varepsilon(f_\pm(q_u, p_u, \varepsilon), g_\pm(q_u, p_u, \varepsilon)),$$

where $f_\pm$ and $g_\pm$ are smooth functions. By (4.6) $f_\pm(\mu q_u, 0, 0) = \mu f_\pm(q_u, 0, 0)$. We define $\nu_\pm : \mathbb{R} \to \mathbb{R}$ by the equation

$$f_\pm(q_u, 0, 0) = q_u \frac{\nu_\pm(\log|q_u|/\log\mu)}{\alpha_\pm^2}.$$

Then the functions $\nu_\pm$ are smooth and periodic with period 1. Hence,

$$\begin{aligned} \mathscr{G}_\varepsilon^\pm(q_u, p_u) &= \mathscr{G}_0^\pm(q_u, p_u) + \varepsilon\left(q_u \frac{\nu_\pm(\log|q_u|/\log\mu)}{\alpha_\pm^2} + O(p_u), g_\pm(q_u, p_u, \varepsilon)\right) \\ &= \left(q_u \frac{q_u p_u + \varepsilon\nu_\pm(\log|q_u|/\log\mu)}{\alpha_\pm^2}, \frac{\alpha_\pm^2}{q_u}\right) \\ &\quad + (O_2(q_u p_u, \varepsilon), O(q_u p_u, \varepsilon)). \end{aligned}$$

This completes the proof of Proposition 4.3. □

Convenient coordinates on $\Delta_\varepsilon^\pm$. On the domains $\Delta_\varepsilon^\pm$ we have two coordinate systems (q_s, p_s) and (q_u, p_u) given by the maps $\mathscr{N}_\varepsilon|_{\Delta_\varepsilon^{s\pm}}$ and $\mathscr{N}_\varepsilon|_{\Delta_\varepsilon^{u\pm}}$ respectively. In the latter coordinate system the separatrix map by definition has the form

$$\mathscr{SM}_\varepsilon = \mathscr{T}_\varepsilon^{n_r} \circ \mathscr{G}_\varepsilon, \tag{4.12}$$

where, we recall, $n_r(z)$ stands for the number of iterations of the point $z \in \Delta_\varepsilon^r$ under the map T_ε until the first return to Δ_ε. In the normal coordinates $n_r = n_r(q, p)$, $(q, p) \in \Delta_\varepsilon^{s\pm}$, is the number of iterations of the map $\mathscr{T}_\varepsilon$ until the trajectory falls into the domain $\Delta_\varepsilon^{u\pm}$.

Away from the stable separatrix $\{q = 0\}$, the following variables are defined:

$$\begin{aligned} x &= \frac{\log|q|}{\log \mathscr{M}_\varepsilon(qp)} = \frac{\log|q|}{\lambda}(1 + O(qp, \varepsilon)), \\ y &= \frac{1}{\varepsilon}\int_0^{qp} \log \mathscr{M}_\varepsilon(\xi)\, d\xi = \frac{1}{\varepsilon} qp(\lambda + O(qp, \varepsilon)). \end{aligned} \tag{4.13}$$

Then $dy \wedge dx = \varepsilon^{-1} dp \wedge dq$. In the variables (x, y) the hyperbolic rotation $\mathscr{T}_\varepsilon$ becomes

$$\mathscr{T}_\varepsilon(x, y) = (x + 1, y). \tag{4.14}$$

Therefore, the map $\mathcal{N}_\varepsilon|_{\Delta_\varepsilon^{u\pm}}(x, y)$ defines smooth coordinates (x mod 1, y) on the cylinders $\overline{\Delta}_\varepsilon^{\pm}/\sim$. To recover the information about the sign of the variable q (lost in (4.13)), we adjoin the sign $s = \operatorname{sign}(q)$ to the coordinate system (x, y). If $s = +$ ($s = -$) then $(x, y) \in \Delta_\varepsilon^{u+}$ (respectively, $(x, y) \in \Delta_\varepsilon^{u-}$). The separatrix map is represented in the coordinates (x mod 1, y, s).

Assuming that x is defined modulo 1, we see from (4.12) and (4.14) that the separatrix map $\mathcal{SM}_\varepsilon$ is determined solely by $\mathcal{G}_\varepsilon$. It remains to represent the gluing map in the variables (x, y). Inverting equations (4.13), we obtain

$$qp = \varepsilon y\left(\frac{1}{\lambda} + O(\varepsilon)\right), \qquad q = s e^{\lambda x(1+O(\varepsilon))}.$$

Let $(x_u, y_u) \in \overline{\Delta}_\varepsilon^{u\pm}$ and $(x_s, y_s) = \mathcal{G}_\varepsilon^{\pm}(x_u, y_u)$. By (4.11) we obtain:

$$\begin{aligned}
q_s p_s &= q_u p_u + \varepsilon \nu_\pm\left(\frac{\log|q_u|}{\log\mu}\right) + O_2(q_u p_u, \varepsilon) = \frac{1}{\lambda}\varepsilon y_u + \varepsilon\nu_\pm(x_u) + O_2(\varepsilon),\\
y_s &= \varepsilon^{-1} q_s p_s (\lambda + O(q_s p_s, \varepsilon)) = y_u + \lambda\nu_\pm(x_u) + O(\varepsilon),\\
x_s &= \frac{-\log|p_s| + \log|q_s p_s|}{\lambda}(1 + O(q_s p_s, \varepsilon))\\
&= \frac{1}{\lambda}\left(\log\left|\frac{q_u}{\alpha_\pm^2} + O(\varepsilon)\right| + \log\frac{\varepsilon|y_s|}{\lambda + O(\varepsilon)}\right)(1 + O(\varepsilon))\\
&= x_u + \frac{1 + O(\varepsilon)}{\lambda}\left(\log\frac{\varepsilon}{\alpha_\pm^2\lambda} + \log|y_s| + O(\varepsilon)\right),\\
\operatorname{sign}(q_s) &= \operatorname{sign}(y_s)\cdot\operatorname{sign}(q_u).
\end{aligned}$$

Finally, writing $\mathcal{SM}_\varepsilon(x, y, s) = (x^+, y^+, s^+)$, we obtain (4.1).

4.3 Poincaré–Melnikov Integral

In this section we obtain formulas for the parameters $\alpha_\pm$ and the functions $\nu_\pm$.

Let F be a first integral of the map T_0, $F(\sigma) = 0$. We suppose that the critical point σ of F is non-degenerate. Let $\Gamma^\pm : \mathbb{R} \to \{z \in D : F(z) = 0\}$ be the upper and lower loops of the unperturbed separatrix parametrized in the natural way:

$$T_0 \circ \Gamma^\pm(t) = \Gamma^\pm(t+1). \tag{4.15}$$

In a neighborhood of the point σ there are symplectic coordinates (u, v) such that the curves $\Gamma^\pm(t)$ satisfy the equations

$$\begin{aligned}
(u, v) &= (c_u^\pm \mu^t + O(\mu^{2t}), O(\mu^{2t})) \quad \text{as } t \to -\infty,\\
(u, v) &= (O(\mu^{-2t}), c_s^\pm \mu^{-t} + O(\mu^{-2t})) \quad \text{as } t \to +\infty,
\end{aligned} \tag{4.16}$$

where $c_s^+, c_u^+ > 0$ and $c_s^-, c_u^- < 0$. These coordinates can be obtained by a linear change of variables from any symplectic coordinates defined in a neighborhood of the point σ. The constants $c_s^\pm$ and $c_u^\pm$ are not uniquely determined; however, only one of these constants can be chosen arbitrarily.

Proposition 4.4.

(a) *The quantities $c_s^+ c_u^+$, $c_s^- c_u^-$, and c_s^+/c_s^- do not depend on the choice of the variables (u, v).*
(b) *The quantities $c_s^+ c_u^+$ and $c_s^- c_u^-$ are preserved under changes of the parametrizations $t \mapsto t + \hat{c}^\pm$ on the curves $\Gamma^\pm$.*
(c) $\alpha_\pm^2 = c_s^\pm c_u^\pm$.

Corollary 4.4. *The constants $\alpha_\pm$ can be computed if the natural parametrization of the unperturbed separatrices is known.*

Proof (of Proposition 4.4). Let (u', v') be another coordinate system satisfying (4.16) (may be, with other constants $c_{s,u}^\pm$). Then

$$u' = uc' + O(u^2 + v^2), \qquad v' = v/c' + O(u^2 + v^2)$$

with some positive c'. These equations imply assertion (a).

Changing the parameterization of Γ^+ (or Γ^-), we multiply c_s^+ (respectively, c_s^-) by a positive constant and divide c_u^+ (respectively, c_u^-) by the same constant. This implies assertion (b).

To prove assertion (c) we note that the normal coordinates (q, p) also satisfy equations (4.16). Moreover, in normal coordinates

$$\begin{aligned} &\text{(i)} \quad \Gamma^\pm = (c_u^\pm \mu^t, 0), \\ &\text{(ii)} \quad \Gamma^\pm = (0, c_s^\pm \mu^{-t}). \end{aligned}$$

The gluing maps $\mathscr{G}_0^\pm$ (4.11)$|_{\varepsilon=0}$ transform parameterizations (i) into (ii):

$$(0, c_s^\pm \mu^{-t}) = \mathscr{G}_0^\pm(c_u^\pm \mu^t, 0) = (0, \alpha_\pm^2/(c_u^\pm \mu^t)).$$

These equations imply assertion (c). □

Remark 4.2. Given normal coordinates (q, p) there exist unique natural parameterizations on $\Gamma^\pm$ such that these curves are as follows:

$$\begin{aligned} &\text{(i)} \quad \Gamma^\pm = (\pm\alpha_\pm \mu^t, 0), \\ &\text{(ii)} \quad \Gamma^\pm = (0, \pm\alpha_\pm \mu^{-t}). \end{aligned} \tag{4.17}$$

The functions $\nu_\pm$ can be obtained from a discrete analogue of the Poincaré–Melnikov integral. We put

$$w(T_0(z)) = \frac{d}{d\varepsilon}\bigg|_{\varepsilon=0} T_\varepsilon(z). \tag{4.18}$$

Then w is a Hamiltonian vector field on D which controls the perturbation T_ε in the first approximation in ε. We define

$$v_\pm^*(t) = \sum_{n=-\infty}^{\infty} \langle dF, w(\Gamma^\pm(t+n))\rangle, \tag{4.19}$$

where the expression $\langle dF, w(A)\rangle$ denotes application of the covector dF to the vector w at the point $A \in D$. Since the critical point σ is non-degenerate, we have

$$F(u,v) = F_0 uv + O_3(u,v), \qquad F_0 \neq 0.$$

Proposition 4.5. *The following equations hold:*

$$v_\pm(t) = \frac{1}{F_0} v_\pm^* \left(t - \frac{\log \alpha_\pm}{\log \mu}\right).$$

Proof. Let A_ε, $-\varepsilon_0 < \varepsilon < \varepsilon_0$ be a smooth curve on D. Then

$$\left\langle dF, \frac{d}{d\varepsilon}\bigg|_{\varepsilon=0} T_\varepsilon^n(A_\varepsilon)\right\rangle - \left\langle dF, \frac{d}{d\varepsilon}\bigg|_{\varepsilon=0} A_\varepsilon\right\rangle = \operatorname{sign} n \cdot \sum_k \langle dF, w(T_0^k(A_0))\rangle, \tag{4.20}$$

where $k \in \{1, \dots, n\}$ if $n > 0$, and $k \in \{0, -1, \dots, n+1\}$ if $n < 0$. This equation can be obtained by a direct calculation with the help of the identities

$$\begin{gathered} \frac{d}{d\varepsilon}\bigg|_{\varepsilon=0} T_\varepsilon^n(A_\varepsilon) = dT_0^n \frac{d}{d\varepsilon}\bigg|_{\varepsilon=0} A_\varepsilon + \operatorname{sign}(n) \cdot \sum_k dT_0^{n-k} w(T_0^k(A_0)), \\ \text{where} \begin{cases} 1 \le k \le n & \text{if } n > 0, \\ n+1 \le k \le 0 & \text{if } n < 0, \end{cases} \\ \langle dF, dT_0^m v\rangle = \langle dF, v\rangle, \quad \text{where } v \text{ is a vector at a point } A \in D. \end{gathered} \tag{4.21}$$

Identity (4.21) can be obtained by differentiation of $F \circ T_0^m = F$ along v.

Now suppose that the point A_ε^{s+} (respectively, A_ε^{u+}) lies on the upper stable separatrix Γ_ε^{s+} (respectively, on the upper unstable one Γ_ε^{u+}) and $A_0^{s+} = A_0^{u+} = A_0$. By using (4.20), we have for any natural n:

$$\left\langle dF, \frac{d}{d\varepsilon}\bigg|_{\varepsilon=0} A_\varepsilon^{s+}\right\rangle = \left\langle dF, \frac{d}{d\varepsilon}\bigg|_{\varepsilon=0} T_\varepsilon^n(A_\varepsilon^{s+})\right\rangle - \sum_{k=1}^{n} \langle dF, w(T_0^k(A_0))\rangle,$$

$$\left\langle dF, \frac{d}{d\varepsilon}\bigg|_{\varepsilon=0} A_\varepsilon^{u+}\right\rangle = \left\langle dF, \frac{d}{d\varepsilon}\bigg|_{\varepsilon=0} T_\varepsilon^{-n}(A_\varepsilon^{u+})\right\rangle + \sum_{k=1-n}^{0} \langle dF, w(T_0^k(A_0))\rangle.$$

Hence,

$$\left\langle dF, \left(\frac{d}{d\varepsilon}\bigg|_{\varepsilon=0} A_\varepsilon^{u+} - \frac{d}{d\varepsilon}\bigg|_{\varepsilon=0} (A_\varepsilon^{s+})\right)\right\rangle = \sum_{k=-\infty}^{+\infty} \langle dF, w(T_0^k(A_0))\rangle. \tag{4.22}$$

Here we have used the equation

$$\lim_{n\to\infty} \frac{d}{d\varepsilon}\bigg|_{\varepsilon=0} T_\varepsilon^n(A_\varepsilon^{s+}) = \lim_{n\to\infty} \frac{d}{d\varepsilon}\bigg|_{\varepsilon=0} T_\varepsilon^{-n}(A_\varepsilon^{u+}).$$

Now let us compute the left-hand side of (4.22) in another way. In the normal coordinates $(q, p) = (q(\varepsilon), p(\varepsilon))$ the curves A_ε^{s+} and A_ε^{u+} have the following form:

$$\begin{aligned} A_\varepsilon^{u+} &= (j_u(\varepsilon), 0), \qquad & j_u(\varepsilon) > 0, \\ A_\varepsilon^{s+} &= (0, j_s(\varepsilon)), \qquad & j_s(\varepsilon) > 0. \end{aligned}$$

These equations give coordinate presentations of A_ε^{u+} and A_ε^{s+} in different domains: Δ_ε^{u+} and Δ_ε^{s+} respectively (see Fig. 4.3). To compare these two curves, we use the coordinate presentation of A_ε^{u+} in Δ_ε^{s+} by using explicit form of $\mathcal{G}_\varepsilon^+$ (see (4.8)):

$$A_\varepsilon^{u+} = \mathcal{G}_\varepsilon^+(j_u(\varepsilon), 0) = \left(\frac{\varepsilon j_u(0)}{\alpha_+^2}\, \nu_+\left(\frac{\log j_u(u)}{\log \mu}\right) + O(\varepsilon^2),\ \frac{\alpha_+^2}{j_u(0)} + O(\varepsilon)\right).$$

Since $A_0^{s+} = A_0^{u+}$, we have: $j_u(0) j_s(0) = \alpha_+^2$.

The normal coordinates (q, p) depend on ε. We must take into account this fact when evaluating in these coordinates any of the vectors $dA_\varepsilon^{u,s+}/d\varepsilon$. However, we can forget about this when evaluating the difference

$$\frac{d}{d\varepsilon}\bigg|_{\varepsilon=0} A_\varepsilon^{u+} - \frac{d}{d\varepsilon}\bigg|_{\varepsilon=0} A_\varepsilon^{s+}.$$

The difference equals

$$\begin{aligned} &\frac{d}{d\varepsilon}\bigg|_{\varepsilon=0}\left(\frac{\varepsilon j_u(0)}{\alpha_+^2}\, \nu_+\left(\frac{\log j_u(u)}{\log \mu}\right) + O(\varepsilon^2),\ \frac{\alpha_+^2}{j_u(0)} + O(\varepsilon) - j_s(\varepsilon)\right) \\ &\quad = \left(\frac{j_u(0)}{\alpha_+^2}\, \nu_+\left(\frac{\log j_u(u)}{\log \mu}\right), \hat{j}\right), \end{aligned} \tag{4.23}$$

where the value of $\hat{j}$ is not essential.

In coordinates $(q, p) = (q(\varepsilon), p(\varepsilon))$ the first integral F satisfies the equation

$$F(q, p) = F_0 qp + O(\varepsilon) + O(q^2 p^2).$$

Hence, according to (4.23),

$$\left\langle dF, \left(\frac{d}{d\varepsilon}\bigg|_{\varepsilon=0} A^{u+}_\varepsilon - \frac{d}{d\varepsilon}\bigg|_{\varepsilon=0} A^{s+}_\varepsilon \right) \right\rangle$$
$$= F_0 \left\langle (j_s(0), 0), \left(\frac{j_u(0)}{\alpha_+^2} \nu_+ \left(\frac{\log j_u(u)}{\log \mu} \right), \hat{j} \right) \right\rangle$$
$$= F_0\, \nu_+(\log j_u(0)/\log \mu). \tag{4.24}$$

We take $A_0 = \Gamma^+(t)$. Then

$$T_0^m(A_0) = \Gamma^+(t+m), \qquad m \in \mathbb{Z}. \tag{4.25}$$

According to (4.17),

$$j_u(0) = \alpha_+ \mu^t, \quad j_s(0) = \alpha_+ \mu^{-t}. \tag{4.26}$$

By using (4.22)–(4.26) we obtain

$$F_0\, \nu_+(t + \log \alpha_+ / \log \mu) = \sum_{k=-\infty}^{+\infty} \langle dF \cdot w(\Gamma^+(t+k)) \rangle = \nu_+^*(t).$$

Analogously we get $F_0\, \nu_-(t + \log \alpha_- / \log \mu) = \nu_-^*(t)$. □

4.4 Hamiltonian System

Here we derive Theorem 4.2 from Theorem 4.1.

Let T_ε be the period-1 map in the system with Hamiltonian $H(z, t, \varepsilon)$. The normal coordinates $(q, p) = (q_\varepsilon, p_\varepsilon)$ of the map T_ε in Theorem 3.1 generate normal coordinates for the Hamiltonian system (see Theorem 3.2). Indeed, it suffices to let the normal coordinates of $g_\varepsilon^t(z)$ be $(q\mathscr{M}_\varepsilon^t, p\mathscr{M}_\varepsilon^{-t})$, where (q, p) are normal coordinates of the point z and g_ε^t is the phase flow on the extended phase space $D \times \mathbb{T}_t$. In the normal coordinates the phase flow has the form

$$g_\varepsilon^t(q, p) = (qe^{\mathscr{H}_\varepsilon'(pq)t}, pe^{-\mathscr{H}_\varepsilon'(pq)t}). \tag{4.27}$$

Comparing $g_\varepsilon^t|_{t=1}$ and (3.1), we see that the functions $\mathscr{M}_\varepsilon(pq)$ and $\mathscr{H}_\varepsilon(pq)$ in Theorems 3.1 and 3.2 are connected as follows:

$$\frac{d\mathscr{H}_\varepsilon(\xi)}{d\xi} = \log \mathscr{M}_\varepsilon(\xi). \tag{4.28}$$

Since the unperturbed Hamiltonian H_0 does not depend on the time, the normalizing change of coordinates depends on time only in terms of order ε. Therefore, the new Hamiltonian is $\mathscr{H}_\varepsilon = H_\varepsilon + O(\varepsilon)$.

Away from the stable separatrix $\{q = 0\}$ on $\mathbb{R}^2_{q,p}$ we introduce the "energy-time" coordinates (h, t) in the following way. We take $t = 0$ for $q = \pm 1$. Assuming that

$\mathscr{H}_\varepsilon(0) = 0$, we see from (4.28) and (4.27) that

$$t = \frac{\log |q_\varepsilon|}{\log \mathscr{M}_\varepsilon}, \qquad h = \varepsilon^{-1}\mathscr{H}_\varepsilon = \varepsilon^{-1}\int_0^{q_\varepsilon p_\varepsilon} \log \mathscr{M}_\varepsilon(\xi)\, d\xi.$$

Thus, the coordinates (h, t) coincide with the coordinates (y, x) (see (4.13)). Therefore, it follows from (4.1) that $\mathscr{S}\mathscr{M}_\varepsilon(h, t, s) = (h^+, t^+, s^+)$, with

$$\begin{cases} h^+ = h + \lambda v_s(t) + O(\varepsilon), \\ t^+ = t + \frac{1+O(\varepsilon)}{\lambda}\left(\log \frac{\varepsilon}{\alpha_s^2\lambda} + \log |h^+|\right), \\ s^+ = s \cdot \operatorname{sign}(h^+). \end{cases} \tag{4.29}$$

Proposition 4.6. *In equations* (4.29) $v_\pm(t) = \frac{1}{\lambda} v_\pm^*(t - \frac{\log \alpha_\pm}{\log \mu})$, *where*

$$-v_\pm^*(t) = \int_{-\infty}^{+\infty} \{H_0, H_1|_{\varepsilon=0}\}(\Gamma^\mp(t+\xi), \xi)\, d\xi. \tag{4.30}$$

Proof. Let $z_\varepsilon(t) = z_0(t) + \varepsilon z'(t) + O(\varepsilon^2)$ be a solution of the system with Hamiltonian H and initial conditions $z_\varepsilon(0) = (u, v)$. Then

$$T_\varepsilon(u, v) = z_\varepsilon(1). \tag{4.31}$$

Below we assume that the point (u, v) does not depend on ε. Differentiating (4.31) with respect to ε at the point $\varepsilon = 0$, we obtain

$$w(T_0(u, v)) = z'(1). \tag{4.32}$$

Differentiating the equation $\frac{d}{dt} H_0(z_\varepsilon(t)) = \varepsilon\{H_1, H_0\}(z_\varepsilon(t), t, \varepsilon)$ with respect to ε and setting $\varepsilon = 0$, we get

$$\frac{d}{dt}\langle dH_0(z_0(t)), z'(t)\rangle = \{H_1, H_0\}(z_0(t)). \tag{4.33}$$

Integrating (4.33) with $z_0(t) = \Gamma^+(t_0 + t)$, we have

$$\begin{aligned} &\langle dH_0(\Gamma^+(t_0+1)), z'(1)\rangle - \langle dH_0(\Gamma^+(t_0)), z'(0)\rangle \\ &= \int_0^1 \{H_1, H_0\}(\Gamma^+(t_0+t), t)\, dt. \end{aligned}$$

Using equations (4.32) and $z'(0) = 0$, we get

$$dH_0 \cdot w(\Gamma^+(t_0+1)) = \int_0^1 \{H_1, H_0\}(\Gamma^+(t_0+\xi), \xi)\, d\xi.$$

The corresponding equation for Γ^- can be found in a similar way. Proposition 4.6 follows from these equations and the definition of the function ν_+^* in (4.19). □

4.5 Separatrix Map for a Pendulum

As an example we consider a pendulum whose suspension point oscillates periodically along the vertical. The Hamiltonian of the system has the form

$$H(\hat{u}, \hat{v}, t, \varepsilon) = \hat{v}^2/2 + \Omega^2 \cos \hat{u} + \varepsilon\theta(\omega t) \cos \hat{u}. \tag{4.34}$$

Here $\hat{u} = \hat{u} \bmod 2\pi$ is the angle between the pendulum and the vertical, $\hat{v}$ is the corresponding momentum, $\Omega > 0$ is the "intrinsic frequency" of the system (Ω^2 is equal to the gravitational acceleration divided by the length of the pendulum), ω is the frequency of oscillation of the suspension point, and the parameter ε is proportional to the amplitude of the oscillations multiplied by ω^2. The law of oscillation of the suspension point is determined by the 2π-periodic function θ. We suppose that the parameter ε is small and the other parameters in the system are of order 1.

We make the system 1-periodic in time by a symplectic change of variables. Let $\tau = \omega t/(2\pi)$ be the new time and $\mathscr{H} = 2\pi H/\omega$ the new Hamiltonian. The Poincaré map in this system has the hyperbolic fixed point $\hat{u} = \hat{v} = 0$. One can easily compute the multiplier: $\mu = e^{2\pi\Omega/\omega}$. In this case, $\lambda = \log \mu = 2\pi\Omega/\omega$. The separatrices $\Gamma^\pm$ have the form

$$(\hat{u}, \hat{v}) = \left(4 \arctan e^{\pm 2\pi\Omega\tau/\omega}, \pm\frac{2\Omega}{\cosh(2\pi\Omega\tau/\omega)}\right).$$

The variables (u, v) (see (4.16)) are as follows:

$$u = \hat{u}\sqrt{\Omega/2} + \hat{v}/\sqrt{2\Omega}, \quad v = -\hat{u}\sqrt{\Omega/2} + \hat{v}/\sqrt{2\Omega}.$$

We get: $c_s^\pm = c_u^\pm = \pm 4\sqrt{2\Omega}$. Hence by Proposition 4.4 $\alpha_\pm^2 = 32\Omega$.

Using Proposition 4.6, we obtain

$$\nu_\pm^*(\tau) = \left(\frac{2\pi}{\omega}\right)^2 \int_{-\infty}^{+\infty} \theta(2\pi\xi)(\hat{v}\sin\hat{u})\big|_{\Gamma^\pm(\tau+\xi)}\, d\xi.$$

Direct computations lead to the equation

$$\nu_\pm^*(\tau) = -\frac{4\pi^2 i\omega}{\Omega^2} \sum_{m\in\mathbb{Z}\setminus\{0\}} \frac{m^2\theta_m}{\sinh(\pi m\omega/(2\Omega))} e^{-2\pi i m\tau}, \tag{4.35}$$

where the Fourier expansion of the function θ has the form

$$\theta(s) = \sum_{m \in \mathbb{Z}} \theta_m e^{ims}.$$

In particular, $\nu_\pm^* \not\equiv$ const for $\theta \not\equiv$ const. Hence (Corollary 4.1) for $\theta \not\equiv$ const. the perturbed system has no analytic first integral for any small $\varepsilon \neq 0$.

The equation $\nu_+^* = \nu_-^*$ appears because the Hamiltonian is invariant with respect to the symmetry

$$(\hat{u}, \hat{v}) \mapsto (-\hat{u}, -\hat{v})$$

which exchanges the upper and lower separatrix loops.

If we increase in (4.35) the value of the parameter ω/Ω (the case of a large frequency of the oscillation of the suspension point), the functions $\nu_\pm^*$ tend to zero exponentially fast. This phenomenon of exponentially small separatrix splitting was discovered by Poincaré. However this observation is somewhat formal because, if we put $\omega/\Omega = 1/\varepsilon^\alpha$, $\alpha > 0$, the above proof of the fact that $\nu_\pm^*$ are responsible for the rate and the form of the separatrix splitting is no longer correct. Analysis of exponentially small separatrix splitting requires a much more complicated argument. This problem is discussed in Chap. 6.

4.6 Some Generalizations

The aim of this section is to indicate a unified scheme for constructing separatrix maps in the systems which satisfy conditions I–IV (Sect. 4.1), as well as in the systems T_ε for which T_0 is the identity map, and therefore there is no hyperbolic fixed point for $\varepsilon = 0$. A system of the second type appears when a resonance curve of an integrable map is disintegrated by a perturbation. Another example is a pendulum with a rapidly oscillating suspension point.

In systems of the second type the multiplier μ_ε of the hyperbolic fixed point σ_ε tends to zero as $\varepsilon \searrow 0$, and the separatrix splitting is exponentially small[2] with respect to ε. Using methods of averaging theory, one can approximate the perturbed system by an integrable map which also has a hyperbolic fixed point. Moreover, in the case of analytic symplectic maps, one can make the difference between these systems exponentially small with respect to ε. This difference is characterized below by a small parameter $\delta = \delta_\varepsilon$. Hence, if one regards the perturbed system as a δ-deformation of an integrable system, then the construction of the separatrix map is similar to the case presented above.

Let T_ε, $0 < \varepsilon < \varepsilon_0$, be a smooth family of two-dimensional symplectic maps close to an integrable map T_0. Suppose that

(A) *there is a smooth family of integrable maps* $\widetilde{T}_\varepsilon$ *such that, for any* $\varepsilon \in (0, \varepsilon_0)$, *the map* $\hat{T} = \widetilde{T}_\varepsilon$ *satisfies conditions* I–IV *and* $|T_\varepsilon - \widetilde{T}_\varepsilon| \leq \delta_\varepsilon$, *where* $\delta_\varepsilon \to 0$ *as* $\varepsilon \to 0$.

---

[2] Provided the system is real-analytic.

If the map $\hat{T} = T_0$ itself satisfies conditions I–IV, one can take $\widetilde{T}_\varepsilon \equiv T_0$ and $\delta_\varepsilon = \varepsilon$.

Since the maps $\widetilde{T}_\varepsilon$ satisfy conditions I–IV, the gluing maps for $\widetilde{T}_\varepsilon$ satisfy (4.8), where $\alpha_\pm = \alpha_\pm(\varepsilon)$. Then the gluing maps for T_ε have the form

$$\begin{cases} q_s = \frac{q_u^2 p_u}{\alpha_\pm^2(\varepsilon)} + \delta_\varepsilon q_u f_\pm(q_u, p_u, \varepsilon) + O_2(q_u p_u), \\ p_s = \frac{\alpha_\pm^2(\varepsilon)}{q_u} + \delta_\varepsilon p_u g_\pm(q_u, p_u, \varepsilon) + O_1(q_u p_u). \end{cases} \tag{4.36}$$

Here $(q, p) = (q(\varepsilon), p(\varepsilon))$ are normal coordinates for T_ε in a neighborhood of the hyperbolic fixed point σ_ε, and as before (q_u, p_u) and (q_s, p_s) are the extended coordinates along the unstable and stable separatrices of σ_ε, respectively.

Let $\mu_\varepsilon = \mathcal{M}_\varepsilon(0)$ be a multiplier of the hyperbolic fixed point σ_ε of the map T_ε and let $\mathcal{M}_\varepsilon(qp)$ be the function defining T_ε in the normal coordinates (see Theorem 3.1). We put

$$f_\pm(q, 0, \varepsilon) = \alpha_\pm^{-2} \hat{\nu}_\pm\left(\frac{\log |q|}{\log \mu_\varepsilon}, \varepsilon\right). \tag{4.37}$$

The functions $\hat{\nu}_\pm$ are periodic with respect to the first argument, with period 1.

Below we assume the condition

(B) $\lim_{\varepsilon \searrow 0} \frac{\delta_\varepsilon}{\log \mu_\varepsilon} = 0$.

Introducing the symplectic coordinates (x, y),

$$\begin{aligned} x &= \frac{\log |q|}{\log \mathcal{M}_\varepsilon(qp)} = \frac{\log |q|}{\log \mu_\varepsilon}(1 + O_1(qp)), \\ y &= \delta_\varepsilon^{-1} \int_0^{qp} \mathcal{M}_\varepsilon(\xi)\, d\xi = \delta_\varepsilon^{-1} qp \log \mu_\varepsilon (1 + O_1(qp)), \end{aligned} \tag{4.38}$$

and adjoining to them the index $s = \operatorname{sign}(q)$, we obtain the following formulas for the separatrix map: $\mathcal{SM}_\varepsilon(x, y, s) = (x^+, y^+, s^+)$, where

$$\begin{cases} x^+ = x + \frac{1}{\log \mu_\varepsilon}\left[\log \frac{\delta_\varepsilon}{\alpha_s^2 \log \mu_\varepsilon} + \log\left|y^+ + \delta_\varepsilon \log \mu_\varepsilon \cdot O\left(1 + \frac{y}{\log \mu_\varepsilon}\right)^2\right|\right], \\ y^+ = y + \log \mu_\varepsilon \cdot \hat{\nu}_s(x, \varepsilon) + \delta_\varepsilon \log \mu_\varepsilon \cdot O\left(1 + \frac{y}{\log \mu_\varepsilon}\right)^2, \\ s^+ = s \cdot \operatorname{sign}(y^+). \end{cases} \tag{4.39}$$

We also assume that

(C) *the functions* $\hat{\nu}_\pm$ *have the form*

$$\hat{\nu}_\pm(\xi, \varepsilon) = \nu_\pm(\xi) + o(1) \quad \textit{as } \varepsilon \to 0, \tag{4.40}$$

where $\nu_\pm$ *are not identically zero.*

As was to be expected, formulas (4.39) coincide with (4.1) in the "standard" situation $\mu_0 > 1$, $\delta_\varepsilon = \varepsilon$, $y \sim 1$.

Chapter 5
Width of the Stochastic Layer

In this chapter we use the separatrix map to study the stochastic layer, appearing in the vicinity of separatrices of near-integrable systems.

5.1 Definitions and Results

This section can be regarded as an informal introduction. Here we present a definition of the stochastic layer and the main results on its size.

Let T be an exact symplectic near-integrable self-map of a two-dimensional domain. We assume that T has a hyperbolic fixed point σ whose asymptotic curves (separatrices) look as shown in Fig. 5.1. Three invariant curves $\gamma_\pm$ and γ_0 closest to the separatrices form the boundary of the stochastic layer. The width w of the stochastic layer is one of important quantities characterizing chaotic properties of T in the vicinity of the separatrices.

Below in this chapter we show that under some natural assumptions

$$w/d \sim 1/\log\mu. \tag{5.1}$$

Here d is the width of a lobe domain $\mathscr{D}$ bounded by segments of the separatrices and $\mu > 1$ is the larger multiplier at the hyperbolic fixed point σ. The symbol $\sim$ means that, if we have a smooth family T_ε of analytic symplectic maps, where T_0 is integrable, then for sufficiently small ε

$$C_1/\log\mu(\varepsilon) < w(\varepsilon)/d(\varepsilon) < C_2/\log\mu(\varepsilon),$$

where C_1 and C_2 are positive constants. The constants C_1 and C_2 can be estimated (see Sect. 5.2).

If $\log\mu(\varepsilon) \sim 1$ as $\varepsilon \to 0$, we see that w and d are of the same order. However, if the stochastic layer appears when a resonant invariant curve of an integrable map disintegrates, the multiplier μ is close to 1. Hence, in this case w is much greater than d. This situation is typical for exponentially small separatrix splitting.

D. Treschev, O. Zubelevich, *Introduction to the Perturbation Theory of Hamiltonian Systems*, Springer Monographs in Mathematics,
DOI 10.1007/978-3-642-03028-4_5,

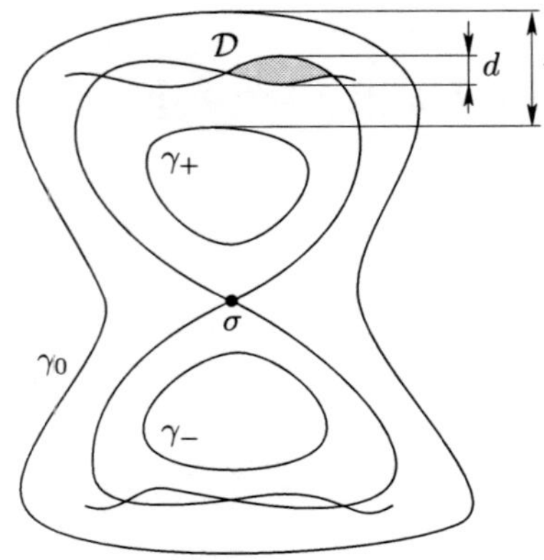

Fig. 5.1 A neighborhood of the separatrices of a hyperbolic point σ: d is the width of the lobe and w is the width of the stochastic layer.

Consider for example a resonant invariant circle $\{y = \text{const.}\}$ of an integrable self-map

$$T_0(x, y) = (x + \nu(y), y)$$

of a cylinder $Z = \{x \bmod 2\pi, y\}$. Take for simplicity the simplest resonance, i.e., the circle determined by the condition $y = y_0$, $\nu(y_0) = 0$. Consider also the perturbed map

$$T_\varepsilon(x, y) = \big(x + \nu(y) + \varepsilon f(x, y), y + \varepsilon g(x, y)\big) + O(\varepsilon^2).$$

Without loss of generality we assume that $\varepsilon \geq 0$. In the variables $(x \bmod 2\pi, \eta)$, $y = y_0 + \sqrt{\varepsilon}\eta$ the map T_ε has the form

$$(x, \eta) \mapsto \big(x + \sqrt{\varepsilon}\nu_0', \eta - \sqrt{\varepsilon}u(x)\big) + O(\varepsilon), \tag{5.2}$$

where $\nu_0' = d\nu/dy|_{y=y_0}$ and $u = -g(x, y_0)$. The condition of exact symplecticity implies that the average of the function u with respect to x vanishes. Therefore, $u(x) = \partial V(x)/\partial x$ for a certain 2π-periodic $V(x)$. Map (5.2) differs from the time-$\sqrt{\varepsilon}$ shift along solutions of the Hamiltonian system with Hamiltonian $\nu_0'\eta^2/2+V(x)$ only in terms of order $O(\varepsilon)$.

Let $\nu_0' > 0$ and let x_0 be a nondegenerate local maximum of the potential V. Then (5.2) has the hyperbolic fixed point $(x, \eta) = (x_0, 0) + O(\sqrt{\varepsilon})$. The multiplier $\mu > 1$ at this point is as follows:

$$\mu = \exp\Big(\sqrt{-\nu_0' V_0''\varepsilon} + O(\varepsilon)\Big), \qquad V_0'' = \partial^2 V/\partial x^2\big|_{x=x_0} < 0.$$

Hence, in this situation $\log\mu \sim \sqrt{\varepsilon}$ and w is much greater than d.

It turns out that in the case of multipliers close to unity, the width of the stochastic layer can be estimated more sharply. In particular, if T satisfies a certain symmetry condition (Condition (S) below) and Conjecture 5.2 about properties of the standard map is true, the following estimate holds:

$$\lim_{\varepsilon\to 0} \frac{w(\varepsilon)\log\mu(\varepsilon)}{d(\varepsilon)} = \frac{4\pi}{k_0}, \tag{5.3}$$

where

$$k_0 = \inf\Bigg\{k' : \text{for all } k > k' \text{ the standard map}$$
$$\begin{pmatrix} I \\ \varphi \bmod 2\pi \end{pmatrix} \mapsto \begin{pmatrix} J \\ \psi \bmod 2\pi \end{pmatrix} = \begin{pmatrix} I + \frac{k}{2\pi}\sin(2\pi\varphi) \\ \varphi + J \end{pmatrix}$$
$$\text{has no invariant curve homotopic to the circle } I = 0\Bigg\}.$$

The constant $k_0 = 0.971635\ldots$ was evaluated numerically in [57, 84, 101].

It is possible to present another version of (5.1). Let $\mathscr{A}$ be the symplectic area of the stochastic layer $\mathscr{S}\mathscr{L}$ and $\mathscr{A}_{\mathscr{D}}$ the area of a lobe.[1] Then

$$\frac{\mathscr{A}}{\mathscr{A}_{\mathscr{D}}\log\mathscr{A}_{\mathscr{D}}^{-1}} \sim \frac{1}{\log^2\mu}. \tag{5.4}$$

The corresponding analog of (5.3) (under the same assumptions) is

$$\lim_{\varepsilon\to 0} \frac{\mathscr{A}\log^2\mu}{\mathscr{A}_{\mathscr{D}}\log\mathscr{A}_{\mathscr{D}}^{-1}} = \frac{3\pi^2}{k_0}. \tag{5.5}$$

In the general (non-symmetric) case the fractions

$$\frac{w(\varepsilon)\log\mu(\varepsilon)}{d(\varepsilon)}, \quad \frac{\mathscr{A}(\varepsilon)\log^2\mu(\varepsilon)}{\mathscr{A}_{\mathscr{D}}(\varepsilon)\log\mathscr{A}_{\mathscr{D}}^{-1}(\varepsilon)}$$

generically do not have limits as $\varepsilon\to 0$ but oscillate between two positive constants.

Formulas (5.1), (5.3)–(5.5) can be regarded as relations between w, $\mathscr{A}$ and the quantities μ, d, $\mathscr{A}_{\mathscr{D}}$. The latter are standard to evaluate: to obtain the functions $d(\varepsilon)$ and $\mathscr{A}_{\mathscr{D}}(\varepsilon)$ one can use the Poincaré–Melnikov theory or its generalizations to the case of exponentially small splitting; $\mu(\varepsilon)$ in the main approximation is usually evaluated easily.

Below in this chapter we formulate conditions under which estimates (5.1), (5.3)–(5.5) hold. Asymptotic formulas (5.1) and (5.4) follow from Theorem 5.1. Equations (5.3) and (5.5) follow from Corollaries 5.1–5.4 provided Conjecture 5.2 is valid.

Particular cases of (5.1) were discovered in [32, 49, 147], but no rigorous proofs were presented. The estimate $w/d \le$ const. is proved in [43] provided $\log\mu \sim 1$.

[1] Below we regard the quantities $\mathscr{A}$ and $\mathscr{A}_{\mathscr{D}}$ as positive, i.e., $\mathscr{A} = |\int_{\mathscr{S}\mathscr{L}}\omega|$ and $\mathscr{A}_{\mathscr{D}} = |\int_{\mathscr{D}}\omega|$, where ω is the symplectic structure.

In [3] the estimate $w/d \geq$ const was established in the same case. Lazutkin [79] has obtained estimate (5.1) for separatrices of the standard map. Our exposition below follows [138].

5.2 Theorems on the Stochastic Layer

Consider a smooth family T_ε, $0 < \varepsilon < \varepsilon_0$, of smooth symplectic maps for which conditions (A)–(C) (Sect. 4.6) are satisfied. In particular, formulas (4.39) hold for these maps.

Below we sometimes need the following additional symmetry condition:

(S) *The map* $\mathscr{S}\mathscr{M}_\varepsilon$ *commutes with the "central symmetry"* $(x, y, s) \mapsto (x, y, -s)$.

Theorems 5.1, 5.2, and 5.3 and also Lemma 5.1 formulated in this section will be proved in Sect. 5.4.

By the *stochastic layer* we mean the domain $\mathscr{S}\mathscr{L}(\varepsilon) \subset D$ bounded by the continuous invariant curves $\gamma_\pm$ and γ_0 (see Fig. 5.1) which are the closest to the hyperbolic point σ_ε. The width of $\mathscr{S}\mathscr{L}(\varepsilon)$ is measured in terms of the function qp, where, as usual, $(q, p) = (q(\varepsilon), p(\varepsilon))$ are normal coordinates corresponding to σ_ε. More precisely, we introduce the following quantities:

$$\underline{w}_* = \min_{(q,p)\in\gamma_*} |qp|, \qquad \overline{w}_* = \max_{(q,p)\in\gamma_*} |qp|,$$

where $*$ stands for a symbol $+$, $-$, or 0. It is natural to say that the width w of the stochastic layer is contained in the interval $(\underline{w}_\bullet + \underline{w}_0, \overline{w}_\bullet + \overline{w}_0)$, where

$$\underline{w}_\bullet = \min\{\underline{w}_+, \underline{w}_-\}, \qquad \overline{w}_\bullet = \max\{\overline{w}_+, \overline{w}_-\}.$$

The width d of the domain $\mathscr{D}$ (see Fig. 5.1) is defined by

$$d = \max_{\mathscr{D}} qp - \min_{\mathscr{D}} qp. \tag{5.6}$$

Since the function qp does not depend on the choice of normal coordinates, it follows that the quantities $\underline{w}_*$, $\overline{w}_*$, and d are well defined.

Lemma 5.1. *For* $\varepsilon \searrow 0$ *(and therefore,* $\delta = \delta(\varepsilon) \searrow 0$*) the following estimates hold:*

(1) $d \sim \delta$, $\mathscr{A}_{\mathscr{D}} \sim \lambda\delta$;
(2) *if the domain* $\mathscr{D}$ *is located on the upper (lower) loop and* $\nu_+(\xi) = a\sin(2\pi(\xi - \xi_+))$ $(\nu_-(\xi) = a\sin(2\pi(\xi - \xi_-))$*, respectively), then* $d = \delta(a + o(1))$ *and* $\mathscr{A}_{\mathscr{D}} = \lambda\delta(a/\pi + o(1))$;
(3) $\underline{\mathscr{A}} \leq \mathscr{A} \leq \overline{\mathscr{A}}$*, where*

$$\underline{\mathscr{A}} = 2|\underline{w}_0 \log \underline{w}_0^{-1} + O(\underline{w}_0)| + \sum_{*\in\{+,-\}} |\underline{w}_* \log \underline{w}_*^{-1} + O(\underline{w}_*)|,$$

$$\overline{\mathscr{A}} = 2|\overline{w}_0 \log \overline{w}_0^{-1} + O(\overline{w}_0)| + \sum_{*\in\{+,-\}} |\overline{w}_* \log \overline{w}_*^{-1} + O(\overline{w}_*)|.$$

Theorem 5.1. *If T_ε satisfies conditions* (A)–(C), *then there exist constants $c_1, c_2 > 0$ such that for small $\varepsilon > 0$*

$$c_1\delta(\varepsilon)/\log\mu(\varepsilon) \le \underline{w}_\pm \le \overline{w}_\pm \le c_2\delta(\varepsilon)/\log\mu(\varepsilon),$$
$$\underline{w}_0 \le \overline{w}_0 \le c_2\delta(\varepsilon)/\log\mu(\varepsilon).$$

Moreover, if T_ε satisfies Condition (S), *then*

$$\underline{w}_0 \ge c_1\delta(\varepsilon)/\log\mu(\varepsilon). \tag{5.7}$$

Remark 5.1. By Theorem 5.1, if conditions (A)–(C) hold, the width of the stochastic layer varies in the interval $[c_1\delta/\log\mu, 2c_2\delta/\log\mu]$. Since by Lemma 5.1 the width d of the domain $\mathscr{D}$ is of order δ, we obtain (5.1).

Conjecture 5.1. Estimate (5.7) remains valid without assumption (S).

Remark 5.2. By Lemma 5.1 and Theorem 5.1, for some positive constants $\tilde{c}_1$ and $\tilde{c}_2$, we have

$$\frac{\tilde{c}_1\delta(\varepsilon)}{\log\mu(\varepsilon)}\log\frac{\log\mu(\varepsilon)}{\delta(\varepsilon)} \le \mathscr{A} \le \frac{\tilde{c}_2\delta(\varepsilon)}{\log\mu(\varepsilon)}\log\frac{\log\mu(\varepsilon)}{\delta(\varepsilon)}.$$

Consider the case $\lim_{\varepsilon\to 0}\mu(\varepsilon) = 1$ (i.e., $T_0 = \mathrm{id}$) in more detail. The computation of the function $\nu_\pm$ in this case is complicated. However, the analysis of the separatrix map turns out to be simpler for the following two reasons. First, the new small parameter $\log\mu$ appears. Second, the experience in the investigation of exponentially small separatrix splitting shows that, in the leading approximation, the 1-periodic functions $\nu_\pm$ contain only the lower harmonics, that is,

$$\hat{\nu}_\pm(\xi,\varepsilon) = \nu_\pm(\xi) + O(\varepsilon), \qquad \nu_\pm(\xi) = a_\pm \sin(2\pi(\xi - \xi_\pm)), \tag{5.8}$$

where $a_\pm > 0$ and $\xi_\pm \in \mathbb{R}/\mathbb{Z}$ are constants.[2]

Theorem 5.2. *Suppose that conditions* (A)–(C) *hold and*

$$\lim_{\varepsilon\to 0}\mu(\varepsilon) = 1, \qquad \lim_{\varepsilon\to 0}\delta(\varepsilon)/\log^2\mu(\varepsilon) = 0.$$

Then

$$\overline{w}_* - \underline{w}_* = O(\delta(\varepsilon)), \qquad * \in \{+,-\}.$$

Moreover, if condition (S) *holds, then $\overline{w}_0 - \underline{w}_0 = O(\delta)$.*

[2] There is good reason to expect that equations (5.8) hold for generic systems with exponentially small separatrix splitting.

Corollary 5.1. *If conditions* (A)–(C) *and* (S) *hold, then by Theorem* 5.1 $\overline{w}_*, \underline{w}_* \sim O(\delta/\log\mu)$*, and hence* $\overline{w}_*/\underline{w}_* = 1 + O(1/\log\mu)$*,* $* \in \{+, -, 0\}$*.*

Below we need the following definitions. Let $C_{a,\omega,\lambda}$ be the symplectic map

$$C_{a,\omega,\lambda}(x, y) = (x^+, y^+) = (x + \omega + y^+, y + \lambda^{-1} a \sin(2\pi x))$$

of the cylinder (x mod 1, y) onto itself. The map $C_{1,0,\varepsilon^{-1}}$ coincides with the standard map.

A closed curve on the cylinder is said to be *topologically horizontal* (a *TH-curve*) if it is homotopic to the circle $\{y = 0\}$.[3] A closed curve γ invariant with respect to a symplectic self-map Q of the cylinder is said to be *rigid* if every smooth symplectic map close to Q has an invariant curve close to γ. We introduce three functions,

$$\overline{m}(a, \omega), \quad \underline{m}(a, \omega), \quad \overline{m}_0(a', a'', \omega', \omega''),$$

as follows:

$$\begin{aligned}
\overline{m} &= \inf\{\lambda_0 > 0 : \text{ for any } \lambda \geq \lambda_0 \text{ the map } C_{a,\omega,\lambda} \\
&\qquad \text{has an invariant rigid TH-curve}\}, \\
\underline{m} &= \sup\{\lambda_0 > 0 : \text{ for any } 0 < \lambda < \lambda_0 \text{ the map } C_{a,\omega,\lambda} \\
&\qquad \text{has no invariant TH-curve}\}, \\
\overline{m}_0 &= \inf\{\lambda_0 > 0 : \text{ for any } \lambda \geq \lambda_0 \text{ the map } C_{a',\omega',\lambda} \circ C_{a'',\omega'',\lambda} \\
&\qquad \text{has an invariant rigid TH-curve}\}.
\end{aligned}$$

Since ω can be eliminated from $C_{a,\omega,\lambda}$ by a change of the variable y, the functions $\underline{m}$ and $\overline{m}$ do not depend on ω. Moreover, it is clear that

$$\underline{m} = \underline{c}a, \quad \overline{m} = \overline{c}a$$

for some constants $\underline{c}$ and $\overline{c}$, $0 < \underline{c} \leq \overline{c}$.

The sharpest estimates for the constant $\underline{c}$ were obtained numerically in [57, 81, 101]. The "last" TH-curve[4] is believed to have the golden mean as the rotation number. In our notation, the result is as follows:

$$\underline{c} = 2\pi/0.971635\ldots.$$

Conjecture 5.2. The constants $\underline{c}$ and $\overline{c}$ coincide.

The function $\overline{m}_0$ also depends on its arguments in a special way. One can readily verify the following assertions.

---

[3] The word *horizontal* is chosen here for the following reason: one can imagine the cylinder embedded in the space $\mathscr{R}^3$ in such a way that its axis is vertical. Then the circle $\{y = 0\}$ belongs to the horizontal plane.

[4] I.e., the one which exists for λ arbitrarily close to $\overline{m}$.

(1) $0 < \overline{m}_0 < \infty$.
(2) $\overline{m}_0(a, a, \omega, \omega) = \overline{c}a$ *for any* a *and* ω.
(3) *The function* $\overline{m}_0$ *can be presented in the form*

$$\overline{m}_0(a', a'', \omega', \omega'') = a\overline{f}(\eta, \omega' - \omega''),$$
$$a' = a\cos(2\pi\eta), \quad a'' = a\sin(2\pi\eta),$$

where $\overline{f}$ *is* 1*-periodic in both arguments and satisfies the identities*

$$\overline{f}(\eta, \omega) = \overline{f}(-\eta, \omega) = \overline{f}(1/2 - \eta, \omega) = \overline{f}(1/4 - \eta, -\omega), \qquad \eta, \omega \in \mathbb{T}.$$

Theorem 5.3. *Suppose that conditions* (A)–(C) *are satisfied,*

$$\lim_{\varepsilon\to 0} \mu(\varepsilon) = 1, \qquad \lim_{\varepsilon\to 0} \delta_\varepsilon / \log^2 \mu_\varepsilon = 0,$$

and equations (5.8) *hold. Then*

(a) $\underline{w}_\pm$ *and* $\overline{w}_\pm$ *satisfy the inequalities*

$$a_\pm\underline{c}(1 + o(1)) \le \log\mu \cdot \underline{w}_\pm/\delta \le \log\mu \cdot \overline{w}_\pm/\delta \le a_\pm\overline{c}(1 + o(1)). \tag{5.9}$$

(b) $\log\mu \cdot \underline{w}_0/\delta \le \log\mu \cdot \overline{w}_0/\delta \le \sup_{\omega',\omega''} \overline{m}_0(a_+, a_-, \omega', \omega'')(1 + o(1))$;
(c) *if condition* (S) *holds, then* $\underline{w}_0$ *and* $\overline{w}_0$ *have estimates similar to* (5.9):

$$a_+\underline{c}(1 + o(1)) \le \log\mu \cdot \underline{w}_0/\delta \le \log\mu \cdot \overline{w}_0/\delta \le a_+\overline{c}(1 + o(1)). \tag{5.10}$$

Corollary 5.2. *If conditions* (A)–(C) *and* (S) *hold, then the upper bound* $\overline{w}$ *and the lower bound* $\underline{w}$ *for the width of the stochastic layer,*

$$\overline{w} = \max\{\overline{w}_+ + \overline{w}_0, \overline{w}_- + \overline{w}_0\}, \quad \underline{w} = \min\{\underline{w}_+ + \underline{w}_0, \underline{w}_- + \underline{w}_0\},$$

satisfy the inequalities

$$2a_+\delta\underline{c}(1 + o(1)) \le \underline{w}\log\mu \le \overline{w}\log\mu \le 2a_+\delta\overline{c}(1 + o(1)). \tag{5.11}$$

Corollary 5.3. *If Conjecture* 5.2 *holds, then* (5.9) *and* (5.10) *imply that*

$$\lim_{\varepsilon\to 0} (\log\mu \cdot w_*/\delta) = a_+\underline{c} = a_+\overline{c}, \qquad * \in \{+, -, 0\}. \tag{5.12}$$

Estimates (5.11) and (5.12) imply (5.3). Indeed, $w(\varepsilon)$ and $d(\varepsilon)$ have the form

$$w(\varepsilon) - \underline{w} = o(1), \quad \overline{w} - w(\varepsilon) = o(1), \quad d(\varepsilon) = \delta a_+ = \delta a_-.$$

The last two equations follow from Lemma 5.1. Lemma 5.1 also implies the following corollary.

Corollary 5.4. *If conditions* (A)–(C) *and* (S) *are satisfied and Conjecture* 5.2 *holds, then*

$$\mathscr{A} = \frac{4\underline{c}a_+\delta_\varepsilon}{\log \mu_\varepsilon} \log \delta^{-1}(\varepsilon)(1 + o(1)), \tag{5.13}$$

$$\mathscr{A}_\mathscr{D} = \frac{a_+\delta_\varepsilon \log \mu_\varepsilon}{\pi}(1 + o(1)). \tag{5.14}$$

Estimate (5.5) follows from (5.13) and (5.14).

5.3 Stochastic Layer for a Pendulum

Consider the Poincaré map for the pendulum with periodically oscillating suspension point. The quantities $\alpha_\pm$ and $\nu_\pm$ were computed in Sect. 4.5. The separatrix map satisfies the symmetry condition (S), because the involution $(\hat{u}, \hat{v}) \mapsto (-\hat{u}, -\hat{v})$ preserves the Hamiltonian (4.34). By Theorem 5.1, the width of the stochastic layer in a neighborhood of the separatrix figure-eight with center at the point $\hat{u} = \hat{v} = 0$ is of order ε provided $\theta, \Omega, \omega \sim 1$ and θ is non-constant. Under the same assumptions, the area $\mathscr{A}$ of the stochastic layer is of order $\varepsilon \log \varepsilon^{-1}$.

Equation (4.35) holds for $\omega/\Omega \sim 1$. In the case $\omega/\Omega \sim \varepsilon^{-1}$ the standard Poincaré–Melnikov theory is no longer valid, and one has to use methods suitable for studying exponentially small effects.

The first non-trivial rigorous results quantitatively describing exponentially small separatrix splitting were obtained in [78]. At present, there are at least two parallel theories that are effective in the investigation of a broad circle of problems of this type. The first theory was created by Lazutkin and his co-authors [51–53]. The second theory was developed in [134, 137].

Consider a system with Hamiltonian (4.34) under the following assumptions:[5]

$$\Omega = 1, \quad \omega = 1/\varepsilon, \quad \theta(s) = 2\varepsilon^{-1} B \cos s. \tag{5.15}$$

The leading multiplier of the hyperbolic fixed point $\hat{u} = \hat{v} = 0$ of the corresponding Poincaré map equals $\mu = e^{2\pi\varepsilon + O(\varepsilon^2)}$.

The separatrix map satisfies condition (S). Following [137], one can show that there is a symplectic change of coordinates $\hat{u}, \hat{v} \mapsto \widetilde{u}, \widetilde{v}$ which is

(i) close to the identity,
(ii) 2π-periodic in time t,
(iii) real-analytic in a complex neighborhood of the separatrices $\Gamma^\pm$ of the system with Hamiltonian $H_0 = \hat{v}^2/2 + \cos \hat{u}$,
(iv) such that the new Hamiltonian function takes the form

$$\widetilde{H}(\widetilde{u}, \widetilde{v}, t, \varepsilon) = H_0(\widetilde{u}, \widetilde{v}) + \varepsilon \widetilde{H}_1(\widetilde{u}, \widetilde{v}, \varepsilon) + \exp(-\widetilde{c}/\varepsilon)\widetilde{H}_2(\widetilde{u}, \widetilde{v}, t, \varepsilon).$$

[5] Equations (5.15) mean that the amplitude and the period of the oscillations of the suspension point are of order ε^2 and ε respectively.

Here $H_0 = \tilde{v}^2/2 + \cos\tilde{u}$, the constant $\tilde{c} \in [0, \pi/2)$ is arbitrary, the functions $\widetilde{H}_1$ and $\widetilde{H}_2$ are real-analytic with respect to $\tilde{u}$ and $\tilde{v}$ in a neighborhood of $\Gamma^\pm$ and smooth with respect to $\varepsilon > 0$, and the function $\widetilde{H}_2$ is 2π-periodic in t.

For any positive $\tilde{c} \in (\pi/4, \pi/2)$ the usual methods of the Poincaré–Melnikov theory applied to this system give the correct asymptotic behavior of the separatrix splitting because the error term[6] $\sim e^{-2\tilde{c}/\varepsilon}$ is much less than the answer ($\sim\varepsilon^{-2} \times e^{-\pi\varepsilon^{-1}/2}$). By this method it is possible to prove that

$$\nu_\pm(t) = -2\pi\varepsilon^{-2}e^{-\pi\varepsilon^{-1}/2}Bf(B^2)\sin(2\pi t),$$

where f is an entire real-analytic function and $f(0) = 2$ [137]. Numerical analysis of this function gives $f(z) = \sum_0^\infty f_n z^n$, where $f_0 = 2$ and

$$\begin{aligned} f_1 &= 0.65856738\ldots, & f_2 &= 6.651741\ldots \times 10^{-2},\\ f_3 &= 3.21010\ldots \times 10^{-3}, & f_4 &= 9.03367\ldots \times 10^{-5},\\ f_5 &= 1.6620\ldots \times 10^{-6}, & f_6 &= 2.1534\ldots \times 10^{-8},\\ f_7 &= 2.070\ldots \times 10^{-10}, & f_8 &= 1.53\ldots \times 10^{-12}. \end{aligned}$$

According to Theorem 5.3, the width of the stochastic layer near the separatrices $\Gamma^\pm$ satisfies estimate (5.11) with

$$a_+\delta = 2\pi\varepsilon^{-2}e^{-\pi\varepsilon^{-1}/2}Bf(B^2), \qquad \log\mu = 2\pi\varepsilon.$$

The area of the stochastic layer can be estimated as follows:

$$\frac{2\pi c}{\varepsilon^4}e^{-\pi\varepsilon^{-1}/2}Bf(B^2)(1+o(1)) \le \mathscr{A} \le \frac{2\pi\overline{c}}{\varepsilon^4}e^{-\pi\varepsilon^{-1}/2}Bf(B^2)(1+o(1)).$$

5.4 KAM Theory and the Birkhoff Theorem

In this section we prove Theorems 5.1, 5.2, and Lemma 5.1.

Proof (of Lemma 5.1). By (4.39), the unstable separatrix is given by the equation $y = 0$ and the stable separatrix by $y_+ = y_+(x, y) = 0$. The last equation can be reduced to the form

$$y = -(\hat{\nu}_s(x,\varepsilon) + O(\delta))\log\mu = -(\nu_s(x) + o(1))\log\mu \quad \text{as } \varepsilon \to 0.$$

Thus, simple zeros of the functions $\nu_\pm(x)$ correspond to homoclinic points. The domain $\mathscr{D}$ is determined by two neighboring simple zeros x_1 and x_2 of ν_s.

Suppose that x_1 and x_2 determine $\mathscr{D}$. The function ν_s preserves its sign on the interval (x_1, x_2). By (5.6) and (4.38), we get

[6] By the Poincaré–Melnikov theory its order is the square of the perturbation.

$$d = \delta\Big(\max_{x_1 \le x \le x_2} \nu_s(x) + o(1)\Big) \quad \text{as } \varepsilon \to 0. \tag{5.16}$$

The area $\mathscr{A}_{\mathscr{D}} = \mathscr{A}_{\mathscr{D}}^s$, $s = \pm$, is defined as the absolute value of the integral $\int_{\mathscr{D}} dp \wedge dq = \delta \int_{\mathscr{D}} dy \wedge dx$. Hence, we have

$$\mathscr{A}_{\mathscr{D}}^s = \delta \log \mu \left| \int_{x_1}^{x_2} \nu_s(x)\, dx + o(1) \right|. \tag{5.17}$$

Assertions (1) and (2) of Lemma 5.1 follow from the equations (5.16) and (5.17).

Let us prove assertion (3). In the normal coordinates q, p the stochastic layer occupies a domain $\mathscr{L}$ such that $\underline{\mathscr{D}} \subset \mathscr{L} \subset \overline{\mathscr{D}}$, where

$$\underline{\mathscr{D}} = \left\{ (q,p) : q^2 + p^2 < \underline{r}^2, \quad |qp| < \begin{cases} \underline{w}_+ & \text{if } q > 0 \text{ and } p > 0, \\ \underline{w}_- & \text{if } q < 0 \text{ and } p < 0, \\ \underline{w}_0 & \text{if } qp < 0 \end{cases} \right\},$$

$$\overline{\mathscr{D}} = \left\{ (q,p) : q^2 + p^2 < \overline{r}^2, \quad |qp| < \begin{cases} \overline{w}_+ & \text{if } q > 0 \text{ and } p > 0, \\ \overline{w}_- & \text{if } q < 0 \text{ and } p < 0, \\ \overline{w}_0 & \text{if } qp < 0 \end{cases} \right\}$$

and $\underline{r}$ and $\overline{r}$, $0 < \underline{r} \le \overline{r}$, do not depend on ε. The desired estimate for $\mathscr{A}$ follows from the inequalities area$(\underline{\mathscr{D}}) \ge \underline{\mathscr{A}}$, area$(\overline{\mathscr{D}}) \le \overline{\mathscr{A}}$. This completes the proof of Lemma 5.1. □

In view of (4.38), Theorems 5.1 and 5.2 result from the following assertion.

Theorem 5.4. *Suppose that $\varepsilon > 0$ is sufficiently small, $1 < \mu < \mu_0$, and*

$$\lim_{\varepsilon \to 0} \delta(\varepsilon)/\log^2 \mu(\varepsilon) = 0.$$

Then there are two constants $0 < y_{\min} < y_{\max}$ (depending only on the functions $\nu_\pm$) such that the following assertions hold.

(1) $|y| < y_{\max}$ *on the invariant curves $\gamma_\pm$ and γ_0.*
(2) *The curves γ_+ and γ_- do not enter the domain $|y| \le y_{\min}$.*
(3) $y_{\max} - y_{\min} < 2 y_{\min} \mu^2 \log \mu$.
(4) *If condition* (S) *holds, the curve γ_0 also does not enter the domain $|y| \le y_{\min}$.*

Proof (of Theorem 5.4). 1. The curves γ_+, γ_-, and γ_0 are located in the domains $\{y > 0, s > 0\}$, $\{y > 0, s < 0\}$, and $\{y < 0\}$, respectively. It is convenient to introduce the new variables u and u_+ determined by the formulas

$$y = y_0(1 + u \log \mu), \quad y_+ = y_0(1 + u_+ \log \mu), \qquad y_0 = \text{const}.$$

The separatrix map (4.39) takes the form

$$
\begin{aligned}
x_+ &= x + \omega_s + \frac{1}{\log \mu} \log \left| 1 + u_+ \log \mu + y_0^{-1} \delta \log \mu \cdot O\left(1 + \frac{y_0}{\log \mu}\right)^2 \right|, \\
u_+ &= u + y_0^{-1} \hat{v}_s(x, \varepsilon) + y_0^{-1} \delta \cdot O\left(1 + \frac{y_0}{\log \mu}\right)^2 \\
s_+ &= s \cdot \operatorname{sign}(y_0) \cdot \operatorname{sign}\left(1 + u_+ \log \mu + y_0^{-1} \delta \log \mu \cdot O\left(1 + \frac{y_0}{\log \mu}\right)^2\right),
\end{aligned}
\tag{5.18}
$$

where the quantities ω_s are given by

$$
\omega_s = \frac{1}{\log \mu} \log \frac{|y_0| \hat{\sigma}}{\alpha_s^2 \log \mu}.
$$

The main parameters in the system (5.18) are δ and y_0^{-1}. The first one is small. The parameter y_0 (the "unperturbed action") is responsible for the type of the dynamics (regular or chaotic). Formally setting $y_0^{-1} = \delta \log \mu \cdot O(1 + y/\log \mu) = 0$, we obtain the integrable map

$$
\begin{aligned}
x_+ &= x + \omega_s + \frac{1}{\log \mu} \log |1 - u_+ \log \mu|, \\
u_+ &= u, \\
s_+ &= s \cdot \operatorname{sign}(I_0).
\end{aligned}
$$

This map is non-degenerate, because $\partial x_+(x, u)/\partial u = (1 + u \log \mu)^{-1} \neq 0$.

Suppose that[7]

$$
|y_0^{-1}| \le \hat{c}_1^{-1} \ll 1, \quad \delta \le \hat{c}_2^{-1} \ll \frac{\hat{c}_1}{\log \mu \cdot (1 + \hat{c}_1/\log \mu)^2}.
$$

Then by KAM theory system (5.18) has an invariant curve $u = u(x) = u_0 + O(y_0^{-1})$, where u_0 is such that the "unperturbed" frequency $\omega_s + (\log \mu)^{-1} \log |1 + u \log \mu|$ is Diophantine. (Obviously, one can take $|u_0| < 1/(2 \log \mu)$.) In fact we obtain 3 families of invariant curves: one for $y_0 < 0$ and two for $y_0 > 0$. This completes the proof of assertion (1) because we can set $y_{max} = \max_x |y_0|(1 + u(x) \log \mu)$.

2. We prove assertion (2) for the curve γ_+. (The case of the curve γ_- is analogous.) For small $\varepsilon > 0$ (5.18) is a twist map. This means that $\partial x_+(x, y)/\partial y > 0$. Following Lazutkin's idea [79], we use the following result of Birkhoff [15] (see also [58]).

Lemma 5.2. *Let γ be a continuous invariant curve of a twist symplectic self-map of the cylinder $\{x \bmod 1, y\}$. If γ is homotopic to the curve $\{y = 0\}$, then γ is the graph of a Lipschitz function $y = r(x)$.*

---

[7] Here the relation $a \ll b$ means that $|a/b| < c$, where $c > 0$ is a sufficiently small constant.

The Lipschitz constant of the function $r(x)$ can be estimated as follows. Let $Q = \partial(x_+, y_+)/\partial(x, y)$ be the Jacobi matrix of the map T_ε. Since T_ε is a twist map, there are constants $\eta_1, \eta_2 > 0$ such that, for any point (x, y) on the curve,

$$(1,0)\, Q \begin{pmatrix} -1 \\ \eta_1 \end{pmatrix} > 0, \quad (1,0)\, Q^{-1} \begin{pmatrix} 1 \\ \eta_2 \end{pmatrix} < 0.$$

Lemma 5.3 ([58]). *For any two points x_1 and x_2 such that $0 \le x_2 - x_1 < 1/2$*

$$-\eta_1(x_2 - x_1) \le r(x_2) - r(x_1) \le \eta_2(x_2 - x_1). \tag{5.19}$$

Indeed, assuming that at least one of the inequalities (5.19) fails, where $0 \le x_2 - x_1 < 1/2$, we have the inequality $\widetilde{x}_2 < \widetilde{x}_1$ for the images $(\widetilde{x}_1, r(\widetilde{x}_1))$ and $(\widetilde{x}_2, r(\widetilde{x}_2))$ of the points $(x_1, r(x_1))$ and $(x_2, r(x_2))$ under the map T_ε (or T_ε^{-1}). This contradicts Lemma 5.2.

Let $\gamma_+ = \{(x, y) : y = r_+(x)\}$. We show that the quantity

$$r_{\max} = \max_x r_+(x)$$

is bounded below by a positive constant (Proposition 5.1) and, moreover, $r_+(x) > r_{\max}\mu^{-2}$ (Proposition 5.2).

Propositions 5.1 and 5.2 imply assertion (2) in Theorem 5.4 for the curve γ_+. Assertion (3) also follows from Proposition 5.2. Indeed, we have: $y_{\max} < y_{\min}\mu^2$. Using the inequality

$$\mu^2 < 1 + 2\mu^2 \log \mu, \qquad \mu > 1,$$

we obtain the desired estimate.

The function r_+ is positive, 1-periodic, and Lipschitz. Let us estimate the Lipschitz constant for small values of the parameter ε by using Lemma 5.3. We first put $\varepsilon = 0$ (therefore $\delta = 0$). The Jacobi matrix Q takes the form

$$Q = \begin{pmatrix} 1 + x_+^{-1} \nu_+'(x) & (x_+ \log \mu)^{-1} \\ \log \mu \cdot \nu_+'(x) & 1 \end{pmatrix},$$

where $x_+ = r_+(x) + \log \mu \cdot \nu_+'(x) = r_+(x_+)$ and $\nu_+' = d\nu_+/dx$. By direct computations we obtain $\eta_2 \ge r_{\max} \log \mu$. The estimate is slightly weaker for small non-zero values of δ. One can put

$$\eta_2 = 2 r_{\max} \log \mu. \tag{5.20}$$

We define the constant

$$\nu'_{\min +} = \min_x d\nu_+/dx.$$

Since ν_+ is periodic and non-constant (assumption (B)), we have $\nu'_{\min +} < 0$.

Proposition 5.1. *Suppose that $\varepsilon > 0$ is sufficiently small. Then*

$$r_{\max} \geq -\nu'_{\min +}/7. \tag{5.21}$$

Proof (of Proposition 5.1). We first put $\varepsilon = 0$. Suppose that (5.18) maps the points $(x_j, r_+(x_j))$ into $(\widetilde{x}_j, r_+(\widetilde{x}_j))$, $j = 1, 2$.

Suppose that $x_1 < x_2$ are close points. Then $\widetilde{x}_1$ and $\widetilde{x}_2$ are also close to each other and $\widetilde{x}_1 < \widetilde{x}_2$. Using equations (5.18), we obtain

$$\Delta\widetilde{x} = \Delta x + \frac{1}{\log \mu} \log\left(1 + \frac{\Delta r_+ + \log \mu \cdot \Delta \nu_+}{r_+(x_1) + \log \mu \cdot \nu_+(x_1)}\right), \tag{5.22}$$

where

$$\Delta\widetilde{x} = \widetilde{x}_2 - \widetilde{x}_1, \quad \Delta x = x_2 - x_1 > 0,$$
$$\Delta r_+ = r_+(x_2) - r_+(x_1), \quad \Delta \nu_+ = \nu_+(x_2) - \nu_+(x_1).$$

We take x_1 such that $\nu'_+(x_1) = \nu'_{\min +}$. For small positive values of Δx we have the estimates

$$\Delta r_+ \leq \eta_2 \Delta x \leq 2 r_{\max} \log \mu \cdot \Delta x, \quad \Delta \nu_+ \leq \nu'_{\min +} \Delta x/2 < 0.$$

Here we have used (5.19) and (5.20). Suppose now that $r_{\max} < -\nu'_{\min -}/6$. Then equation (5.22) implies the estimate

$$\Delta\widetilde{x} < \Delta x + \frac{1}{\log \mu} \log\left(1 + \frac{2r_{\max} + \nu'_{\min +}/2}{r_{\max}} \Delta x \log \mu\right)$$
$$< \Delta x + \frac{(2r_{\max} + \nu'_{\min +}/2)\Delta x}{r_{\max}} < \frac{3r_{\max} + \nu'_{\min +}/2}{\Delta x / r_{\max}} < 0.$$

This contradicts the inequality $\Delta\widetilde{x} > 0$. The estimates are preserved for small values of ε (possibly with slightly different constants). This is why we took $\nu'_{\min +}/7$ instead of $\nu'_{\min +}/6$ in (5.21). Proposition 5.1 is proved. □

Proposition 5.2. *For any $x \in \mathbb{T}$, $r_{\max}/r(x) < \mu^2$.*

Proof (of Proposition 5.2). We put $\varepsilon = 0$ and assume that there is a point $\hat{x} \in \mathbb{T}$ such that $r_{\max} \geq r(\hat{x})\mu$. We can assume that $0 < \hat{x} - x_0 < 1$, where x_0 is such that $r(x_0) = r_{\max}$. Let $\theta_0, \hat{\theta} \in \mathbb{T}$ be a pair of points such that $(5.18)|_{\varepsilon=0}$ maps the points $(\theta_0, r(\theta_0))$ and $(r(\hat{\theta}), \hat{\theta})$ to $(r(x_0), x_0)$ and $(r(\hat{x}), \hat{x})$, respectively.

Let us show that $\hat{\theta} - \theta_0 > 1$ in the sense that the image of the segment

$$\{(y, x) : y = r(x), \ \hat{\theta} \leq x \leq \theta_0\}$$

covers the entire circle under the natural projection $(y, x) \mapsto x$ onto $\mathbb{T}$. After proving this fact, we arrive at a contradiction to Lemma 5.2, and this will prove Proposition 5.2.

The quantity $\hat{\theta} - \theta_0$ is estimated by $(5.18)|_{\varepsilon=0}$ as follows:

$$\hat{x} - x_0 = \hat{\theta} - \theta_0 + \frac{1}{\log \mu} \log \frac{r(\hat{x})}{r(x_0)}.$$

Hence,

$$\hat{\theta} - \theta_0 = \hat{x} - x_0 - \frac{1}{\log \mu} \log \frac{r(\hat{x})}{r_{\max}} > 1.$$

The estimates are somewhat weaker in the case of small non-zero values of ε. This is why we assumed in Proposition 5.2 that $r_{\max}/r(x) < \mu^2$. Proposition 5.2 is proved. □

A similar proof of assertion (4) in Theorem 5.4 can be obtained with the help of condition (S). □

Chapter 6
The Continuous Averaging Method

6.1 Description of the Method

There are several problems in perturbation theory, where standard methods do not lead to satisfactory results. We mention as examples the problem of an inclusion of a diffeomorphism into a flow in the analytic set up, and the problem of quantitative description of exponentially small effects in dynamical systems. In these cases one possible approach is based on the continuous averaging. The method appeared as an extension of the Neishtadt averaging procedure [95], effectively working in the presence of exponentially small effects.

To present the general idea let us transform the system of ordinary differential equations

$$\dot{z} = \hat{u}(z), \tag{6.1}$$

by using the change of variables

$$z \mapsto Z(z, \Delta). \tag{6.2}$$

Here z is a point of the manifold M, $\hat{u}$ is a smooth vector field on M, Δ is a non-negative parameter, and change (6.2) is defined as a shift along solutions of the equation[1]

$$Z' = f(Z, \delta), \qquad Z(z, 0) = z, \quad 0 \le \delta \le \Delta, \tag{6.3}$$

where the prime denotes the derivative with respect to δ.

Let the change $z \mapsto Z$ transform (6.1) to the following system:

$$\dot{Z} = u(Z, \delta). \tag{6.4}$$

Differentiating equation (6.4) with respect to δ, we have

$$\dot{f}(Z, \delta) = u_\delta(Z, \delta) + \partial_f u(Z, \delta) \quad \text{or} \quad u_\delta = [u, f].$$

[1] Such a construction for a change of variables is called the Lie method. The corresponding Hamiltonian version is called the Deprit–Hori method.

D. Treschev, O. Zubelevich, *Introduction to the Perturbation Theory of Hamiltonian Systems*, Springer Monographs in Mathematics,
DOI 10.1007/978-3-642-03028-4_6, © Springer-Verlag Berlin Heidelberg 2010

Here ∂_f is the differential operator on M corresponding to the vector field f, the subscript δ denotes the partial derivative, and $[\cdot,\cdot]$ is the vector commutator: $[u_1, u_2] = \partial_{u_1} u_2 - \partial_{u_2} u_1$. Putting $f = \xi u$, where ξ is some fixed linear operator, we obtain the Cauchy problem

$$u_\delta = -[\xi u, u], \qquad u|_{\delta=0} = \hat{u}. \tag{6.5}$$

The equation $f = \xi u$ is crucial for our method. Traditionally the vector field f in the Lie method is constructed as a series in the small parameter. The choice of the operator ξ depends on the form to which we want to transform the initial equations. We call (6.5) an averaging system.

If (6.1) is a Hamiltonian system with Hamiltonian $\hat{H} = \hat{H}(z)$ and the symplectic structure ω, it is natural to search for the change (6.2) among symplectic ones, and to regard equation (6.3) as Hamiltonian with some Hamiltonian function $F(z, \delta)$. Under these assumptions systems (6.4) are also Hamiltonian. Their Hamiltonian functions H satisfy the equation $H(Z, \delta) = \hat{H}(z)$. Differentiating this equation with respect to δ, we get

$$H_\delta(Z, \delta) + \partial_{f(Z,\delta)} H(Z, \delta) = 0 \quad \text{or} \quad H_\delta = -\{F, H\}$$

because $\partial_f H = \{F, H\}$. Putting $F = \xi H$ for some linear operator ξ, we obtain

$$H_\delta = -\{\xi H, H\}, \qquad H|_{\delta=0} = \hat{H}. \tag{6.6}$$

Now let us present a nonautonomous analog of (6.6). To obtain such an analog, assume that the functions $\hat{H}$ and F depend explicitly on time. Then the Hamiltonian H also depends explicitly on t. We obtain an equation for H by the reduction to the autonomous case. Let E be a variable, canonically conjugate to time t. Consider the autonomous system with Hamiltonian $H + E$ and the symplectic structure $\omega + dE \wedge dt$. Let $\{\,,\,\}_*$ be the new Poisson bracket and $F = \xi H$. Then (6.6) takes the form:

$$(H + E)_\delta = -\{\xi H, H + E\}_*, \qquad H|_{\delta=0} = \hat{H}.$$

It is equivalent to the following one:

$$H_\delta = (\xi H)_t - \{\xi H, H\}, \qquad H|_{\delta=0} = \hat{H}(z, t). \tag{6.7}$$

This is a nonautonomous analog of system (6.6).

Analogously a nonautonomous analog of (6.5) can be constructed:

$$u_\delta = (\xi u)_t - [\xi u, u], \qquad u|_{\delta=0} = \hat{u}(z, t). \tag{6.8}$$

Properties of the averaging system can be illustrated by the following example. Consider the Hamiltonian system with one and a half degrees of freedom

$$\dot{y} = -\varepsilon \partial \hat{H}/\partial x, \quad \dot{x} = \varepsilon \partial \hat{H}/\partial y, \qquad \hat{H} = \hat{H}(x, y, t). \tag{6.9}$$

Here $y \in \mathbb{R}$, $x \in \mathbb{T}$; the function $\hat{H}$ is real-analytic in x, y, and 2π-periodic in t. The system contains two slow variables x and y ($\dot{x} = O(\varepsilon)$, $\dot{y} = O(\varepsilon)$) and one fast variable $t \in \mathbb{T}$.

Let us try to weaken the dependence of $\hat{H}$ on time with the help of the canonical change

$$(x, y) \mapsto (X(x, y, t, \Delta), Y(x, y, t, \Delta)), \qquad \Delta > 0,$$

where

$$X' = \partial F/\partial Y, \quad Y' = -\partial F/\partial X, \qquad F = F(X, Y, t, \delta) = \xi H. \tag{6.10}$$

We put[2]

$$\xi H(x, y, t, \delta) = \sum_{k \in \mathbb{Z}} i\sigma_k H^k(x, y, \delta)e^{ikt}, \qquad \sigma_k = \operatorname{sign} k, \tag{6.11}$$

where H^k are the Fourier coefficients in the expansion

$$H(x, y, t, \delta) = \sum_{k \in \mathbb{Z}} H^k(x, y, \delta)e^{ikt}.$$

Equation (6.7) takes the form

$$H^k_\delta = -|k|H^k - \varepsilon\{\xi H, H\}^k. \qquad k \in \mathbb{Z}. \tag{6.12}$$

Here $\{\,,\,\}^k$ denotes the Fourier coefficient corresponding to the number k. The more detailed form of system (6.12) is as follows:

$$H^k_\delta = -|k|H^k + i\varepsilon\sigma_k\{H^0, H^k\} - 2i\varepsilon \sum_{l+m=k,\, m<0<l} \{H^l, H^m\},$$

$$H^k|_{\delta=0} = \hat{H}^k, \qquad k \in \mathbb{Z}.$$

The terms $\varepsilon\{\xi H, H\}^k$ in (6.12) are proportional to the small parameter. Therefore, in zero approximation they are negligible. By putting $\varepsilon = 0$, we obtain

$$H^k_\delta = -|k|H^k.$$

For $k \neq 0$ their solutions rapidly tend to zero as $\delta \to \infty$.

A more precise approximation for system (6.12) is obtained if we take into account the terms $i\varepsilon\sigma_k\{H^0, H^k\}$. We have the following system:

$$H^k_\delta = -|k|H^k + i\varepsilon\sigma_k\{H^0, H^k\}.$$

The solution has the form

$$H^k = e^{-|k|\delta}\hat{H}^k \circ g^{i\varepsilon\sigma_k\delta}, \tag{6.13}$$

[2] Such an operator ξ is called the Hilbert transform.

where g^s is the time-s shift $(x, y)|_{t=0} \mapsto (x, y)|_{t=s}$ along solutions of the system

$$\dot{x} = \partial H^0/\partial y, \quad \dot{y} = -\partial H^0/\partial x.$$

The complex singularities of the functions $\hat{H}^k \circ g^s$ of the complex variable s prevent an unbounded continuation of solutions (6.13) to all the set of positive δ. Nevertheless, the functions (6.13) can be made exponentially small in ε since δ can be chosen of order $\sim 1/\varepsilon$.

Certainly, these arguments cannot be regarded as a proof of the fact that systems of the type (6.12) can be used for an averaging. The rigorous statements and estimates are presented below in this chapter.

6.2 Majorants

The main tool we use in the analysis of equations (6.5)–(6.8) is the majorant method. This section contains some simple properties of majorants.

Let two functions $f(z), g(z)$ $z = (z_1, \ldots, z_m) \in \mathbb{C}^m$ be analytic at the point $z = 0$:

$$f(z) = \sum_{\beta} f_\beta z^\beta, \quad g(z) = \sum_{\beta} g_\beta z^\beta,$$

$$\beta = (\beta_1, \ldots, \beta_m), \qquad \beta_j \geq 0, \qquad z^\beta = z_1^{\beta_1} \cdots z_m^{\beta_m}.$$

Then g is said to be a majorant for f ($f \ll g$) if for any multi-index β we have $g_\beta \geq |f_\beta|$. In the case when f and g are vector-valued functions we say that $f \ll g$ if any component of the vector g is a majorant for the corresponding component of f.

Lemma 6.1. *The relation $\ll$ satisfies the following properties*:

(1) *If $f_1 \ll g_1$ and $f_2 \ll g_2$ then $f_1 + f_2 \ll g_1 + g_2$ and $f_1 f_2 \ll g_1 g_2$.*
(2) *If $f \ll g$ then $\partial f/\partial z_j \ll \partial g/\partial z_j$ for any $j = 1, \ldots, m$.*
(3) *If $f(z, \lambda) \ll g(z, \lambda)$ for any value of the parameter $\lambda \in [a, b]$ then*

$$\int_a^b f(z, \lambda)\, d\lambda \ll \int_a^b g(z, \lambda)\, d\lambda.$$

(4) *Let $|f(z)| \leq c$ in the domain*

$$\{z = (z_1, \ldots, z_m) : |z_j| \leq b, \ j = 1, \ldots, m\}.$$

Then $f(z) \ll c/w$, where $w = b^{-m}(b - z_1) \cdots (b - z_m)$.

Assertions (1)–(3) of Lemma 6.1 are obvious. Assertion (4) follows from the Cauchy formula

$$f(z) = \frac{1}{(2\pi i)^m} \oint d\zeta_1 \cdots \oint d\zeta_{m-1} \oint \frac{f(\zeta) d\zeta_m}{(\zeta_1 - z_1) \cdots (\zeta_m - z_m)},$$

where the integration is performed along the circles $\{|z_j| = b\}$. Indeed, we have the estimates

$$\left| \frac{\partial^{\beta_1 + \cdots + \beta_m} f}{\partial z_1^{\beta_1} \ldots \partial z_m^{\beta_m}}(0) \right| = \left| \frac{1}{(2\pi i)^m} \oint d\zeta_1 \cdots \oint d\zeta_{m-1} \oint \frac{\beta_1! \cdots \beta_m! f(\zeta) d\zeta_m}{\zeta_1^{\beta_1+1} \cdots \zeta_m^{\beta_m+1}} \right|$$
$$\leq \frac{\beta_1! \cdots \beta_m! c}{b^{\beta_1 + \cdots + \beta_m}} = \frac{\partial^{\beta_1 + \cdots + \beta_m}}{\partial z_1^{\beta_1} \ldots \partial z_m^{\beta_m}} \bigg|_{z=0} \frac{c}{w}.$$

Below we use majorants to estimate solutions of initial value problems. The main idea is as follows. Let

$$f = f(z, \delta) = (\ldots, f^{-1}(z, \delta), f^0(z, \delta), f^1(z, \delta), \ldots), \qquad z \in \mathbb{C}^n, \quad \delta \in \mathbb{R}$$

be an infinite-dimensional vector-function, where each component f^k, $k \in \mathbb{Z}$, is analytic in z and takes values in $\mathbb{C}^m$. Consider the Cauchy problem for the system of ordinary differential equations

$$f^k_\delta(z, \delta) = F^k(f(z, \delta), z, \delta), \qquad f^k(z, 0) = \hat{f}^k(z), \tag{6.14}$$

with some known functionals F^k and initial data $\hat{f}^k$.

We call the system

$$\mathbf{f}^k_\delta(z, \delta) = \mathbf{F}^k(\mathbf{f}(z, \delta), z, \delta), \qquad \mathbf{f}^k(z, 0) = \hat{\mathbf{f}}^k(z), \tag{6.15}$$

a majorant system associated with (6.14) if

(a) $\hat{f}(z) \ll \hat{\mathbf{f}}(z)$ for any $k \in \mathbb{Z}$ and
(b) $F^k(g(z), z, \delta) \ll \mathbf{F}^k(\mathbf{g}(z), z, \delta)$ for any $k \in \mathbb{Z}$, $\delta \geq 0$, and $g \ll \mathbf{g}$.

Below we use a version of the following principle.

Majorant principle. Suppose that $\mathbf{f}(z, \delta)$, $0 \leq \delta \leq \delta_0$ is a solution of a majorant system associated with (6.14). Then system (6.14) has a solution and

$$f^k(z, \delta) \ll \mathbf{f}^k(z, \delta) \quad \text{for any } \delta \in [0, \delta_0], \quad k \in \mathbb{Z}.$$

Moreover, in (6.15) it is possible to replace "=" by "≫".

The same principle is true if we rewrite systems (6.14)–(6.15) in the integral form

$$f^k(z, \delta) = \hat{f}^k(z) + \int_0^\delta F^k(f(z, s), z, s)\, ds,$$
$$\mathbf{f}^k(z, \delta) = \hat{\mathbf{f}}^k(z) + \int_0^\delta \mathbf{F}^k(\mathbf{f}(z, s), z, s)\, ds,$$

where again in the majorant equations "=" can be replaced by "≫".

These statements look very natural: however their rigorous proof is not straightforward, see [150]. The proofs are based on some statements which generalize the classical Cauchy–Kovalevskaya theorem from the space of analytic functions to the general scales of Banach spaces. Now we formulate a version of the Nirenberg–Nishida theorem [97, 98] which gives such a generalization. Readers unacquainted with the concept of a scale of Banach spaces can find the necessary definitions in Sect. 9.5.

The Nirenberg–Nishida theorem. The scales of Banach spaces arise as the axiomatization of the set of analytic functions. As in the Cauchy theorem for ordinary differential equations the main point of the Nirenberg–Nishida theorem is the Lipschitz type conditions (6.16) for the right-hand side of (6.17).

Let $\{(E_s, \|\cdot\|_s)\}_{0<s<1}$ be a scale of Banach spaces with

$$\|\cdot\|_{s'} \le \|\cdot\|_s, \qquad s' < s.$$

Suppose that for some η, $R > 0$ for all $0 < s' < s < 1$ the mapping

$$f : \{u \in E_s : \|u\|_s \le R\} \times \{\delta \in \mathbb{R} \mid 0 \le \delta < \eta\} \to E_{s'}$$

is continuous. Furthermore, suppose that for $\|u\|_s$, $\|v\|_s < R$ and for all $\delta \in [0, \eta)$

$$\|f(u,\delta) - f(v,\delta)\|_{s'} \le C\frac{\|u - v\|_s}{s - s'}. \tag{6.16}$$

The function $f(0, \cdot) : [0, \eta) \to E_s$ is continuous for all $0 < s < 1$ and

$$\|f(0,\delta)\|_s \le \frac{K}{1 - s}$$

with some constant $K > 0$. Consider the following Cauchy problem:

$$u_\delta(\delta) = f(u(\delta), \delta), \quad u(0) = 0. \tag{6.17}$$

Theorem 6.1. *Under the above assumptions, problem* (6.17) *has a unique solution*

$$u(\delta) \in C^1\big([0, a(1 - s)), E_s\big), \qquad 0 < s < 1$$

with some constant $a > 0$, and this solution satisfies the inequality $\|u(\delta)\|_s \le R$.

Remark 6.1. The zero initial condition in problem (6.17) does not restrict the generality: if $u(0) \ne 0$ then one can change the variable $u = u(0) + v$.

6.3 An Inclusion of a Map into a Flow

In this section we present a proof of Theorem 1.10 from Chap. 1 on an inclusion of a real-analytic map into a real-analytic flow in the case when no additional structures are presented (the map and the flow are general). The main idea is a smoothing of the dependence of the initial isotopy g^t on time with the help of the continuous averaging method.

Here we present a general plan of the argument. The absent technical details are contained in Sect. 9.5.

Recall that T is the time-2π map corresponding to a vector field $\hat{u}(\hat{z}, t)$ which is real-analytic in $\hat{z}$, C^2-smooth and 2π-periodic in t (see Lemmas 1.2, 1.3). The system generated by $U(z, t)$ (see the statement of Theorem 1.10) will be a result of the application of the averaging procedure to the system generated by $\hat{u}(\hat{z}, t)$. Thus the new vector will satisfy

$$U(z, t) = u(z, t, \Delta),$$

where $u(z, t, \delta)$ is a solution of the Cauchy problem (6.8) and Δ is positive. We choose the operator ξ in the following way:

$$\begin{aligned} \xi u &= i \sum_{k\in\mathbb{Z}} \sigma_k u^k(z, \delta) e^{ikt}, \qquad \sigma_k = \operatorname{sign} k, \\ u &= \sum_{n\in\mathbb{Z}} u^n(z, \delta) e^{int}. \end{aligned} \tag{6.18}$$

Putting $v^j = u^j e^{|j|\delta}$ we transform (6.8) into the infinite system

$$v^k_\delta = i\sigma_k[v^0, v^k] - 2i \sum_{l,n} [v^l, v^n] e^{-(|l|+|n|)\delta}, \tag{6.19}$$

$$v^k(z, 0) = \hat{u}^k(z), \qquad k \in \mathbb{Z}, \tag{6.20}$$

where $\hat{u}^k(z)$ are the Fourier coefficients of the function $\hat{u}(z, t)$, and the indices l, n in the sum satisfy the condition

$$l + n = k, \qquad n < 0 < l. \tag{6.21}$$

Problem (6.19)–(6.20) is the Cauchy–Kovalevskaya type problem. But unlike the classical Cauchy-Kovalevskaya problem, (6.19)–(6.20) consists of an infinite number of equations and an infinite number of unknown functions v^k.

Our next goal is to establish the existence theorem for (6.19)–(6.20).

By truncation of the system (6.19):

$$v^k_\delta = i\sigma_k[v^0, v^k],$$

it is easy to obtain informal evidence that one should expect the existence of a solution to (6.19) on some interval $\delta \in (0, \Delta)$, $\Delta > 0$. Then $u^j(z, \Delta)$ would decrease exponentially as $|j| \to \infty$.

Since the system is infinite dimensional, the classical technique which has been employed by Kovalevskaya does not suit, so we have to use some modern abstract version of the Kovalevskaya theorem, namely the Nirenberg–Nishida theorem. By using the Nirenberg–Nishida theorem we prove that the system (6.19) has a solution and therefore provides smoothing with respect to t. (The precise formulation of the existence result is contained in Lemma 6.2.)

Now we turn to the technical part of our argument. Choose a positive number R such that all the functions $\hat{u}^k(z)$ are analytic in the polydisk

$$D(2R) = \{z \in \mathbb{C}^m \mid \|z\| < 2R\}.$$

For the domains D_s we take the following polydisks:

$$D_s = D(sR).$$

Full details of the construction built below are given in Sect. 9.5.

Let H_s be a set of functions $f : D_s \to \mathbb{C}$ analytic in D_s and continuous in the closure $\overline{D}_s$. Being endowed with the norm

$$\|f\|_s^H = \sup_{z \in D_s} |f(z)|,$$

the set H_s becomes a Banach space. The spaces H_s, $0 < s < 1$, form a scale of Banach spaces.

Now let us introduce a scale $(F_s, \|\cdot\|_s^F)$. The Banach spaces F_s consist of the sequences $u = \{u_k(x)\}_{k \in \mathbb{Z}}$, $u_k \in H_s$, and the norms are

$$\|u\|_s^F = \sqrt{\sum_{k \in \mathbb{Z}} (1 + |k|^2)(\|u_k\|_s^H)^2}.$$

Lemma 6.2. *There exist positive constants Δ, s such that the Cauchy problem* (6.19)–(6.20) *has a unique solution*

$$v_j(z, \delta) = \{v_j^k(z, \delta)\}_{k \in \mathbb{Z}} \in C^1([0, \Delta], F_s), \qquad j = 1, \dots, m, \quad \delta \in [0, \Delta].$$

Corollary 6.1. *The vector field $U(z, t) = u(z, t, \Delta)$ is real-analytic in both z and t.*

Remark 6.2. Let us take any point $\tilde{z} \in M$ with a complex chart $U_{\tilde{z}}$ such that $\tilde{z}$ has zero coordinates and $U_{\tilde{z}} = D_s$ (see Remark 1.3). Lemma 6.2 states the local existence theorem for (6.19)–(6.20) in every chart $U_{\tilde{z}}$, $\tilde{z} \in M$. Due to the compactness of M we have a finite cover of M with such charts. Equations (6.19)–(6.20) are defined independently on coordinate systems, and thus the solutions from all the charts form a global solution on the whole manifold M.

Now let us prove Lemma 6.2. The function $\hat{u}(z, t)$ is C^2-smooth in t, and thus for all $z \in D(R)$ we have

$$|\hat{u}^n(z)| \le c_1(|n| + 1)^{-2}, \qquad n \in \mathbb{Z}. \tag{6.22}$$

This is a standard estimate for the Fourier coefficients of a C^2-smooth periodic function. Moreover, due to the compactness of the set $\overline{D(R)}$, the constant c_1 can be chosen independently of z. By (6.22)

$$\{\hat{u}_j^k(z)\}_{k\in\mathbb{Z}} \in F_s, \qquad j = 1, \dots, m, \quad 0 < s < 1.$$

Lemma 6.2 is a direct consequence of the Nirenberg–Nishida theorem. Formally speaking, to apply this theorem, following Remark 6.1, we present the system (6.19)–(6.20) in the form (6.17), i.e., the initial conditions (6.20) become zero.

The change of variables from Remark 6.1 does not bring any serious problems in our argument. So for simplicity's sake we do not make this change but verify the conditions of the Nirenberg–Nishida theorem for the right-hand side of (6.19).

There is only one nontrivial condition to check. Let us verify that the right-hand side of the system (6.19) satisfies (6.16). This right-hand side contains terms of the type

$$\sum_{l,n} v_j^l \frac{\partial v_r^n}{\partial z_j} e^{-(|l|+|n|)\delta}, \qquad r = 1, \dots, m.$$

It is sufficient to check that these terms satisfy (6.16). But this follows from Corollary 9.14 if, in Proposition 9.9, we put $b_{l,n} = e^{-(|l|+|n|)\delta}$ for $n < 0 < l$ and $b_{l,n} = 0$ otherwise.

To show that the right-hand side of (6.19) is continuous in δ and v, introduce the notation

$$g(\delta, v) = \left\{\sum_{l,n}[v^l, v^n]e^{-(|l|+|n|)\delta}\right\}.$$

Take $\delta, \delta_0 \geq 0$. Then we have

$$\begin{aligned}\left(\|g(\delta, v) - g(\delta_0, v)\|_s^F\right)^2 \\ = \sum_{k\in\mathbb{Z}}(1+|k|^2)\left(\left\|\sum_{l,n}[v^l, v^n](e^{-(|l|+|n|)\delta} - e^{-(|l|+|n|)\delta_0})\right\|_s^H\right)^2. \qquad (6.23)\end{aligned}$$

By the Cauchy estimate, the right-hand side of this formula is estimated from above by

$$\begin{aligned}&\sum_{k\in\mathbb{Z}}(1+|k|^2)\left(\sum_{l,n}\|[v^l, v^n]\|_s^H\right)^2 \\ &\quad\leq \frac{c}{(s'-s)^2}\sum_{k\in\mathbb{Z}}(1+|k|^2)\left(\sum_{l,n}\|v^l\|_{s'}^H\|v^n\|_{s'}^H\right)^2. \qquad (6.24)\end{aligned}$$

Here c is an inessential positive constant and $s < s' < 1$. By formula (9.59) the last term of (6.24) is estimated as follows

$$\frac{c}{(s'-s)^2}(\|v\|_{s'}^F)^4.$$

Consequently the sum (6.23) converges uniformly in δ and we can take the termwise limit as $\delta \to \delta_0$. Thus

$$\|g(\delta, v) - g(\delta_0, v)\|_s^F \to 0 \tag{6.25}$$

as $\delta \to \delta_0$.

As we have already stated, the right-hand side of (6.19) satisfies the inequality (6.16) and this inequality is uniform in δ. By this observation and formula (6.25) we obtain the desired continuity.

Lemma 6.2 is proved.

6.4 The Case of Arbitrary Algebra χ

In this section we finish the proof of Theorem 1.10, i.e., we consider the case of an arbitrary algebra χ, and prove Remarks 1.5–1.6 from Chap. 1.

The domains D_s are the same as in Sect. 6.3 and thus the spaces H_s, F_s are also the same. Introduce the following locally convex spaces (for the details see Sect. 9.5):

$$H = \bigcap_{0<s<1} H_s, \quad F = \bigcap_{0<s<1} F_s.$$

Let L be a closed subspace of F. Suppose that for all $\delta \geq 0$ the right-hand side of the system (6.19) takes the space L to itself and the initial conditions (6.20) also belong to the space L. In other words, the system (6.19)–(6.20) is defined on L.

By Theorem 9.15 the space L is associated with the scale $\{L_s\}$. Therefore, applying the Nirenberg–Nishida theorem to the system (6.19)–(6.20) in the scale $\{L_s\}$, we obtain the same assertion as in Lemma 6.2 where the spaces F_s are replaced with L_s.

Below we verify that the role of the space L can in particular be taken by the space X_I or the algebra χ. That is, the right-hand side of the system (6.19)–(6.20) is defined on the corresponding space.

Lemma 6.3. *Suppose that $\hat{u} \in \chi$. Then the vector field $u(z, t, \delta)$ also lies in χ. Moreover, if $\hat{u}$ is I-reversible, u is also I-reversible.*

The lemma follows from the next proposition.

Proposition 6.1. *If $u(z, t, \delta) \in \chi$, the right-hand side of* (6.8) *also belongs to χ. Moreover, if $\hat{u}$ is I-reversible, the right-hand side of* (6.8) *is also I-reversible.*

Proof (of Proposition 6.1). First, consider the case when no reversibility is assumed. Note that $u(z, t) \in \chi$ for any $t \in \mathbb{T}$ if and only if any Fourier coefficient $u^k(z)$ belongs to χ. This implies that for $u \in \chi$ we have $\xi u \in \chi$ and $[\xi u, u] \in \chi$.

Now consider the reversible case ($T \in X_I$). Let us call a vector field $v(z, t)$ on M symmetric with respect to the involution I (or I-symmetric) if

$$v(z, t) = dI\, v(Iz, -t) \tag{6.26}$$

(cf. with (1.11), Chap. 1). Let S be the set of I-symmetric vector fields on M. (Recall that the set of I-reversible fields is denoted by R.)

Proposition 6.2. *For any two vector fields $v \in R$ and $w \in S$*

$$\xi v \in S, \quad w_t \in R, \tag{6.27}$$

$$[v, w] \in R. \tag{6.28}$$

This proposition implies Proposition 6.1 in the reversible case. To prove Proposition 6.2, we note that for any $v \in R$ and $w \in S$ with Fourier series

$$v(z,t) = \sum_{n\in\mathbb{Z}} v^n(z)e^{int}, \quad w(z,t) = \sum_{n\in\mathbb{Z}} w^n(z)e^{int},$$

the Fourier coefficients v^n and w^l satisfy the equations

$$v^n(z) = -dI\, v^{-n}(Iz), \quad w^l(z) = dI\, w^{-l}(Iz). \tag{6.29}$$

These equations can be obtained from Fourier expansions of equations (1.11), Chap. 1, and (6.26) of the present chapter. Equations (6.27) easily follow from (6.29).

Obviously, it is sufficient to prove (6.28) for the case

$$v = v^n(z)e^{int} + v^{-n}(z)e^{-int}, \quad w = w^l(z)e^{ilt} + w^{-l}(z)e^{-ilt}.$$

We have

$$\begin{aligned}[v, w] &= [v^n, w^l]e^{i(n+l)t} + [v^{-n}, w^{-l}]e^{-i(n+l)t} \\ &\quad + [v^n, w^{-l}]e^{i(n-l)t} + [v^{-n}, w^l]e^{-i(n-l)t}.\end{aligned}$$

By using (6.29), we get

$$\begin{aligned}dI\,[v^n, w^l](z) &= [dI\, v^n(z), dI\, w^l(z)] \\ &= [-v^{-n}(Iz), w^{-l}(Iz)] = -[v^{-n}, w^{-l}](Iz).\end{aligned}$$

The equation $dI[v^n, w^{-l}](z) = -[v^{-n}, w^l](Iz)$ can be checked analogously. Proposition 6.2 is proved.

Remark 1.5 follows from the observation that, if u is Hamiltonian with a single-valued Hamiltonian function, the same is true for the vector fields ξu and $[\xi u, u]$.

Finally we discuss the proof of Remark 1.6. Let the map T be a perturbation of order ε of the map T_0. Then the map $Q = T \circ T_0^{-1}$ is such that

(1) in some Riemannian metric $\mathrm{dist}(Q(z), z) \le (\mathrm{const.})\varepsilon$,
(2) $T = Q \circ T_0$.

According to Proposition 1.3, Q is smoothly isotopic to the identity. Let the isotopies g_0^t and g_Q^t, $t \in [0, 2\pi]$, link the identity map with T_0 and Q respectively. It is possible to assume that the vector fields

$$\hat{u}_0 = \left(\frac{d}{dt}g_0^t\right) \circ g_0^{-t} \quad \text{and} \quad \hat{u}_Q = \left(\frac{d}{dt}g_Q^t\right) \circ g_Q^{-t}$$

are 2π-periodic and C^2-smooth in t, analytic in z, and $\hat{u}_Q = O(\varepsilon)$ on M. The isotopy $g^t = g_Q^t \circ g_0^t$ links the identity map with T. The vector field $\hat{u} = \frac{dg^t}{dt} \circ g^{-t}$ is smooth and $O(\varepsilon)$-close to $\hat{u}_0$:

$$\hat{u}(z,t) = \hat{u}_0(z,t) + \varepsilon \hat{u}_*(z,t,\varepsilon).$$

Let us put

$$u(z,t,\delta) = \hat{u}_0(z,t) + \varepsilon u_*(z,t,\delta), \qquad u|_{\delta=0} = \hat{u}, \quad u_*|_{\delta=0} = \hat{u}_*.$$

We assume that u satisfies the averaging system (6.8) with the operator ξ of the form

$$\xi u = \varepsilon i \sum_{k\in\mathbb{Z}} \sigma_k u^k(z) e^{ikt}.$$

By using the Nirenberg–Nishida theorem, it is possible to verify that the averaging system has a solution for small values of $\delta \geq 0$. For any small positive δ the vector field u is analytic in both z and t. Moreover, $u = \hat{u}_0 + O(\varepsilon)$. □

6.5 Fast Phase Averaging

Many physical models are represented by systems of ODE which contain an angular variable changing much faster than the other variables. Taking the fast phase as a new time, we can rewrite the equations in the form

$$\dot{z} = \varepsilon \hat{v}(z,t,\varepsilon), \qquad z \in M, \tag{6.30}$$

where M is the m-dimensional phase space of the system, and ε is a small parameter. Here ε is the ratio of a typical velocity of slow variable change to a typical velocity of phase rotation. The vector field $\hat{v}$ is assumed to be smooth[3] and to depend on time 2π-periodically.

It is well known that, by a change of the variables, it is possible to weaken the dependence of the system (6.30) on time. In particular, by using the standard averaging method (see for example [17]), it is easy to construct a 2π-periodic in t change of the variables $z \mapsto z_*$ such that the equations (6.30) take the form

$$\dot{z}_* = \varepsilon \bar{v}^0(z_*) + \varepsilon^2 \hat{v}_*(z_*,\varepsilon) + \varepsilon \tilde{v}(z_*,t,\varepsilon). \tag{6.31}$$

[3] Actually, in the averaging problem only smoothness in z is important. It is enough to have continuity in t and sometimes even this requirement can be weakened.

Here the only term on the right-hand side depending explicitly on time is $\varepsilon\tilde{v} = O(\varepsilon^K)$. The natural number K is arbitrary and

$$\bar{v}^0(z) = \frac{1}{2\pi}\int_0^{2\pi} \hat{v}(z, t, 0)\,dt$$

is the time average of $\hat{v}(z, t, 0)$.

Now suppose that the functions $\hat{v}$ depend analytically on the phase variables. In this case Poincaré noted the existence and divergence of power series in the small parameter presenting a change of variables eliminating time from the equations: terms at ε^k in these series are of order $k!$. In a general situation this statement has been proved in [118].

Neishtadt [11, 95] noted that, in the case of functions $\hat{v}$ which are analytic in phase variables, it is possible to obtain in (6.31)

$$\tilde{v} = O(e^{-\alpha/\varepsilon}), \tag{6.32}$$

where $\alpha > 0$ is some constant (the parameter ε is assumed to be nonnegative). Hence, explicit dependence of the equations on time can be made exponentially small. The method Neishtadt used to prove this assertion is based on a large (of order $1/\varepsilon$) number of successive changes of variables. These changes gradually weaken the dependence of the equations on time. Ramis and Schafke [111] obtained analogous results analyzing diverging series, produced by the standard averaging method.

It is also known that in general a constant $A > \alpha$ exists such that it is impossible to construct a change of the variables $z \mapsto z_*$ which is 2π-periodic in t and such that $\tilde{v} = O(e^{-A/\varepsilon})$. This statement follows, for example, from an estimate of the separatrix splitting rate in Hamiltonian systems of type (6.30) with one and a half degrees of freedom. In this section, following paper [136], we obtain realistic estimates for a "maximal" α for which the estimate (6.32) is possible.

Let g^t be the phase flow of the averaged system[4]

$$\dot{z} = \bar{v}^0(z). \tag{6.33}$$

Suppose that the manifold M is real-analytic. We fix its complex neighborhood $M_{\mathbb{C}}$. Let Q be compact in M and $V_Q \subset M_{\mathbb{C}}$ its complex neighborhood. For example one may assume that $M = \mathbb{R}^m$ and Q is a closure of some bounded domain. Then it is natural to take

$$M_{\mathbb{C}} = \{z \in \mathbb{C}^m : z = x + ia,\ x \in \mathbb{R}^m,\ a \in \mathbb{R}^m,\ |a| < c\},$$
$$V_Q = \{z \in \mathbb{C}^m : z = x + b,\ x \in Q,\ b \in \mathbb{C}^m,\ |b| < \tilde{c}\}.$$

It is reasonable to assume that $\tilde{c}$ is small.

Suppose that, for any real $s \in (-\alpha, \alpha)$ and for any point $z \in V_Q$, the map g^{is} is analytic at the point z and moreover $g^{is}(z) \in M_{\mathbb{C}}$. We define the set

[4] It would be more correct to call (6.33) the first approximation averaged system in the fast time.

$$U_{Q,\alpha} = \bigcup_{-\alpha<s<\alpha} g^{is}(V_Q).$$

Theorem 6.2. *Let the positive constants α, ρ, ε_0 be such that*

(1) $U_{Q,\alpha} \subset M_{\mathbb{C}}$.
(2) *The vector field $\hat{v}$ is real-analytic in z and C^2-smooth in t, ε on $U_{Q,\alpha} \times \mathbb{T} \times [0, \varepsilon_0]$.*

Then for sufficiently small ε_0, there exists an open set $V'_Q \subset V_Q$ and a map

$$F : V'_Q \times \mathbb{T} \times (0, \varepsilon_0) \to M_{\mathbb{C}}, \qquad Q \subset V'_Q,$$

which is 2π-periodic in t real-analytic in z and such that

(a) $F(z, t, \varepsilon) = z_* = z + O(\varepsilon)$.
(b) *The map $F \times id_t$ transforms the vector field $\binom{\varepsilon\hat{v}}{1}$ on the extended phase space $M \times \mathbb{T}_t$ into $\binom{\varepsilon v_*}{1}$, where*

$$v_*(z, t, \varepsilon) = \bar{v}^0(z) + \varepsilon\hat{v}_*(z, \varepsilon) + \tilde{v}(z, t, \varepsilon). \tag{6.34}$$

(c) *The time-dependent remainder $\tilde{v}$ is estimated as follows:*

$$|\tilde{v}(z, t, \varepsilon)| \le Ce^{-\alpha/\varepsilon}, \qquad z \in V'_Q, \quad t \in \Sigma_\rho, \quad \varepsilon \in [0, \varepsilon_0). \tag{6.35}$$

Theorem 6.2 means in particular that, in the case when components of $\hat{v}$ are entire functions of z, the quantity α in (6.35) can be an arbitrary positive number such that for all $s \in [-\alpha, \alpha]$ the maps $z \mapsto g^{is}(z)$ are holomorphic at any point $z \in Q$.

Theorem 6.2 will be obtained below as a corollary of its local version (Theorem 6.3). To formulate the latter we consider instead of the compact Q a point $z^0 \in M$. Let $V \subset M_{\mathbb{C}}$ be its neighborhood.

Suppose that, for any real s such that $s \in (-\alpha, \alpha)$ and for any point $z \in V$, the map g^{is} is analytic at the point z and moreover $g^{is}(z) \in M_{\mathbb{C}}$. We define the set

$$U_\alpha = \bigcup_{-\alpha<s<\alpha} g^{is}(V).$$

Theorem 6.3. *Let the positive constants α, ρ, ε_0 be such that*

(1) $U_\alpha \subset M_{\mathbb{C}}$.
(2) *$\hat{v}$ is analytic in z and C^2-smooth in t, ε on the set $U_\alpha \times \mathbb{T} \times [0, \varepsilon_0]$.*

Then for sufficiently small ε_0 there exists a map $f : V' \times \mathbb{T} \times (0, \varepsilon_0) \to V$ which is 2π-periodic in t and real-analytic in z, where $V' \subset V$ is a neighborhood of the point z^0, and the following assertions hold.

(a) $f(t, z, \varepsilon) = z_* = z + O(\varepsilon)$.
(b) *In coordinates z_*, t the vector field $\varepsilon\hat{v}$ takes the form εv_* (see (6.34)), where for some constant C_0*

$$|\tilde{v}(z,t,\varepsilon)| \le C_0 e^{-\alpha/\varepsilon}, \qquad z \in V', \quad t \in \mathbb{T}, \quad \varepsilon \in [0,\varepsilon_0). \tag{6.36}$$

Theorem 6.2 can be deduced from Theorem 6.3 if we take into account the fact that the averaging system we use in the proof of Theorem 6.3 is defined independently of the neighborhood V and of the point z^0. Hence, the maps f we construct in Theorem 6.3, pasted together, produce the map F. We can put $V'_Q = \bigcup_z V'(z)$, where $z \in Q$ and the neighborhoods $V'(z)$ are defined in Theorem 6.3. Due to the compactness of Q we can assume that the union is taken over a finite system of sets. The constant C in (6.35) can be taken as the maximum among the constants C_0 corresponding to the sets $V'(z)$ from this system.

The proof of Theorem 6.3 is based on the continuous averaging. Namely, we solve the Cauchy problem (6.8), where instead of u and $\hat{u}$ we write εv and $\varepsilon\hat{v}$:

$$v_\delta = (\xi v)_t - \varepsilon[\xi v, v], \qquad v|_{\delta=0} = \hat{v}(z,t,\varepsilon). \tag{6.37}$$

The operator ξ is defined as before. Let

$$v(z,t,\varepsilon,\delta) = \sum_{k\in\mathbb{Z}} v^k(z,\varepsilon,\delta)e^{ikt}.$$

Then we put

$$\xi v(z,t,\varepsilon,\delta) = \sum_{k\in\mathbb{Z}} i \operatorname{sign} k\, v^k(z,\varepsilon,\delta)e^{ikt} \tag{6.38}$$

(cf. (6.11)). According to an informal argument analogous to the those presented in Sect. 6.1, one can hope that the system (6.37)–(6.38) really performs an averaging in time t. The required change of variables corresponds to the value $\delta = \alpha/\varepsilon$. An analysis of a solution for the system (6.37) on the interval $\delta \in [0, \alpha/\varepsilon]$ is contained in the next section.

Remark 6.3. According to the definition, the averaging procedure possesses the following important property. Suppose that the vector field $\hat{v}$ for all fixed t and ε belongs to a certain subalgebra χ in the Lie algebra of vector fields on M. Then for fixed t, ε and δ the vector field $v(z,t,\varepsilon,\delta)$ also lies in χ. Therefore, the diffeomorphism f belongs to the corresponding Lie group. Moreover, if $\hat{v}$ is reversible with respect to some involution $I : M \to M$, the vector field $v(z,t,\varepsilon,\delta)$ is also I-reversible. In particular, if the initial vector field $\hat{v}$ is Hamiltonian then $v_* = v(z,t,\varepsilon,\alpha/\varepsilon)$ is also Hamiltonian, and the corresponding change of variables F is symplectic.

As an example consider a pendulum with a periodically vertically oscillating suspension point. The Hamiltonian of the system is presented in Sect. 4.5. Let the oscillation period of the suspension point be small. More precisely, equations (5.15) from Sect. 5.3 hold. Then the Hamiltonian function is as follows:

$$H = y^2/2 + \cos x + 2B\cos(t/\varepsilon)\cos x. \tag{6.39}$$

By using the change $t = \varepsilon\tau$, $H = \varepsilon^{-1}\mathscr{H}$, we get

$$\mathscr{H} = \varepsilon(\mathscr{H}_0 + \mathscr{H}_1), \qquad \mathscr{H}_0 = y^2/2 + \cos x, \quad \mathscr{H}_1 = 2B\cos x\cos\tau.$$

The system with the Hamiltonian $\mathscr{H}$ is of form (6.30), where the dot denotes the derivative with respect to the new time τ. The corresponding first approximation averaged system (6.33) is also Hamiltonian: the Hamiltonian function is $\mathscr{H}_0$.

Solutions of the averaged system are elliptic functions of τ. The positions of their complex singularities can be expressed in terms of elliptic integrals. Consider the energy level $L_1 = \{(x, y) \in \mathbb{R}^2 : \mathscr{H}_0(x, y) = 1\}$, corresponding to separatrices of the unperturbed pendulum. The distance from the real axis $\operatorname{Re}\tau$ to a singularity of a solution with initial conditions, lying on L_1, equals $\pi/2$. Hence, according to Theorem 6.2, if Q lies in a small neighborhood of the separatrices, the constant α in the exponential estimate of the perturbation $\tilde{v}$ can be chosen close to $\pi/2$.

Analogous arguments have been used to solve the problem of separatrix splitting in the system with Hamiltonian (6.39), see [134, 137].

6.6 Analytic Properties of the Averaging Procedure

We divide the analysis of the problem (6.37)–(6.38) into two parts. First we construct the majorant estimates for the solution to this problem as if this solution exists. Such estimates are called a priori estimates. It is a heuristic part of the proof.

In the second part we give a formal proof of Theorem 6.3.

We use the spaces of functions which are analytic in polydiscs. These local observations are justified by the same argument as in Remark 6.2.

Our method is a version of the method of a priori estimates on the space of countably dimensional vectors of analytic functions. This technique is a generalization of the classical majorant method which goes back to Cauchy, Weierstrass and Kovalevskaya.

Below for brevity we do not write ε among arguments of the functions we deal with.

A priori estimates. Putting $\bar{v} = \hat{v}^0|_{\varepsilon=0}$,

$$v^0(z,\delta) = w^0(z,\delta) + \overline{v}(z), \quad v^k(z,\delta) = w^k(z,\delta)e^{-|k|\delta}, \qquad k \neq 0,$$

we write the system (6.37)–(6.38) in the form

$$\begin{aligned} w^0_\delta &= -2i\varepsilon \sum_{l>0} [w^l, w^{-l}]e^{-2l\delta}, \\ w^k_\delta &= i\sigma_k\varepsilon[\overline{v} + w^0, w^k] - 2i\varepsilon \sum_{l,n} [w^l, w^n]e^{(|k|-|l|-|n|)\delta}, \end{aligned} \tag{6.40}$$

$$w^0(z,0) = \hat{v}^0(z) - \overline{v}(z), \quad w^k(z,0) = \hat{v}^k(z), \qquad k \neq 0, \tag{6.41}$$

where l, n in the sum $\sum_{l,n}$ satisfy (6.21).

For any vector field $u(z)$ on M we put

$$\mathbf{g}^s u(z) = Dg^s u \circ g^{-s}(z), \qquad s \in \mathbb{C}, \tag{6.42}$$

where Dg^s is the differential of the phase flow g^s with respect to z. For any $z \in M$, $\mathbf{g}^s : T_z M \to T_z M$ is a linear operator. Hence $u \mapsto \mathbf{g}^s u$ is a linear operator on the space of smooth (or analytic) vector fields on M.

The operators $\mathbf{g}^s$ form a one-parametric group: $\mathbf{g}^{s_1}\mathbf{g}^{s_2} = \mathbf{g}^{s_1+s_2}$. Moreover,

$$\mathbf{g}^s[u_1, u_2] = [\mathbf{g}^s u_1, \mathbf{g}^s u_2].$$

Proposition 6.3. *Let the vector field $w(z, \delta)$ on M be the solution of the initial value problem*

$$w_\delta = i\sigma\varepsilon[\overline{v}, w] + \eta(z, \delta), \qquad w(z, 0) = \hat{w}(z)$$

with some known η and $\hat{w}$. Then

$$w(z, \delta) = \mathbf{g}^{i\sigma\varepsilon\delta}\hat{w}(z) + \int_0^\delta \mathbf{g}^{i\sigma\varepsilon(\delta-s)}\eta(z, s)\, ds.$$

The proof follows from the identity

$$\frac{d}{ds}\mathbf{g}^s u = [\bar{v}, \mathbf{g}^s u]. \tag{6.43}$$

According to Proposition 6.3, (6.40) is equivalent to the system

$$\begin{aligned} w^0(z, \delta) &= \hat{v}^0(z) - \overline{v}(z) + i\varepsilon \int_0^\delta \eta^0(z, s)\, ds, \\ w^k(z, \delta) &= \mathbf{g}^{i\sigma_k\varepsilon\delta}\hat{v}^k(z) + i\varepsilon \int_0^\delta \mathbf{g}^{i\sigma_k\varepsilon(\delta-s)}\eta^k(z, s)\, ds, \end{aligned} \tag{6.44}$$

$$\begin{aligned} \eta^0(z, s) &= -2\sum_{l>0}[w^l, w^{-l}](z, s)e^{-2ls}, \\ \eta^k(z, s) &= \sigma_k[w^0, w^k] - 2\sum_{l,n}[w^l, w^n](z, s)e^{(|k|-|l|-|n|)s}. \end{aligned} \tag{6.45}$$

Lemma 6.4. *For any real $s \in [-\alpha, \alpha]$ the Fourier coefficients $\hat{v}^k$ satisfy the majorant estimates*

$$\mathbf{g}^{is}(\hat{v}^0(z) - \overline{v}(z)) \ll \frac{\varepsilon\beta\kappa}{(\kappa - \zeta)}\mathbf{1}, \quad \mathbf{g}^{is}\hat{v}^k(z) \ll \frac{\beta\kappa}{k^2(\kappa - \zeta)}\mathbf{1}, \qquad k \neq 0,$$

where $\zeta = z_1 + \cdots + z_m$, $\mathbf{1} = (1, \ldots, 1)^T \in \mathbb{R}^m$ and β, κ are positive constants.

Proof. According to assumption (2) of Theorem 6.3, for any small ε and $s \in [-\alpha, \alpha]$ the vector field $\mathbf{g}^{is}\hat{v}(z, t)$ is analytic in $z \in V$ and C^2-smooth in $t \in \mathbb{T}$. We put $V' = \{z \in \mathbb{C} : |z| \leq \kappa\} \subset V$ for some positive κ. Let

$$\beta' = \max_{s\in[-\alpha,\alpha],(z,t)\in V'\times\mathbb{T}} \left| \frac{d^2}{dt^2} \mathbf{g}^{is}\hat{v}(z,t) \right|.$$

Then obviously

$$\max_{s\in[-\alpha,\alpha],z\in V'} \left|k^2 \mathbf{g}^{is}\hat{v}^k(z)\right| \le \beta', \qquad k \ne 0.$$

By using assertion (4) of Lemma 6.1, for any $s \in [-\alpha, \alpha]$ and $k \ne 0$ we obtain the estimate

$$\mathbf{g}^{is}\hat{v}^k(z) \ll \frac{\beta'\kappa^m}{k^2(\kappa - z_1)\cdots(\kappa - z_m)}\mathbf{1} \ll \frac{\beta'\kappa}{k^2(\kappa - \zeta)}\mathbf{1}. \tag{6.46}$$

The last inequality follows from the obvious estimate

$$\frac{\kappa^m}{(\kappa - z_1)\cdots(\kappa - z_m)} \ll \frac{\kappa}{\kappa - z_1 - \cdots - z_m}.$$

Note that $\hat{v}^0 - \overline{v}(z) = O(\varepsilon)$. Therefore for some positive β''

$$\max_{s\in[-\alpha,\alpha],z\in V_0} \left|\mathbf{g}^{is}(\hat{v}^0(z) - \overline{v}(z))\right| \le \varepsilon\beta''.$$

This implies that

$$\mathbf{g}^{is}(\hat{v}^k(z) - \overline{v}(z)) \ll \frac{\varepsilon\beta''\kappa}{\kappa - \zeta}\mathbf{1}. \tag{6.47}$$

Lemma 6.4 follows from the estimates (6.46)–(6.47). □

Define the function

$$\varphi(\delta) = 32m\beta\kappa \int_0^\delta (\varepsilon^2 K + K^{-1}e^{-2s})\,ds.$$

It depends on the parameters β, κ, K.

Lemma 6.5. *Let the positive constant K be such that*

$$\varphi(\delta) < \kappa^2 \quad \textit{for any } \delta \in [0, \alpha/\varepsilon]. \tag{6.48}$$

Then for any $\delta \in [0, \alpha/\varepsilon]$ and $\tau \in [-\alpha + \varepsilon\delta, \alpha - \varepsilon\delta]$

$$\mathbf{g}^{i\tau} w^0(z,\delta) \ll \varepsilon K \mathbf{w}(\zeta,\delta)\mathbf{1}, \quad \mathbf{g}^{i\tau} w^k(z,\delta) \ll k^{-2}\mathbf{w}(\zeta,\delta)\mathbf{1}, \qquad k \ne 0, \tag{6.49}$$

where the scalar function $\mathbf{w}$ satisfies the initial value problem

$$\mathbf{w}_\delta = 8m(\varepsilon^2 K + K^{-1}e^{-2\delta})\mathbf{w}\mathbf{w}_\zeta, \qquad \mathbf{w}(\zeta, 0) = \frac{\beta\kappa}{\kappa - \zeta}. \tag{6.50}$$

Lemma 6.5 implies Theorem 6.3. Indeed, the function $\mathbf{w}$ can be found explicitly:

$$\mathbf{w}(\zeta,\delta) = \frac{2\beta\kappa}{\kappa - \zeta + \sqrt{(\kappa-\zeta)^2 - \varphi(\delta)}}.$$

Due to (6.48), $\mathbf{w}$ is analytic at the point $z = 0$ for $\delta \in [0, \alpha/\varepsilon]$. Hence

$$v^0(z,\alpha/\varepsilon) - \overline{v}(z) \ll \varepsilon K \mathbf{w}(\zeta,\alpha/\varepsilon)\mathbf{1}, \quad v^k(z,\alpha/\varepsilon) \ll e^{-|k|\alpha/\varepsilon}\mathbf{w}(\zeta,\alpha/\varepsilon)\mathbf{1}.$$

These estimates imply Theorem 6.3.

Now let us prove Lemma 6.5. First, note that according to (6.44) the functions

$$\mathbf{g}^{i\tau} w^n(z,\delta), \qquad 0 \le \delta \le \alpha/\varepsilon, \quad -\alpha + \varepsilon\delta \le \tau \le \alpha - \varepsilon\delta, \quad n \in \mathbb{Z}$$

satisfy the equations

$$\begin{aligned} \mathbf{g}^{i\tau} w^0(z,\delta) &= \mathbf{g}^{i\tau} w^0(z,0) + i\varepsilon \int_0^\delta \mathbf{g}^{i\tau}\eta^0(z,s)\,ds, \\ \mathbf{g}^{i\tau} w^k(z,\delta) &= \mathbf{g}^{i(\tau+\sigma_k\varepsilon\delta)}\hat{v}^k(z) + i\varepsilon \int_0^\delta \mathbf{g}^{i(\tau+\sigma_k\varepsilon(\delta-s))}\eta^k(z,s)\,ds, \end{aligned} \tag{6.51}$$

where the η^n are defined by (6.45) and $\sigma_k = \operatorname{sign} k$. Hence, the vector fields on the right-hand sides are combinations of the vector fields

$$\mathbf{g}^{i\tau'} w^{n'}(z,\delta'), \qquad 0 \le \delta' \le \alpha/\varepsilon, \quad -\alpha + \varepsilon\delta' \le \tau' \le \alpha - \varepsilon\delta', \quad n' \in \mathbb{Z}.$$

Consider the system

$$\begin{aligned} \mathbf{w}^0(\zeta,\delta) &= \mathbf{w}^0(\zeta,0) + \varepsilon \int_0^\delta \rho^0(\zeta,s)\,ds, \\ \mathbf{w}^k(\zeta,\delta) &= \mathbf{w}^k(\zeta,0) + \varepsilon \int_0^\delta \rho^k(\zeta,s)\,ds, \qquad k \neq 0, \end{aligned} \tag{6.52}$$

$$\begin{aligned} \rho^k(\zeta,s) &= 2m \sum_{n<0<l,\, l+n=k} (\mathbf{w}^l_\zeta \mathbf{w}^n + \mathbf{w}^l \mathbf{w}^n_\zeta)(\zeta,s)\,ds, \\ \mathbf{w}^0(\zeta,0) &= \varepsilon K \frac{\beta\kappa}{\kappa-\zeta}, \quad \mathbf{w}^k(\zeta,0) = \frac{1}{k^2}\frac{\beta\kappa}{\kappa-\zeta}, \qquad k \neq 0. \end{aligned} \tag{6.53}$$

Now, assuming that the solutions to both systems exist, we give an informal explanation why the solution to the system (6.52), (6.53) majorates the solution to the system (6.44), (6.41). In this sense we say that the system (6.52), (6.53) is a majorant system for the system (6.44), (6.41). The reader should regard all the argument of this section only as a heuristic method to obtain the system (6.52), (6.53) from (6.44), (6.41). For now we summarize the argument above as the following

Proposition 6.4. *The system* (6.52), (6.53) *is a majorant system for* (6.44), (6.41).

Now let us check the majorant estimates (6.49). First, we note that according to Lemma 6.4, they are satisfied for $\delta = 0$. We replace the left- and right-hand sides of (6.51) by the corresponding majorants according to the assertions of Lemma 6.5 and show that after this operation (6.51) remain true if we replace "=" by "$\gg$".

We use the simple remark that, for any two vector fields u', u'' such that $u'(z) \ll \mathbf{w}'(\zeta)\mathbf{1}$ and $u''(z) \ll \mathbf{w}''(\zeta)\mathbf{1}$,

$$[u', u''](z) \ll m\big(\mathbf{w}'_\zeta(\zeta)\mathbf{w}''(\zeta) + \mathbf{w}'(\zeta)\mathbf{w}''_\zeta(\zeta)\big)\mathbf{1}.$$

We begin with majorants for

$$Q^0 = \mathbf{g}^{i\tau}\eta^0(z, s) \quad \text{and} \quad Q^k = \mathbf{g}^{i(\tau+\sigma_k\varepsilon(\delta-s))}\eta^k(z, s).$$

Lemma 6.5 predicts the estimate

$$\begin{aligned} Q^0 &\ll \sum_{l>0} \frac{4me^{-2ls}}{l^4}\mathbf{w}(\zeta, s)\mathbf{w}_\zeta(\zeta, s)\mathbf{1} \ll 4me^{-2s}\mathbf{w}(\zeta, s)\mathbf{w}_\zeta(\zeta, s)\sum_{l>0}\frac{1}{l^4}\mathbf{1} \\ &\ll 8me^{-2s}\mathbf{w}(\zeta, s)\mathbf{w}_\zeta(\zeta, s)\mathbf{1} = \frac{e^{-2s}\mathbf{w}_s(\zeta, s)}{\varepsilon^2 K + K^{-1}e^{-2s}}\mathbf{1} \ll K\mathbf{w}_s(\zeta, s)\mathbf{1}. \end{aligned}$$

Analogously

$$\begin{aligned} Q^k &\ll \frac{2m\varepsilon K}{k^2}\mathbf{w}(\zeta, 0)\mathbf{w}_\zeta(\zeta, 0)\mathbf{1} + \sum_{l,n}\frac{4me^{(|k|-|l|-|n|)s}}{l^2n^2}\mathbf{w}(\zeta, s)\mathbf{w}_\zeta(\zeta, s)\mathbf{1} \\ &\ll \frac{2m}{k^2}\mathbf{w}(\zeta, s)\mathbf{w}_\zeta(\zeta, s)\left(\varepsilon K + 2e^{-2s}\sum_{l,n}\frac{k^2}{l^2n^2}\right)\mathbf{1}. \end{aligned}$$

Here we used $\mathbf{w}(z, 0) \ll \mathbf{w}(z, s)$ for any $s \in [0, \alpha/\varepsilon]$. Recall that l, n in the sum satisfy (6.21). Hence, $\sum_{l,n} k^2 l^{-2} n^{-2} < \sum_{j=1}^\infty j^{-2} = \pi^2/6$ and

$$\begin{aligned} Q^k &\ll \frac{2m}{k^2}\mathbf{w}(\zeta, s)\mathbf{w}_\zeta(\zeta, s)(\varepsilon K + \pi^2 e^{-2s}/3)\mathbf{1} \\ &= \frac{\varepsilon K + \pi^2 e^{-2s}/3}{4(\varepsilon^2 K + K^{-1}e^{-2s})}\frac{\mathbf{w}_s(\zeta, s)}{k^2}\mathbf{1} \ll \frac{\mathbf{w}_s(\zeta, s)}{\varepsilon k^2}\mathbf{1}. \end{aligned}$$

Now according to our plan we check the majorant analogs of (6.51):

$$\begin{aligned} \varepsilon K\mathbf{w}(\zeta, \delta) &\gg \varepsilon K\mathbf{w}(\zeta, 0) + \varepsilon\int_0^\delta K\mathbf{w}_s(\zeta, s)\,ds, \\ k^{-2}\mathbf{w}(\zeta, \delta) &\gg k^{-2}\mathbf{w}(\zeta, 0) + \varepsilon\int_0^\delta \frac{\mathbf{w}_s(\zeta, s)}{\varepsilon k^2}ds. \end{aligned}$$

Both these relations follow from (6.50).

The existence theorem. Let $R > 0$ be a number such that the function $\mathbf{w}(\zeta, \alpha/\varepsilon)$ is analytic in the disk $\{\|\zeta\| < R\}$.

To construct the spaces H_s, G_s, F_s (see Sect. 9.5), we replace the domain D in (9.53) by the polydisk $\|z\| < sR$ with $0 < s < 1$.

We then obtain the following definitions of our spaces.

The space H_s is a set of functions $f : D_s \to \mathbb{C}$ which are analytic in D_s and continuous in the closure $\overline{D}_s$. This space is endowed with the norm

$$\|f\|_s^H = \sup_{z \in D_s} |f(z)|.$$

The space G_s is the space of the sequences $u = \{u_k(x)\}_{k\in\mathbb{Z}}, u_k \in H_s$, with the norm

$$\|u\|_s^G = \sup_{k\in\mathbb{Z}}\{(1 + |k|^2)\|u_k\|_s^H\}.$$

The Banach spaces F_s consist of the sequences $u = \{u_k(x)\}_{k\in\mathbb{Z}},\ u_k \in H_s$, and the norm is

$$\|u\|_s^F = \sqrt{\sum_{k\in\mathbb{Z}}(1 + |k|^2)(\|u_k\|_s^H)^2}.$$

Recall that

$$F = \bigcap_{0<s<1} F_s, \quad G = \bigcap_{0<s<1} G_s.$$

Let V stand for the set of functions

$$w(z, \delta) = \{w^k(z, \delta)\}_{k\in\mathbb{Z}} \in C([0, \alpha/\varepsilon], F)$$

which satisfy the following conditions.

(1) For any $\delta \in [0, \alpha/\varepsilon]$ and $\tau \in [-\alpha + \varepsilon\delta, \alpha - \varepsilon\delta]$, the estimates (6.49) hold.
(2) For any $\delta', \delta'' \in [0, \alpha/\varepsilon]$ and $0 < s < 1$

$$\|w(\cdot, \delta') - w(\cdot, \delta'')\|_s^F \le K_s|\delta' - \delta''|. \tag{6.54}$$

The positive constants $\{K_s\}$ are the same for all $w(z, \delta) \in V$; these constants will be defined in the sequel. Observe that V is a convex set.

Let $C([0, \alpha/\varepsilon], F)$ stand for the space of continuous mappings from the interval $[0, \alpha/\varepsilon]$ to the space F with the topology defined in Sect. 9.5.

Theorem 6.4. *With a suitable choice of the constants* $\{K_s\}_{s\in(0,1)}$, *the system* (6.44)–(6.45) *has a solution in the set* V.

Remark 6.4. If a solution to the integral equations (6.44)–(6.45) belongs to the space $C([0, \alpha/\varepsilon], F)$ then it is a differentiable function in the variable δ and it also solves the system (6.40)–(6.41). This observation follows from Proposition 6.3 and the standard argument on the connection between initial value problems and corresponding integral equations.

We introduce the set W which consists of functions

$$w(z,\delta) = \{w^k(z,\delta)\}_{k\in\mathbb{Z}} \in C([0,\alpha/\varepsilon], F)$$

such that the estimate (6.54) holds and

$$w^0(z,\delta) \ll \varepsilon K \mathbf{w}(\zeta,\delta)\mathbf{1}, \quad w^k(z,\delta) \ll k^{-2}\mathbf{w}(\zeta,\delta)\mathbf{1}, \qquad k \neq 0. \tag{6.55}$$

Since (6.55) is the special case of (6.49) when $\tau = 0$, it follows that

$$V \subseteq W. \tag{6.56}$$

Define a mapping $P[w]$ by the formula

$$P^0[w](z,\delta) = \hat{v}^0(z) - \overline{v}(z) + i\varepsilon \int_0^\delta \eta^0(z,s)\,ds,$$

$$P^k[w](z,\delta) = \mathbf{g}^{i\sigma_k\varepsilon\delta}\hat{v}^k(z) + i\varepsilon \int_0^\delta \mathbf{g}^{i\sigma_k\varepsilon(\delta-s)}\eta^k(z,s)\,ds, \qquad k \neq 0.$$

We describe the domain of $P[w]$ below.

Now we prove Theorem 6.4. The plan is as follows. First we show that the set W is compact in $C([0,\alpha/\varepsilon], F)$. Then observing that the set V is closed, we conclude that it is compact as a closed subspace of W. Finally we show that $P[V] \subseteq V$ and $P \in C(V,V)$. Due to these facts, by the Schauder–Tikhonov theorem, there exists a fixed point $\tilde{w} \in V$ of the mapping P: $P[\tilde{w}] = \tilde{w}$. This finishes the proof.

Lemma 6.6. *The set W is compact in $C([0,\alpha/\varepsilon], F)$.*

Proof. Consider the set $W_u \subset F$ which consists of functions $w(z,\delta) \in W$ with $\delta = u$. By Propositions 9.10, 9.11 and formula (6.55) the sets W_u, $u \in [0,\alpha/\varepsilon]$, are bounded in G and closed in F.

By Corollary 9.13 the sets W_u, $u \in [0,\alpha/\varepsilon]$, are compact in F.

Now by (6.54) the proof of the lemma follows from Theorem 9.13. □

The operator $\mathbf{g}^{i\tau}$, defined by (6.42):

$$\mathbf{g}^{i\tau} w(z,\delta) = \{\mathbf{g}^{i\tau} w^k(z,\delta)\}_{k\in\mathbb{Z}},$$

is a continuous operator from V to W.

Lemma 6.7. *The set V is compact in $C([0,\alpha/\varepsilon], F)$.*

Proof. By Lemma 6.6 and inclusion (6.56) it is sufficient to check that V is closed in $C([0,\alpha/\varepsilon], F)$.

Consider a sequence $\{w_j\} \in V$ such that $w_j \to w$ in $C([0,\alpha/\varepsilon], F)$ as $j \to \infty$ and $w \in C([0,\alpha/\varepsilon], F)$. Then we have

$$\mathbf{g}^{i\tau} w_j^0(z,\delta) \ll \varepsilon K \mathbf{w}(\zeta,\delta)\mathbf{1}, \quad \mathbf{g}^{i\tau} w_j^k(z,\delta) \ll k^{-2}\mathbf{w}(\zeta,\delta)\mathbf{1}, \qquad k \neq 0. \tag{6.57}$$

The convergence in $C([0,\alpha/\varepsilon], F)$ implies pointwise convergence with respect to the variable δ. Therefore in formulas (6.57) we pass to the limit as $j \to \infty$ at every point $\delta \in [0, \alpha/\varepsilon]$ and for every $\tau \in [-\alpha + \varepsilon\delta, \alpha - \varepsilon\delta]$. Then, using the continuity of the operator $\mathbf{g}^{i\tau}$ and Proposition 9.11, we obtain

$$\mathbf{g}^{i\tau} w^0(z, \delta) \ll \varepsilon K \mathbf{w}(\zeta, \delta)\mathbf{1}, \quad \mathbf{g}^{i\tau} w^k(z, \delta) \ll k^{-2}\mathbf{w}(\zeta, \delta)\mathbf{1}, \qquad k \neq 0.$$

Thus $w \in V$. □

Lemma 6.8. *Define a set of constants* $\{K_s\}_{s\in(0,1)}$ *such that*

$$K_s = \frac{c_5}{(s'-s)}\Big(1 + \sup_{\xi\in[0,\alpha/\varepsilon]} (\|\mathbf{w}(\cdot,\xi)\|_{s'}^H)^2\Big), \qquad s' = \frac{1+s}{2}.$$

Then the mapping P takes the set V (see (6.54)*) to itself.*

Proof. From the majorant argument above it follows that if a function $w(z, \delta) \in F$ satisfies the relations (6.49) then $P[w](z, \delta)$ also satisfies these relations. This follows from the formulation and proof of Lemma 6.5 and Proposition 6.4.

So it remains to choose the constants $\{K_s\}_{s\in(0,1)}$ such that for any $w \in V$ the function $P[w]$ satisfies formula (6.54).

Let us estimate the terms P^k with $k \neq 0$. Putting for definiteness $\delta'' > \delta'$, by the Mean Value Theorem we have

$$\|P^k[w](\cdot,\delta') - P^k[w](\cdot,\delta'')\|_s^H \leq \sup_{\xi\in[\delta',\delta'']} \|P^k_\delta[w](\cdot,\xi)\|_s^H \cdot |\delta' - \delta''|, \qquad k \neq 0. \tag{6.58}$$

Formula (6.43) gives

$$P^k_\delta[w](z,\xi) = i\varepsilon\big(\sigma_k[\bar{v}(z), P^k[w](z,\xi)] + \eta^k(z,\xi)\big), \quad k \neq 0. \tag{6.59}$$

We know that $P[w] \in W$. Thus, due to the properties of the relation "$\ll$" (see Lemma 9.10) and by the Cauchy estimate (9.54), it follows that

$$\|[\bar{v}, P^k[w]](\cdot,\xi)\|_s^H \leq \frac{c}{k^2(s'-s)}\|\bar{v}\|_{s'}^H \|\mathbf{w}(\cdot,\xi)\|_{s'}^H, \qquad 0 < s < s' < 1. \tag{6.60}$$

The positive constant c is independent of s, s', k and w.

Consider the term η^k. By the same argument as above we have

$$\begin{aligned}\|\eta^k(\cdot,\xi)\|_s^H &\leq \|[w^0, w^k](\cdot,\xi)\|_s^H + 2\sum_{l,n} \|[w^l, w^n](\cdot,\xi)\|_s^H \\ &\leq \frac{c_1(\|\mathbf{w}(\cdot,\xi)\|_{s'}^H)^2}{s'-s}\Big(\frac{1}{k^2} + \sum_{l,n}\frac{1}{n^2 l^2}\Big), \qquad 0 < s < s' < 1.\end{aligned} \tag{6.61}$$

The positive constant c_1 depends only on m. Recall that the summation in this formula is taken over all integers n, l such that $n < 0 < l, l+n = k \neq 0$. Consequently we obtain the estimate

$$\sum_{l,n} \frac{1}{n^2 l^2} \le \frac{1}{k^2} \sum_{l>0,\, l\neq k} \frac{1}{l^2(1-l/k)^2} \le \frac{c_2}{k^2}$$

with some constant $c_2 > 0$. Combining the formulas (6.59), (6.60), (6.61), and substituting them into (6.58) we get

$$\|P^k[w](\cdot,\delta') - P^k[w](\cdot,\delta'')\|_s^H \le \frac{c_3}{k^2(s'-s)}\Big(1 + \sup_{\xi\in[\delta',\delta'']} (\|\mathbf{w}(\cdot,\xi)\|_{s'}^H)^2\Big)\cdot|\delta'-\delta''|, \qquad k\neq 0. \tag{6.62}$$

Analogously

$$\|P^0[w](\cdot,\delta') - P^0[w](\cdot,\delta'')\|_s^H \le \frac{c_4}{(s'-s)} \sup_{\xi\in[\delta',\delta'']} (\|\mathbf{w}(\cdot,\xi)\|_{s'}^H)^2\cdot|\delta'-\delta''|. \tag{6.63}$$

By (6.62) and (6.63) we obtain

$$\|P[w](\cdot,\delta') - P[w](\cdot,\delta'')\|_s^F \le K_s\cdot|\delta'-\delta''| \tag{6.64}$$

with K_s defined above. □

The mapping $P : V \to V$ is a Lipschitz mapping in the following sense

$$\sup_{\xi\in[0,\alpha/\varepsilon]} \|P[w'](\cdot,\xi) - P[w''](\cdot,\xi)\|_s^F \le \frac{c_6}{s'-s} \sup_{\xi\in[0,\alpha/\varepsilon]} \|w'(\cdot,\xi) - w''(\cdot,\xi)\|_{s'}^F,$$
$$w', w'' \in V, \quad 0 < s < s' < 1.$$

As done in Sect. 6.3, this formula is derived with the help of Corollary 9.14. Thus the mapping $P : V \to V$ is continuous and, by the Schauder–Tikhonov theorem (see Sect. 9.5), it has a fixed point $\tilde{w}$:

$$P(\tilde{w}) = \tilde{w}, \qquad \tilde{w} \in V.$$

This fixed point is the desired solution to problem (6.44)–(6.45).

Theorem 6.4 is proved.

Chapter 7
The Anti-Integrable Limit

7.1 Perturbation of the Standard Map

We have seen in the previous chapters that the dynamics in near-integrable Hamiltonian systems and symplectic maps remains quite regular: stochastic regimes exist, but are located in small domains. It is natural to expect that chaotic properties become more pronounced when the "distance to the set of integrable systems" increases.

Consider as an example the standard map (1.5) from Chap. 1, where the parameter ε is not small but, on the contrary, very large. The limit as $\varepsilon \to \infty$ in systems of such a type is called the anti-integrable limit [12].

Let us rewrite the map SM in the "Lagrangian form". To this end suppose that

$$\begin{pmatrix} x_- \\ y_- \end{pmatrix} \overset{\text{SM}}{\mapsto} \begin{pmatrix} x \\ y \end{pmatrix} \overset{\text{SM}}{\mapsto} \begin{pmatrix} x_+ \\ y_+ \end{pmatrix}.$$

Then x_-, x, x_+ satisfy the equation

$$\varepsilon^{-1}(x_+ - 2x + x_-) = \sin x. \tag{7.1}$$

The standard map written in this form is defined on the cylinder $\mathscr{Z} = \mathbb{R}^2_{(x_-,x)}/\sim$, where the equivalence relation $\sim$ is as follows:

$$(x_1', x_2') \sim (x_1'', x_2'') \quad \text{if and only if } x_1' - x_1'' = x_2' - x_2'' \in 2\pi\mathbb{Z}.$$

In the other words, the cylinder $\mathscr{Z}$ is the quotient space of the plane $\mathbb{R}^2_{(x_-,x)}$ with respect to the action of the group of shifts

$$(x_-, x) \mapsto (x_- + 2\pi l, x + 2\pi l), \qquad l \in \mathbb{Z}.$$

The standard map sends the point $(x_-, x) \in \mathscr{Z}$ to $(x, x_+) \in \mathscr{Z}$, where x_-, x, x_+ satisfy (7.1).

D. Treschev, O. Zubelevich, *Introduction to the Perturbation Theory of Hamiltonian Systems*, Springer Monographs in Mathematics,
DOI 10.1007/978-3-642-03028-4_7,

Infinite sequences $\dots, x_{-1}, x_0, x_1, \dots$ such that the triple $(x_-, x, x_+) = (x_{l-1}, x_l, x_{l+1})$ satisfies (7.1) for any integer l, are called trajectories of the standard map.

The Lagrangian form of the standard map admits a variational formulation. Namely, trajectories of the system are extremals of the formal sum

$$\sum_{l=-\infty}^{\infty} L(x_l, x_{l+1}), \qquad L(x', x'') = \frac{1}{2\varepsilon}(x' - x'')^2 - \cos x''. \tag{7.2}$$

The extremality means that, for any trajectory $\dots, x^0_{-1}, x^0_0, x^0_1, \dots$ and any integer n,

$$\frac{\partial}{\partial x_n} \sum_{l=-\infty}^{\infty} L(x_l, x_{l+1}) = 0 \quad \text{for } \dots, x_{-1}, x_0, x_1, \dots = \dots, x^0_{-1}, x^0_0, x^0_1, \dots.$$

If $\varepsilon = \infty$, the standard map is meaningless because x_+ cannot be found in terms of x and x_- from equation $(7.1)|_{\varepsilon^{-1}=0}$. However, the corresponding variational problem is well-defined. Its solutions are sequences of the form

$$\dots, \pi k_{-1}, \pi k_0, \pi k_1, \dots, \qquad k_j \in \mathbb{Z}. \tag{7.3}$$

For large values of the parameter ε the standard map has many trajectories close to sequences (7.3). More precisely the following Theorem 7.1 holds [12]. Let S_K be the set of all sequences (7.3) such that $|k_l - k_{l+1}| \le K$ for any $l \in \mathbb{Z}$.

Let us introduce on the set of infinite sequences the metric ρ, corresponding to the uniform convergence norm, i.e., for any two sequences

$$X' = \dots. x'_{-1}, x'_0, x'_1, \dots, \qquad X'' = \dots, x''_{-1}, x''_0, x''_1, \dots,$$

we put

$$\hat\rho(X', X'') = \sup_{l \in \mathbf{Z}} |x'_l - x''_l|.$$

(For some pairs X', X'' the last expression can be equal to infinity).

Theorem 7.1 ([12]). *For any $K > 0$ and any $\sigma > 0$ there exists a (sufficiently large) $\varepsilon_0 > 0$ such that, for any $X' \in S_K$ and any $\varepsilon > \varepsilon_0$, the standard map has a unique trajectory X'' with $\rho(X', X'') < \sigma$.*

Theorem 7.1 means that for large values of ε some trajectories of the standard map turn out to be in a one-to-one correspondence with elements of the set S_K. Sequences from S_K can be regarded as codes of the corresponding trajectories. This possibility to code trajectories by elements of a sufficiently large set is typical for chaotic systems.

Theorem 7.1 was generalized by MacKay and Meiss [83] to multidimensional symplectic maps with configurational space $\mathbb{T}^m$. Below we present a proof of a theorem on the anti-integrable limit in a more general situation.

7.2 A General Construction

Suppose that a discrete group G acts on an m-dimensional smooth manifold M.[1] Below we assume that any point $x \in M$ has a neighborhood U such that the sets $g(U)$, $g \in G$, do not intersect: $g'(U) \cap g''(U) = \emptyset$ for $g' \neq g''$. This means that the quotient space M/G is a smooth manifold.

The action of G on M generates the "diagonal" action of G on the direct product $M \times M$: for any pair $x, y \in M$ and $g \in G$ we have

$$g(x, y) = (g(x), g(y)).$$

Let the smooth function $L : M \times M \to \mathbb{R}$ be invariant with respect to the action of G:

$$L(x, y) = L(g(x), g(y)), \qquad x, y \in M, \quad g \in G.$$

We assume also that the function L satisfies the following nondegeneracy property: for any $x \in M$ the map $\Theta_x : M \to T_x^* M$ defined by

$$y \mapsto \Theta_x(y) = \frac{\partial L}{\partial x}(x, y)$$

is a diffeomorphism. The functions L satisfying these properties are called discrete Lagrangians.

The assumption that the maps Θ_x are global diffeomorphisms can be weakened. We will frequently assume that the inverse maps Θ_x^{-1} exist not everywhere on $T_x^* M$. In this situation the dynamics in the system is, in general, not defined globally. However this does not mean that such systems are not interesting. It is sufficient to recall that the Poincaré map in an autonomous system almost never defined globally.

The quotient spaces M/G and $(M \times M)/G$ will be said to be respectively the configurational and phase spaces of the discrete Lagrangian system.[2]

The dynamics in the system with Lagrangian L is defined as follows: $(x_-, x) \mapsto (x, x_+)$, where $x_-, x, x_+ \in M$ are such that

$$\frac{\partial}{\partial x}\big(L(x_-, x) + L(x, x_+)\big) = 0. \tag{7.4}$$

Trajectories of a discrete Lagrangian system are extremals of the formal sum (7.2) in the same sense as this was explained in the previous section for the standard map.

A discrete Lagrangian is defined up to a constant multiplier, i.e., the Lagrangians $L(x', x'')$ and $cL(x', x'')$, where $c \neq 0$ is a constant, determine the same dynamical system. The dynamics are also preserved after adding to the Lagrangian a term of the form $f(x') - f(x'')$.

[1] I.e. a homomorphism of the group G to the group of diffeomorphisms of the manifold M is defined. Diffeomorphisms corresponding to elements of the group G are denoted below by the same letters. In the case of the standard map $M = \mathbb{R}$, and G is the group of shifts: $x \mapsto x + 2\pi l$, $l \in \mathbb{Z}$.

[2] In the case of the standard map these spaces are the one-dimensional torus $\mathbb{T}$ and the cylinder $\mathscr{Z}$.

As an example assume that the manifold M is Riemannian and the group G acts on M by isometries. Let $\operatorname{dist}(\cdot\,,\cdot)$ be the distance induced by the Riemannian metric. Then the Lagrangian $L(x', x'') = \operatorname{dist}^2(x', x'')$ is a smooth function for any pair of sufficiently close points x', x''. It is invariant with respect to the diagonal action of G. The corresponding functions Θ_x have inverses in the vicinity of zero of the space T_x^*M. Hence any pair of sufficiently close points x_-, x determines a unique point x_+ satisfying equation (7.4).

The function $x_+ = x_+(x_-, x)$ in this example has a simple geometric meaning. Since the points x_-, x are close to one another, there exists a unique shortest geodesic linking x_- with x. Then x_+ is situated on the same geodesic, x lies (locally) between x_- and x_+ and

$$\operatorname{dist}(x_-, x) = \operatorname{dist}(x, x_+) = \frac{1}{2}\operatorname{dist}(x_-, x_+).$$

Let $L : M \times M \times U \to \mathbb{R}$, where $U \subset \mathbb{R}$ is a neighborhood of zero, be a smooth function invariant with respect to the diagonal action of G on $M \times M$:

$$L(x, y, \mu) = L(gx, gy, \mu), \qquad x, y \in M, \quad \mu \in U, \quad g \in G.$$

Suppose that the maps Θ_x, corresponding to the function $L(x, y, 0)$ are invertible (everywhere or locally in the domains which appear below) and $f : M \to \mathbb{R}$ is a smooth function invariant with respect to the action of G on M:

$$f(x) = f(gx) \quad \text{for all } x \in M, \quad g \in G.$$

Then the functions

$$\mathscr{L}_\lambda = \mathscr{L}_\lambda(x, y) = L(x, y, 1/\lambda) + \lambda f(y), \qquad \lambda \in \mathbb{R}$$

also determine certain discrete Lagrangian systems. We define the anti-integrable limit in these systems as the limit $\lambda \to \infty$.

Suppose that the configurational space M/G is compact and all critical points of the function f are isolated. Let $\mathrm{Cr} \subset M$ be the set of nondegenerate critical points of f. The action of G preserves the set Cr. Moreover, Cr can be regarded as a finite union of disjoint orbits O_l:

$$O_l = \bigcup_{g \in G} g(y_l), \qquad y_l \in \mathrm{Cr}\,.$$

Let S_K, $K > 0$, be the set of all sequences

$$\dots, x_{-1}, x_0, x_1, \dots, \qquad x_j \in \mathrm{Cr}, \quad \rho(x_j, x_{j+1}) \le K, \quad j \in \mathbb{Z},$$

where ρ is the distance in some G-invariant Riemannian metric on M. As in the previous section, let us introduce a metric on the set of infinite sequences

$$\dots, x_{-1}, x_0, x_1, \dots, \qquad x_j \in M.$$

For any two sequences

$$X' = \dots, x'_{-1}, x'_0, x'_1, \dots, \qquad X'' = \dots, x''_{-1}, x''_0, x''_1, \dots,$$

we put

$$\hat{\rho}(X', X'') = \sup_{l \in \mathbb{Z}} \rho(x'_l, x''_l) \le \infty.$$

Theorem 7.2. *For any $K > 0$ and any $\sigma > 0$ there exists $\lambda_0 > 0$ such that, for any $X' \in S_K$ and any $\lambda > \lambda_0$ the discrete Lagrangian system with Lagrangian $\mathscr{L}_\lambda$ has a unique trajectory X'' with $\hat{\rho}(X', X'') < \sigma$.*

The proof of Theorem 7.2 is based on the contraction principle. Let us introduce local coordinates in neighborhoods of points $x \in \mathrm{Cr}$ so that the coordinates of any two points x', x'' such that $x' = g(x'')$, $g \in G$, coincide. To this end, for any orbit O_l we take a point $x \in O_l$ and introduce arbitrarily coordinates in a neighborhood of x. Coordinates in neighborhoods of other points of O_l are obtained as translations of this coordinate system by the maps $g \in G$.

The map $x \mapsto \partial f/\partial x$ is invertible in a neighborhood of any point $x' \in \mathrm{Cr}$. This fact follows from the nondegeneracy of the critical point x' and from the implicit function theorem. The inverse map $\Phi_{x'} : B \to V_{x'}$ acts from a neighborhood B of zero in $\mathbb{R}^m$ to a neighborhood $V_{x'}$ of the point x'. We can assume that the set B does not depend on x'. Hence, due to the invariance of the coordinate systems with respect to the action of the group G, for any two points x', x'' contained in the same orbit O_l, we have $V_{x'} = g(V_{x''})$ for some $g \in G$.

Let $\Omega_\sigma(X')$ be a neighborhood of the sequence $X' \in S_K$ in the set of all sequences on M:

$$\Omega_\sigma(X') = \{X : \hat{\rho}(X, X') \le \sigma\}.$$

Consider the map $\mathscr{F}$ of the metric space $\Omega_\sigma(X')$ to the space of sequences $\dots, x_{-1}, x_0, x_1, \dots$ on M such that the sequence $\hat{X} = \mathscr{F}(X)$ is defined as follows:

$$\hat{x}_j = \Phi_{x'_j}\Big(-\frac{1}{\lambda}\frac{\partial}{\partial x_j}\big(L(x_{j-1}, x_j, 1/\lambda) + L(x_j, x_{j+1}, 1/\lambda)\big)\Big). \tag{7.5}$$

The fixed points of $\mathscr{F}$ are trajectories of the system with Lagrangian L_λ lying in a σ-neighborhood of the code X'. Indeed, for $X = \hat{X}$ equation (7.5) implies that

$$\frac{\partial}{\partial x_j}\big(\mathscr{L}_\lambda(x_{j-1}, x_j) + \mathscr{L}_\lambda(x_j, x_{j+1})\big) = 0.$$

Hence, to prove Theorem 7.2, it is sufficient to check that for large λ the following two statements hold:

(a) $\mathscr{F}(\Omega_\sigma(X')) \subset \Omega_\sigma(X')$,
(b) the map $\mathscr{F}$ is contracting.

For any $y \in \mathrm{Cr}$ let $\mathscr{B}_\sigma(y) \subset V_y$ be the ball

$$\mathscr{B}_\sigma(y) = \{x \in V_y : \rho(x, y) < \sigma\}.$$

The metric ρ, the function f, and the local coordinates we use are invariant with respect to the action of G. Hence the set

$$\bigcap_{y \in \mathrm{Cr}} \frac{\partial f}{\partial x}(\mathscr{B}_\sigma(y)) \subset \mathbb{R}^m \tag{7.6}$$

is open and nonempty. Indeed, since for any $y_1, y_2 \in \mathrm{Cr}$ such that $y_1 = g(y_2)$, $g \in G$, we have

$$\frac{\partial f}{\partial x}(\mathscr{B}_\sigma(y_1)) = \frac{\partial f}{\partial x}(\mathscr{B}_\sigma(y_2)),$$

it follows that (7.6) is an intersection of a finite number of nonempty open sets. In particular, the set (7.6) contains an open ball $B_r \subset \mathbb{R}^m$ centered at zero with a radius $r > 0$. Taking if necessary smaller r, we can assume that $B_r \subset B$. According to the definition, $\Phi_y^{-1}(B_r) \subset \mathscr{B}_\sigma(y)$ for any Cr.

Since $X' \in S_K$, $K < \infty$, for sufficiently small $\mu_0 > 0$ we have

$$\sup_{X \in \Omega_\sigma(X'), |1/\lambda| \le \mu_0} \left| \frac{\partial}{\partial x_j}\big(L(x_{j-1}, x_j, 1/\lambda) + L(x_j, x_{j+1}, 1/\lambda)\big)\right| = C < \infty.$$

Let $\lambda > \max\{C/r, 1/\mu_0\}$. Then for any $X \in \Omega_\sigma(X')$ the argument of the function Φ_{x_j} in (7.5) lies in B_r. Hence, $\hat{X} \in \Omega_\sigma(X')$.

To check that the map $\mathscr{F}$ is contracting, we put

$$c_1 = \sup \frac{\rho(\Phi_y(b_1), \Phi_y(b_2))}{|b_1 - b_2|},$$

where the supremum is taken over all $y \in \mathrm{Cr}$ and $b_1, b_2 \in B_r$, and $|\cdot|$ denotes the standard norm in $\mathbb{R}^m$. The constant c_1 is obviously finite.

For any two sequences $X, \hat{X} \in \Omega_\sigma(X')$ and $|1/\lambda| \le \mu_1$ (the positive constant $\mu_1 \le \mu_0$ is sufficiently small) we have

$$\Big| \frac{\partial}{\partial x_j}\big(L(x_{j-1}, x_j, 1/\lambda) + L(x_j, x_{j+1}, 1/\lambda)\big) - \frac{\partial}{\partial \hat{x}_j}\big(L(\hat{x}_{j-1}, \hat{x}_j, 1/\lambda) + L(\hat{x}_j, \hat{x}_{j+1}, 1/\lambda)\big)\Big| \le c_2\, \hat{\rho}(X, \hat{X}),$$

with some constant c_2. By putting $X^* = \mathscr{F}(X)$, $\hat{X}^* = \mathscr{F}(\hat{X})$, we obtain the estimate

$$\rho(x_j^*, \hat{x}_j^*) \le \frac{c_1 c_2}{\lambda} \hat{\rho}(X, \hat{X}).$$

Hence, the map $\mathscr{F}$ is contracting for $\lambda > \max\{2c_1c_2, 1/\mu_1\}$.

It is easy to show that the set of trajectories constructed in Theorem 7.2 forms a *uniformly hyperbolic set*. Although this set is uncountable, its measure and even the Hausdorff dimension vanish.

The absence of an analytic integral in systems close to the anti-integrable limit is intuitively obvious. However, a formal proof of this fact needs some work.[3]

On the other hand one should not think that an unbounded growth of chaos in the anti-integrable limit leads to ergodicity of the system. An elementary straightforward calculation shows that

For arbitrarily large values of the perturbing parameter the standard map has an elliptic periodic trajectory of period 2.

This observation is a trivial particular case of beautiful results, contained in [45], which show that the number of elliptic periodic points can be arbitrarily large for large values of the perturbing parameter and moreover, for an increasing sequence of the parameter, these periodic points can be asymptotically dense on the phase torus.

Since in a general situation elliptic periodic trajectories are surrounded by stability islands, the standard map does not become ergodic in the anti-integrable limit. However, in principle, the standard map can be ergodic for some large values of ε.

7.3 Further Examples

1. Consider a particle with a small mass μ which moves in the space $\mathbb{R}^m$ in the force field with potential $V(x,t)$. Suppose that the potential is 2π-periodic in the variables $x_1, \dots, x_m$ and t. Moreover, we assume that

$$V(x,t) = \frac{1}{2\pi} v(x)\delta(t), \tag{7.7}$$

where $\delta(t)$ is the periodic δ-function:

$$\delta(t) = \begin{cases} \infty, & t \in 2\pi\mathbb{Z}, \\ 0, & t \in \mathbb{R} \setminus 2\pi\mathbb{Z}, \end{cases} \qquad \int_{2\pi k-\sigma}^{2\pi k+\sigma} \delta(t)\, dt = 1$$

for any $k \in \mathbb{Z}$, $\sigma \in (0, 2\pi)$.

The Hamiltonian of the system has the form

$$H = \frac{|p|^2}{2\mu} + \frac{1}{2\pi} v(x)\delta(t),$$

where $p = (p_1, \dots, p_m)$ is the momentum canonically conjugate to the coordinates x. The Hamiltonian equations read

---

[3] In the case of two-dimensional phase space: nonintegrability follows from the existence of transversal homoclinics to periodic solutions.

$$\dot{p} = -\frac{1}{2\pi}\frac{\partial v}{\partial x}(x)\delta(t), \qquad \dot{x} = p/\mu.$$

Therefore p gets increments equal to $-(2\pi)^{-1}\partial v/\partial x$ at the time moments $2\pi l$, $l \in \mathbb{Z}$. During the remaining time the particle is free.

The Poincaré map can be written down explicitly. For any integer l we put $x(2\pi l - 0) = x_l$, $p(2\pi l - 0) = p_l$. Then

$$\begin{pmatrix} x_l \\ p_l \end{pmatrix} \mapsto \begin{pmatrix} x(2\pi l + 0) \\ p(2\pi l + 0) \end{pmatrix} = \begin{pmatrix} x_l \\ p_l - \frac{1}{2\pi}\frac{\partial v}{\partial x}(x_l) \end{pmatrix}$$
$$\mapsto \begin{pmatrix} x_{l+1} \\ p_{l+1} \end{pmatrix} = \begin{pmatrix} x_l + 2\pi p_{l+1}/\mu \\ p_l - \frac{1}{2\pi}\frac{\partial v}{\partial x}(x_l) \end{pmatrix}.$$

The quantities x_{l-1}, x_l, x_{l+1} satisfy the equation

$$x_{l+1} - 2x_l + x_{l-1} = \frac{1}{\mu}\frac{\partial v}{\partial x}(x_l).$$

The discrete Lagrangian of this system has the form

$$L(x', x'') = \frac{|x' - x''|^2}{2} - \frac{1}{\mu}\frac{\partial v}{\partial x}(x'').$$

In this case $M = \mathbb{R}^m$, $G = \mathbb{Z}^m$ and for any $x \in \mathbb{R}^m$, $k \in \mathbb{Z}^m$, we have $k(x) = x + 2\pi k$. The Lagrangian L is invariant with respect to the action of the group $\mathbb{Z}^m$ by shifts. The limit $\mu \to 0$ is anti-integrable. Therefore for small μ the system has an uncountable hyperbolic set, carrying a chaotic dynamics.

Note that for small μ the system remains close to the anti-integrable limit if the potential V is an ordinary periodic function close (as a distribution) to (7.7).

If a light particle travels in a potential force field, where the potential $V(x, t)$ does not satisfy (7.7), the theory of the anti-integrable limit ($\mu \to 0$) exists, but becomes technically more complicated [22].

2. Consider a plane billiard system in a strip bounded by graphs of two periodic functions. We assume that a particle moves in the domain

$$D = \{(x, y) \in \mathbb{R}^2 : f_1(x) \le y \le f_2(x) + d\},$$

where $f_{1,2}$ are 2π-periodic functions (Fig. 7.1) and the parameter d is large. The motion inside the domain is assumed to be free. Reflections from the boundary are elastic.

This is a discrete Lagrangian system, where the function L is the length of the line segment between two subsequent points of the impact with the boundary. We will consider motions such that the particle collides alternately with the upper and lower walls. In other words, we are interested in trajectories whose links do not deviate much from the verticals $x = \text{const}$.

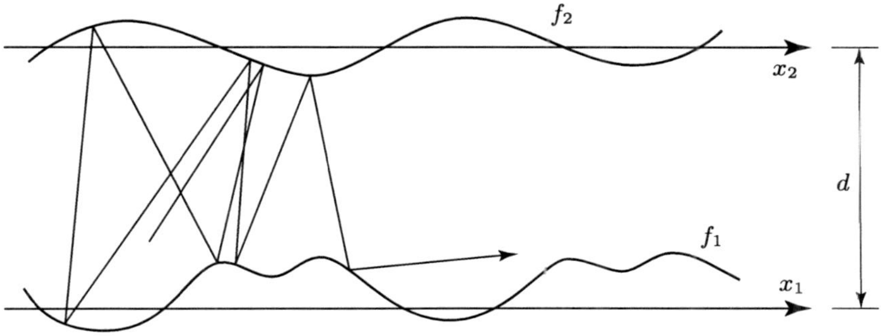

Fig. 7.1 Billiard in a wide strip.

Let x_1 be the coordinate on the lower boundary and x_2 on the upper one. The length of the corresponding line segment is

$$L(x_1, x_2) = \sqrt{(x_2 - x_1)^2 + (d + f_2(x_2) - f_1(x_1))^2}.$$

In this case $M = \mathbb{R}$, $G = \mathbb{Z}$, the group $\mathbb{Z}$ acts on $\mathbb{R}$ by the shifts $x \mapsto x + 2\pi k$, $k \in \mathbb{Z}$. The Lagrangian is invariant with respect to these shifts, and for large d

$$L(x_1, x_2) = d + f_2(x_2) - f_1(x_1) + \frac{1}{2d}(x_2 - x_1)^2 + O(d^{-2}).$$

Thus the limit $d \to \infty$ is anti-integrable provided the functions f_1 and f_2 do not coincide.

It is easy to construct also a multidimensional analog of this system.

3. In conclusion we present a system whose configurational space is not a torus. Suppose that on the Lobachevski plane $\mathscr{L}$ the discrete group of motions G acts, and $\mathscr{L}/G$ is a compact manifold. Recall that in this case $\mathscr{L}/G$ is diffeomorphic to a sphere with $n > 1$ handles. The Lobachevski metric induces on $\mathscr{L}/G$ a metric of a constant negative curvature. For any function $f : \mathscr{L} \to \mathbb{R}$ invariant with respect to the action of the group G, the function $L : \mathscr{L} \times \mathscr{L} \to \mathbb{R}$ defined by

$$L(x', x'') = \mathrm{dist}^2\,(x', x'') + \lambda f(x'')$$

is a discrete Lagrangian. Here the distance is taken in the Lobachevski metric. The limit $\lambda \to \infty$ is anti-integrable.

Note that the system corresponding to the value $\lambda = 0$ has a first integral $\mathrm{dist}(x', x'') = \mathrm{const} \geq 0$. The system is ergodic on nonzero levels of this integral. On the other hand, for large λ, apparently, there is neither integral, nor ergodicity.

7.4 Anti-Integrable Limit in the Separatrix Map

It turns out that, in the vicinity of separatrices area-preserving maps behave similarly to maps near the anti-integrable limit. Here we discuss this phenomenon for model (4.1), Chap. 4

$$\begin{pmatrix} I \\ \varphi \\ \sigma \end{pmatrix} \mapsto \begin{pmatrix} I_+ \\ \varphi_+ \\ \sigma_+ \end{pmatrix}, \qquad \begin{aligned} I_+ &= I + \lambda v_\sigma(\varphi), \\ \varphi_+ &= \varphi + \omega_\sigma + \frac{1}{\lambda}\log|I_+|, \\ \sigma_+ &= \sigma \operatorname{sign} I_+, \qquad \sigma, \sigma_+ \in \{-1, 1\}. \end{aligned} \tag{7.8}$$

Recall that, in the main approximation, the map (7.8) determines the dynamics near separatrices of a hyperbolic fixed point of any area-preserving map close to an integrable one. The functions $v_{\pm 1}$ are periodic and their average vanishes.

The map (7.8) can be represented in the form

$$I = \partial W/\partial\varphi, \quad \varphi_+ = \partial W/\partial I_+, \quad \sigma_+ = \sigma \operatorname{sign} I_+, \tag{7.9}$$

where the generating function

$$W = W(I_+, \varphi, \sigma) = \varphi I_+ + \lambda V_\sigma(\varphi) + (\omega_\sigma + \lambda^{-1}\log|I_+| - \lambda^{-1})\, I_+.$$

Here the "potentials" V_σ are primitives of the functions $-v_\sigma$. The primitives are periodic in φ since the mean values of v_σ vanish. The first two equations (7.9) can be written as follows:

$$\varphi_+ dI_+ + I d\varphi = dW(I_+, \varphi, \sigma).$$

Now let us represent (7.8) in the Lagrangian form. To this end consider another generating function (the Legendre transform of W) $h = \varphi_+ I_+ - W$. By using this function, we write down the separatrix map in the following way:

$$I_+ d\varphi_+ - I d\varphi = dh. \tag{7.10}$$

It is easy to obtain an explicit formula for h:

$$h = h(\varphi, \varphi_+, \sigma, \vartheta_+) = \vartheta_+ \lambda^{-1} e^{\lambda(\varphi_+ - \varphi - \omega_\sigma)} - \lambda V_\sigma(\varphi), \qquad \vartheta_+ = \operatorname{sign} I_+.$$

The Lagrangian form of the separatrix map is as follows:

$$\begin{pmatrix} \varphi_- \\ \varphi \\ \sigma_- \\ \vartheta \end{pmatrix} \mapsto \begin{pmatrix} \varphi \\ \varphi_+ \\ \sigma \\ \vartheta_+ \end{pmatrix}, \qquad \begin{aligned} &\sigma = \sigma_- \vartheta, \\ &\frac{\partial}{\partial\varphi}(h_- + h) = 0, \end{aligned} \tag{7.11}$$

where

$$h_- = h(\varphi_-, \varphi, \sigma_-, \vartheta), \quad h = h(\varphi, \varphi_+, \sigma, \vartheta_+).$$

The first equation (7.11) is generated by the definition of ϑ ($\vartheta = \operatorname{sign} I$) and by the last equation (7.8). The second equation (7.11) follows from (7.10) since, according to (7.10) and to the analogous equation $I d\varphi - I_- d\varphi_- = dh_-$, we have

$$I = -\partial h/\partial\varphi = \partial h_-/\partial\varphi.$$

It is easy to check that the quantities φ_+ σ and ϑ_+ are computed uniquely from (7.11) in terms of φ_-, φ, σ_-, and ϑ.

Now we can present a variational principle for the separatrix map (7.8). Any sequence

$$f = \{f_j\}, \quad f_j = \begin{pmatrix} \varphi_j \\ \sigma_j \\ \vartheta_j \end{pmatrix}, \qquad \sigma_{j+1} = \sigma_j \vartheta_{j+1},$$

is said to be a path. Let Π be the set of all paths.

In general the index j takes all integer values. However, it is possible to consider also semifinite and finite paths. Paths finite from the left begin with a triple f_j, where $\varphi_j = +\infty$. Paths finite from the right end with f_j, where $\varphi_j = -\infty$. Paths finite from the left and from the right are called finite.

The action S is defined as the formal sum

$$S = S(f) = \sum_j h(\varphi_j, \varphi_{j+1}, \sigma_j, \vartheta_{j+1}).$$

The path f^0 is said to be an extremal iff $\partial S/\partial\varphi_j|_{f=f^0} = 0$ for any j.

Proposition 7.1. *The path f is an extremal if and only if the sequence*

$$(\varphi_j, \varphi_{j+1}, \sigma_j, \vartheta_{j+1})^T$$

is a trajectory of (7.11).

The proof is straightforward. Below we identify extremals with Lagrangian trajectories of the separatrix map.

Note that semifinite trajectories belong to separatrices. Finite ones belong to both stable and unstable separatrices, and therefore they are homoclinic trajectories.

We define the distance ρ on Π in the following way. Let f' and f'' be paths, where

$$f'_j = \begin{pmatrix} \varphi'_j \\ \sigma'_j \\ \vartheta'_j \end{pmatrix}, \quad f''_j = \begin{pmatrix} \varphi''_j \\ \sigma''_j \\ \vartheta''_j \end{pmatrix}.$$

We put $\rho(f', f'') = \infty$ if the sequences σ'_j, ϑ'_j do not coincide with $\sigma''_j, \vartheta''_j$ or if for some j only one of the triples is defined. Otherwise we put

$$\rho(f', f'') = \sup_j |\varphi'_j - \varphi''_j|.$$

Here we put $|-\infty - (-\infty)| = |+\infty - (+\infty)| = 0$.

Let $\mathrm{Cr}(\sigma)$ denote the set of nondegenerate critical points of the function V_σ. The set Π contains the subset of simple paths (codes). By the definition a path f is simple if $\varphi_j \in \mathrm{Cr}(\sigma_j)$ for any j.

Theorem 7.3. *Suppose that the sets* $\mathrm{Cr}(\pm 1)$ *are finite and the constants* c_1, $c_2 = c_2(c_1)$ *are sufficiently large. Then, for any simple path* f^* *such that* $\varphi^*_j - \varphi^*_{j+1} > c_1$ *for all* j, *there exists a unique trajectory* $\tilde{f}$ *in the* c_2^{-1}*-neighborhood of* f^*.

Theorem 7.3 is similar to Theorems 7.1–7.2 about the anti-integrable limit. The simple path f^* can be regarded as a code corresponding to the trajectory $\tilde{f}$.

The proof of Theorem 7.3 conceptually coincides with the proof of Theorem 7.2. Here again the main tool is the contraction principle. We only present the contracting operator F having $\tilde{f}$ as a fixed point. All details of the proof can be restored easily. Any path f which belongs to the c_2^{-1}-neighborhood of the simple path f^* is mapped to $\hat{f} = F(f)$, where the sequences σ_j, ϑ_j and $\hat{\sigma}_j$, $\hat{\vartheta}_j$ coincide, and

$$\hat{\varphi}_j = (V'_{\sigma_j})^{-1}\big(\vartheta_j \lambda^{-1} e^{\lambda(\varphi_j - \varphi_{j-1} - \omega_{\sigma_{j-1}})} - \vartheta_{j+1}\lambda^{-1} e^{\lambda(\varphi_{j+1} - \varphi_j - \omega_{\sigma_j})}\big).$$

Here V'_{σ_j} is the derivative of V_{σ_j} and the local inverse $(V'_{\sigma_j})^{-1}$ is assumed to act from a neighborhood of zero to a neighborhood of the point φ^*_j.

Theorem 7.3 establishes a symbolic dynamics in a neighborhood of separatrices of an area-preserving map. Recall that we deal only with the main approximation of the separatrix map. The same result can be easily obtained in the general situation.

The traditional approach to symbolic dynamics near separatrices is presented in [4, 16, 99]. The multidimensional separatrix map and the corresponding symbolic dynamics is discussed in [139, 140].

Chapter 8
Hill's Formula

8.1 General Remarks

In 1886, in his study of stability of the lunar orbit, Hill [61] published a formula which expresses the characteristic polynomial of the monodromy matrix for a second order time periodic equation in terms of the determinant of a certain infinite matrix. Here is a slightly modified version of this result. Consider the Hill equation

$$\ddot{x} = a(t)x, \qquad x \in \mathbb{R}, \quad t \bmod 2\pi,$$

where

$$a = \sum_{k=-\infty}^{+\infty} a_k e^{ikt}, \qquad a_k \in \mathbb{C},$$

is a 2π-periodic function. The stability of the zero solution is determined by the eigenvalues of the monodromy matrix (the multipliers) ρ and ρ^{-1}. Consider the infinite matrix

$$H = (h_{jk}), \qquad h_{jk} = \frac{k^2\delta_{jk} + a_{k-j}}{k^2 + 1}, \qquad j, k \in \mathbb{Z}, \tag{8.1}$$

where δ_{jk} is the Kronecker symbol. Hill showed that

$$\frac{\rho + \rho^{-1} - 2}{e^{2\pi} + e^{-2\pi} - 2} = \det H. \tag{8.2}$$

Hill computed the determinant approximately replacing H by a 3×3 matrix which gave quite a good approximation. He used equation (8.2) to find the multipliers approximately. Astronomical tables obtained by this method are well-known in astronomy.

Hill's argument was not rigorous because he did not prove convergence for the infinite determinant $\det H$. Several years later Poincaré [105] explained an exact meaning of the Hill's infinite determinant and presented a rigorous proof of Hill's

D. Treschev, O. Zubelevich, *Introduction to the Perturbation Theory of Hamiltonian Systems*, Springer Monographs in Mathematics, DOI 10.1007/978-3-642-03028-4_8, © Springer-Verlag Berlin Heidelberg 2010

formula. Hill's beautiful result entered textbooks on differential equations, but was almost forgotten by the dynamical systems community until the end of XXth century when an analog of formula (8.2) appeared for discrete Lagrangian systems in [82] and independently in [131]. Here H was finite and turned out to be the Hessian matrix associated with the action functional at the critical point, generated by the periodic solution. In [19] (see also [73]) a general form of the Hill formula was obtained for a periodic solution of an arbitrary Lagrangian system on a smooth manifold. In this case H has a meaning of a properly regularized Hessian of the action functional at the critical point determined by a periodic solution. Both discrete and continuous versions of Hill's formula give non-trivial information on the dynamical stability of the periodic orbit in terms of its Morse index.

We consider two similar but formally different cases:

- Continuous Lagrangian systems with configuration space M and time periodic Lagrangian $\mathscr{L}(x, v, t)$ on $TM \times \mathbb{R}$. Then τ-periodic trajectories $x(t)$ are critical points of the action functional

$$A[x] = \int_0^\tau \mathscr{L}(x(t), \dot{x}(t), t)\, dt, \qquad x(0) = x(\tau).$$

- Discrete Lagrangian systems with Lagrangian $L(x, y)$ on $M \times M$. Then n-periodic trajectories are sequences $x_j = x_{n+j}$ which are critical points of the action functional

$$A(\mathbf{x}) = \sum_{j=1}^{n} L(x_j, x_{j+1}), \qquad \mathbf{x} = (x_1, \dots, x_n), \qquad x_{n+1} = x_1.$$

Usually one case can be reduced to the other, but this reduction may be cumbersome. Hence it makes sense to consider both cases separately.

All versions of Hill's formula look as follows:

$$\rho^{-m} \det(P - \rho I) = \beta \det \mathbf{H}_\rho, \tag{8.3}$$

where P is the monodromy $2m \times 2m$ matrix of the periodic solution, $\mathbf{H}_\rho$ is the modified Hessian matrix which coincides with the ordinary Hessian for $\rho = 1$, and β is a nonzero scaling factor, usually with known sign.

We start with the discrete case since it is technically simpler.

8.2 Discrete Case

Discrete Lagrangian systems (DLS). Let M be a smooth m-dimensional manifold and $L : M^2 = M \times M \to \mathbb{R}$ a smooth function. Let ∂_1, ∂_2 be the differentials in the first and second variables:

$$\partial_1 L(x, y) = \frac{\partial L(x, y)}{\partial x}, \quad \partial_2 L(x, y) = \frac{\partial L(x, y)}{\partial y}, \tag{8.4}$$

and let

$$B(x, y) = -\partial_1 \partial_2 L(x, y) = -\partial_{12} L(x, y).$$

In local coordinates,

$$B(x, y) = -\left(\frac{\partial^2 L}{\partial y_j \partial x_i}\right) \tag{8.5}$$

is the matrix of mixed partial derivatives of L. In invariant terms, this is a linear operator $B(x, y) : T_x M \to T_y^* M$. We say that L is a discrete Lagrangian if B is nondegenerate, i.e.,

$$\det B(x, y) \neq 0, \qquad x, y \in M. \tag{8.6}$$

Due to condition (8.6) L locally defines a map $T : M^2 \to M^2$, $T(x, y) = (y, z)$, where $z = z(x, y)$ is determined by the equation

$$\frac{\partial}{\partial y}(L(x, y) + L(y, z)) = \partial_2 L(x, y) + \partial_1 L(y, z) = 0. \tag{8.7}$$

In general T is a multivalued map with the graph

$$\Gamma = \{(x, y, y, z) \in M^2 \times M^2 : \partial_2 L(x, y) + \partial_1 L(y, z) = 0\}.$$

The dynamical system determined by T is called the discrete Lagrangian system (DLS) with configurational space M and Lagrangian L.

Remark 8.1. We deal with a small neighborhood of a periodic orbit of T. Hence it is sufficient to assume that condition (8.6) holds in this neighborhood.

It is easy to check (see e.g. [145]) that T is symplectic with respect to the symplectic 2-form $\omega = B(x, y)\, dx \wedge dy$,

$$\begin{aligned} &\omega(\mathbf{u}, \mathbf{v}) = \langle B(x, y)u_1, v_2\rangle - \langle B(x, y)v_1, u_2\rangle, \\ &\mathbf{u} = (u_1, u_2), \quad \mathbf{v} = (v_1, v_2). \end{aligned} \tag{8.8}$$

Remark 8.2. Let us pass to Hamiltonian variables by the map $S : M^2 \to T^*M$, $(x, y) \mapsto (x, p_x)$, $p_x = -\partial_1 L(x, y)$. It is locally invertible and replaces T by a locally defined map $\tilde{T} = STS^{-1} : T^*M \to T^*M$. The map $\tilde{T}$ is symplectic with respect to the standard symplectic form $dp_x \wedge dx$ on T^*M, and L is the generating function of T^*:

$$\tilde{T}(x, p_x) = (y, p_y), \qquad p_x = -\partial_1 L(x, y), \quad p_y = \partial_2 L(x, y).$$

Such a symplectic map T^* is usually called a twist map.

The map T remains the same after multiplication of the Lagrangian by a constant, after addition of a constant to L and after the so-called gauge transformation

$$L(x,y) \mapsto L(x,y) + f(x) - f(y)$$

with an arbitrary smooth $f : M \to \mathbb{R}$.

A typical example of DLS is the standard map with its multidimensional generalizations

$$M = \mathbb{R}^m, \quad L(x,y) = \frac{1}{2}\langle B(x-y), x-y\rangle - \frac{1}{2}(V(x) + V(y)), \tag{8.9}$$

where B is a constant symmetric nondegenerate $m \times m$ matrix. Obviously B satisfies (8.5). Note that one can replace the potential $\frac{1}{2}(V(x)+V(y))$ by $V(x)$ because they are gauge-equivalent.

Consider a billiard system in a domain in $\mathbb{R}^{m+1}$ bounded by a smooth m-dimensional connected oriented hypersurface M. Let $l(x,y) = |x-y|$ be the length of the segment $[x,y]$ joining x, y. Then the billiard system is a DLS with the Lagrangian l. First consider a plane billiard system in a domain $D \subset \mathbb{R}^2$. For simplicity we assume that D is homeomorphic to a disk. Then we can use for a coordinate on the boundary $\partial D = M$ the arc length, counted counter-clockwise. Then it is well known (see e.g. [73]) that

$$\frac{\partial^2 l(x,y)}{\partial x \partial y} = \frac{\sin\alpha \sin\beta}{l} = \frac{(\mathbf{n}(y) - \mathbf{n}(x), \mathbf{e})}{l},$$

where α, β are the angles between $[x,y]$ and the corresponding tangent lines (see Fig. 8.1), $\mathbf{n}$ is the outer unit normal vector, and $\mathbf{e} = \overrightarrow{xy}/l$. Thus

$$\frac{\partial^2 l(x,y)}{\partial x \partial y} > 0. \tag{8.10}$$

It is convenient to replace l by $L(x,y) = -l(x,y)$ and to assume that $B(x,y) = -\partial_{12} L(x,y) : T_x M \to T_y M$ (we use the Euclidean metric to identify TM

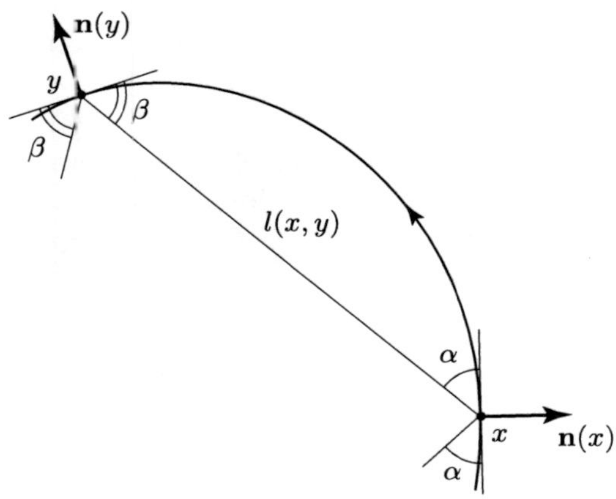

Fig. 8.1 A piece of a billiard trajectory.

with T^*M. Then we obtain

$$\langle B(x,y)v, \Pi v\rangle > 0, \qquad \Pi v := -\Pi_0 v, \tag{8.11}$$

where $\Pi_0 v$ is the parallel transport of v in the ambient space $\mathbb{R}^2$.

In more dimensions, we define a linear isomorphism $\Pi(x,y): T_yM \to T_xM$ as the parallel projection along the segment $[x,y]$ composed with $\iota: v \mapsto -v$.

Proposition 8.1. *$\langle B(x,y)v, \Pi(x,y)v\rangle > 0$ for all $v \in T_xM$. In particular $B(x,y): T_xM \to T_yM$ multiplies orientation by $(-1)^m$, i.e., taking any two local coordinate systems which determine the same orientation on M near x and y, we have:* $\operatorname{sign}(\det B) = (-1)^m$.

Proof. Take two points $x_0, y_0 \in M = \partial D$ and their small neighborhoods U, V in M. Let $\pi: V \to U$ be the projection along lines parallel to $[x_0, y_0]$, and let $\Pi = -D\pi(y_0): T_{y_0}M \to T_{x_0}M$. We claim that the quadratic form $Q(v) = \langle B(x,y)v, \Pi v\rangle$ is positive definite. Indeed, let Γ be the plane containing the segment $[x_0, y_0]$ and v. Then $u = \Pi v$ also lies in this plane. Consider the billiard system in the 2-dimensional domain $D \cap \Gamma$. By the inequality for a planar billiard system, we obtain that $\langle Bv, \Pi v\rangle > 0$. The statement about the determinant is proved as follows: the operator $G = \Pi^*B + B^*\Pi$ is positive definite and hence preserves orientation. Hence Π^*B preserves orientation (the sum of a positive definite and antisymmetric matrix has positive determinant). □

In [145] the reader can find other examples of DLS (mostly integrable) including multivalued ones.

For a continuous Lagrangian system (CLS) with Lagrangian $\mathscr{L}(x,v)$, an analog of the operator $B(x,y)$ is the matrix $\partial_{22}\mathscr{L}(x,v)$ of second partial derivatives. Indeed, consider a DLS with the Lagrangian $L(x,y) = \mathscr{L}(x,(y-x)/\varepsilon)$. Then in the limit $\varepsilon \to 0$, orbits of DLS converge to orbits of CLS with the Lagrangian $\mathscr{L}$. A computation shows that

$$\varepsilon^2 B(x,y) = D_{vv}\mathscr{L}(x,v) + O(\varepsilon), \qquad v = (y-x)/\varepsilon.$$

In particular, for an analog of a positive definite Lagrangian system,

$$\det B(x,y) > 0.$$

Unfortunately for $m \geq 2$ there is no clear discrete analog of positive definite continuous Lagrangian systems, i.e., of the condition $D_{vv}\mathscr{L}(x,v) > 0$. Indeed, in general the matrix B is not symmetric and, moreover, its symmetry does not have an invariant meaning.

The Poincaré and Hesse matrices. Let $(x_i)_{i\in\mathbb{Z}}$ be a periodic trajectory of a DLS, i.e., $T(x_{j-1}, x_j) = (x_j, x_{j+1})$ and $x_{j+n} = x_j$ for all j. The number n is said to be the period of the trajectory. The periodic orbit is determined by a point $\mathbf{x} = (x_1, \ldots, x_n) \in M^n$. A cyclic permutation of $\mathbf{x}$ gives the same periodic trajectory. Thus $\mathbf{x}$ should be viewed as an element of $\mathscr{M} = M^n/\mathbb{Z}_n$. By (8.7),

$$\partial_2 L(x_{j-1}, x_j) + \partial_1 L(x_j, x_{j+1}) = 0, \qquad j = 1, \dots, n, \tag{8.12}$$

where by definition $x_0 = x_n$ and $x_1 = x_{n+1}$. Thus $\mathbf{x}$ is a critical point of the function (the action functional)

$$\mathscr{A}(\mathbf{x}) = L(x_1, x_2) + L(x_2, x_3) + \cdots + L(x_n, x_1), \qquad \mathbf{x} \in M^n.$$

The point $p = (x_1, x_2)$ is a fixed point of the map $T^n : M^2 \to M^2$. The perturbed dynamics near the trajectory is determined by the map T^n in a neighborhood of p. The linear approximation is determined by the linear Poincaré map $P = DT^n(p) : T_pM^2 \to T_pM^2$. In local coordinates, P becomes the monodromy matrix

$$P = \left. \frac{\partial T^n(x, y)}{\partial(x, y)} \right|_{(x,y)=(x_1,x_2)}.$$

It is defined uniquely up to a similarity $P \mapsto S^{-1}PS$. Eigenvalues of P are called multipliers of the periodic orbit $\mathbf{x}$. They determine dynamical properties of the periodic trajectory in the linear approximation.

Let

$$\mathbf{H} = \frac{\partial^2 \mathscr{A}(\mathbf{x})}{\partial \mathbf{x}^2}$$

be the Hessian matrix of $\mathscr{A}$ at the critical point $\mathbf{x}$. Denote

$$B_k = B(x_k, x_{k+1}), \qquad x_{n+1} = x_1.$$

Theorem 8.1 (Hill's formula).

$$\det(P - I) = \frac{(-1)^m \det \mathbf{H}}{\prod_{k=1}^n \det B_k}. \tag{8.13}$$

Invariant version of Hill's formula. The left-hand side of (8.13) obviously does not depend on the choice of local coordinates in M. However an invariant meaning of the right-hand side

$$\Delta = \frac{\det \mathbf{H}}{\prod_{k=1}^n \det B_k}$$

is a priori not clear. To explain why Δ is coordinate independent, note that, for any critical point $\mathbf{x} \in M^n$, the Hessian of $\mathscr{A}$ at $\mathbf{x}$ determines a symmetric bilinear form h on $X = T_\mathbf{x}M^n = T_{x_1}M \times \cdots \times T_{x_n}M$:

$$h(\mathbf{u}, \mathbf{v}) = \sum_{j=1}^n (\langle A_j u_j, v_j\rangle - \langle B_{j-1} u_{j-1}, v_j\rangle - \langle B_j^* u_{j+1}, v_j\rangle),$$
$$u_0 = u_n, \quad u_{n+1} = u_1,$$

where $\mathbf{u} = (u_1, \dots, u_n)$, $\mathbf{v} = (v_1, \dots, v_n)$. Here

$$A_j = \partial_{22}L(x_{j-1}, x_j) + \partial_{11}L(x_j, x_{j+1})$$

is the Hessian of the function $x \mapsto L(x_{j-1}, x) + L(x, x_{j+1})$ at $x = x_j$, i.e., a symmetric operator $A_j : T_{x_j}M \to T^*_{x_j}M$, and $B_j = B(x_j, x_{j+1})$ can be viewed as a linear operator $T_{x_j}M \to T^*_{x_{j+1}}M$. As usual, $\langle\, ,\rangle$ is the canonical pairing of a vector and covector.

The form h is represented by a symmetric operator $\mathbf{H} : X \to X^*$:

$$h(\mathbf{u}, \mathbf{v}) = \langle \mathbf{Hu}, \mathbf{v}\rangle, \qquad \mathbf{u}, \mathbf{v} \in X,$$

where

$$(\mathbf{Hu})_j = A_j u_j - B_{j-1}u_{j-1} - B_j^* u_{j+1}, \qquad j = 1, \dots, n.$$

Remark 8.3. The variational system of the periodic trajectory $\mathbf{x}$ is defined as

$$A_j u_j - B_{j-1}u_{j-1} - B_j^* u_{j+1} = 0, \qquad j \in \mathbb{Z}. \tag{8.14}$$

This is the linear approximation to the system (8.12) near the periodic trajectory. If u_j is any solution of the variational system, then $P(u_j, u_{j+1}) = (u_{j+n}, u_{j+n+1})$.

Define a symmetric linear operator $\mathbf{A} : X \to X^*$ and a nonsymmetric linear operator $\mathbf{B} : X \to X^*$ by

$$(\mathbf{Au})_j = A_j u_j, \quad (\mathbf{Bu})_j = -B_{j-1}u_{j-1}, \quad (\mathbf{B}^*\mathbf{u})_j = -B_j^* u_{j+1}.$$

Then

$$\mathbf{H} = \mathbf{A} + \mathbf{B} + \mathbf{B}^*.$$

If we introduce local coordinates, then $\mathbf{B}$ becomes an $mn \times mn$ matrix, and it is easy to see that

$$\det \mathbf{B} = \prod_{k=1}^{n} \det B_k.$$

By (8.4), $\mathbf{B}$ is non-degenerate. Thus $\mathbf{B}^{-1}\mathbf{H} : X \to X$ and $(\mathbf{B}^*)^{-1}\mathbf{H} : X \to X$ are well-defined operators, and

$$\Delta = \det(\mathbf{B}^{-1}\mathbf{H}) = \det((\mathbf{B}^*)^{-1}\mathbf{H}).$$

Another way to give an invariant meaning to (8.13) is to regard $\det \mathbf{H}$ and $\det \mathbf{B}$ as linear operators

$$\wedge^{mn} X \to \wedge^{mn} X^*.$$

Since these spaces are 1-dimensional, $\det \mathbf{H}/\det \mathbf{B}$ is a well defined scalar.

The third way to give an invariant meaning to Δ is to introduce on M a Riemannian metric $(\, ,)$ which defines a scalar product on $X = T_{\mathbf{x}}M^n$. Then $\mathbf{H}$ is replaced by a self-adjoint operator $H : X \to X$ defined by $(H\mathbf{u}, \mathbf{v}) = \langle \mathbf{Hu}, \mathbf{v}\rangle$. The linear operators $B_j : T_{x_j}M \to T^*_{x_{j+1}}M$ are replaced by $C_j : T_{x_j}M \to T_{x_{j+1}}M$, where $(C_j u, v) = -\langle B_j u, v\rangle$. Define an operator $C : T_{x_1}M \to T_{x_1}M$ by $C =$

$C_n C_{n-1} \cdots C_1$. Then

$$\Delta = \det H / \det C.$$

The numerator and denominator depend on the Riemannian metric, but not the ratio.

Generalized Hill determinant. Let us define a generalization of the Hessian $\mathbf{H}$ as follows. Let $\mathscr{X}$ be the infinite dimensional space of all complex vector fields $(v_j)_{j\in\mathbb{Z}}$, $v_j \in T_{x_j}M$. For any $\rho \in \mathbb{C}$, consider the subspace $X_\rho \subset \mathscr{X}$ of complex quasiperiodic vector fields $(v_j)_{j\in\mathbb{Z}}$ such that $v_{j+n} = \rho v_j$. If $|\rho| = 1$, the Hessian h defines a Hermitian form h_ρ on X_ρ:

$$h_\rho(\mathbf{u}, \bar{\mathbf{v}}) = \sum_{j=1}^{n} (\langle A_j u_j, \bar{v}_j\rangle - \langle B_{j-1}u_{j-1}, \bar{v}_j\rangle - \langle B_j^* u_{j+1}, \bar{v}_j\rangle), \qquad u_{n+j} = \rho u_j.$$

Since $\mathbf{v} = (v_1, \ldots, v_n)$ determines $(v_j)_{j\in\mathbb{Z}}$ uniquely, h_ρ can be viewed as a Hermitian form on $X = X_1$.

We have $h_\rho(\mathbf{u}, \bar{\mathbf{v}}) = \langle \mathbf{H}_\rho \mathbf{u}, \bar{\mathbf{v}}\rangle$, where

$$(\mathbf{H}_\rho \mathbf{u})_j = A_j u_j - B_{j-1}u_{j-1} - B_j^* u_{j+1}, \qquad u_0 = \rho^{-1}u_n, \quad u_{n+1} = \rho u_1.$$

The operator $\mathbf{H}_\rho$ makes sense also for $|\rho| \neq 1$, but then it is non-Hermitian. We usually regard $\mathbf{H}_\rho$ as an operator in X, defining $\mathbf{H}_\rho \mathbf{u}$, $\mathbf{u} = (u_1, \ldots, u_n)$, by setting $u_0 = \rho^{-1}u_n$, $u_{n+1} = \rho u_1$.

In coordinates, $\mathbf{H}_\rho$ is an $mn \times mn$ matrix which coincides with $\mathbf{H}$ with two exceptions: in the upper right $m \times m$ block $-B_n$ is replaced by $-\rho^{-1}B_n$ and in the lower left $m \times m$ block $-B_n^*$ is replaced by $-\rho B_n^*$. In particular, $\mathbf{H}_1 = \mathbf{H}$.

Here is a generalization of Hill's formula:

Theorem 8.2 ([82, 131]). *For any* $\rho \in \mathbb{C}$

$$\rho^{-m} \det(P - \rho I) = (-1)^m \frac{\det \mathbf{H}_\rho}{\det \mathbf{B}}. \tag{8.15}$$

Taking $\rho = 1$ in (8.15), we obtain (8.13).

Proposition 8.2. *Both sides in* (8.15) *are polynomials of degree m in the variable* $\chi = (\rho + \rho^{-1})/2$ *with coefficient at* χ^m *equal* 1.

Indeed, the characteristic polynomial $F(\rho) = \det(P - \rho I)$ of the symplectic operator P is reciprocal: $F(\rho) = \rho^{2m}F(\rho^{-1})$. Therefore

$$G(\rho) = \rho^{-m}\det(P - \rho I) = G(\rho^{-1})$$

is a symmetric polynomial in ρ and ρ^{-1}. Thus G is a function of χ.

It is more natural to represent $\mathbf{H}_\rho$ in a different way. Define an operator $\mathscr{H}$: $\mathscr{X} \to \mathscr{X}$ by

$$(\mathscr{H}\mathbf{u})_j = A_j u_j - B_{j-1}u_{j-1} - B_j^* u_{j+1}, \qquad j \in \mathbb{Z}.$$

Then $\mathbf{H}_\rho$ is the restriction of $\mathscr{H}$ to the finite-dimensional space X_ρ of quasiperiodic vector fields. We identify X_ρ^* with the set of $\mathbf{p} = (p_j) \in E_j^*$, $j \in \mathbb{Z}$, such that $p_{j+n} = \rho^{-1} p_j$. The pairing is

$$\langle \mathbf{p}, \mathbf{u} \rangle = \sum_{j=1}^{n} \langle p_j, u_j \rangle.$$

Then $\mathbf{H}_\rho : X_\rho \to X_{\rho^{-1}}^*$. Let us make a change of variables $\mathbf{C}_\rho : X = X_1 \to X_\rho$, $\mathbf{C}_\rho^* : X_\rho^* \to X_1^*$,

$$(\mathbf{C}_\rho \mathbf{w})_j = \rho^{j/n} w_j, \quad (\mathbf{C}_\rho^* \mathbf{p})_j = \rho^{j/n} p_j.$$

The operator $\mathbf{H}_\rho : X_\rho \to X_{\rho^{-1}}^*$ is replaced by $\hat{\mathbf{H}}_\rho = \mathbf{C}_{\rho^{-1}}^* \mathbf{H}_\rho \mathbf{C}_\rho$, where

$$(\hat{\mathbf{H}}_\rho \mathbf{w})_j = A_j w_j - \rho^{-1/n} B_{j-1} w_{j-1} - \rho^{1/n} B_j^* w_{j+1}, \qquad w_{n+j} = w_j.$$

Hence

$$\hat{\mathbf{H}}_\rho = \mathbf{A} + \rho^{-1/n}\mathbf{B} + \rho^{1/n}\mathbf{B}^*.$$

Thus

$$\lim_{|\rho|\to\infty} \rho^{-1/n}\hat{\mathbf{H}}_\rho = \mathbf{B}^*.$$

We obtain

$$\lim_{|\rho|\to\infty} (\rho^{-m} \det((\mathbf{B}^*)^{-1}\mathbf{H}_\rho)) = 1.$$

Hill's formula (8.15) follows immediately. Indeed, both $\rho^m \det((\mathbf{B}^*)^{-1}\mathbf{H}_\rho)$ and $\det(P-\rho I)$ are polynomials in ρ of order $2m$ with leading coefficient 1. Both vanish precisely when the variational system (8.14) of the periodic trajectory has a nonzero solution such that $u_{j+n} = \rho u_j$.

Applications. Identity (8.13) implies the following corollary.

Corollary 8.1. *The dynamical non-degeneracy of a periodic trajectory* $\det(P-I) \neq 0$ *is equivalent to the geometric non-degeneracy* $\det(\mathbf{H}) \neq 0$.

Below in this section we suppose that $\det B(x, y)$ is positive for all $x, y \in M$. Then (8.13) gives

$$(-1)^m \det(\mathbf{H}) \det(P - I) > 0.$$

Corollary 8.2. *Suppose that m is odd and* $\operatorname{ind}\mathbf{H}$ *is even (for example,* $\operatorname{ind}\mathbf{H} = 0$, *i.e.,* $\mathbf{x}$ *is a non-degenerate local minimum of the action* $\mathscr{A}$*). Then* $\mathbf{x}$ *is dynamically unstable: there is a real multiplier $\rho > 1$. When m is even, the same holds when* $\operatorname{ind}\mathbf{H}$ *is odd.*

Corollary 8.3. *Suppose that m is odd, n is even and* $\mathbf{x}$ *is a non-degenerate local maximum of the action* $\mathscr{A}$. *Then* $\mathbf{x}$ *is dynamically unstable: there is a real multiplier* $\rho > 1$.

Indeed, it is sufficient to use the following result.

Proposition 8.3. *If* $\det(P - I) < 0$, *there is a real multiplier* $\rho > 1$.

Proof. Consider the characteristic polynomial $F(\rho) = \det(P - \rho I)$. Its roots are the multipliers of the periodic solution $\mathbf{x}$. We have $F(+\infty) = +\infty$ and $F(1) = \det(P - I) < 0$. Then there exists a real root $\rho > 1$. □

Corollary 8.4. *If* $\operatorname{ind} \mathbf{H}_{-1}$ *is odd, there is a real multiplier* $\rho < -1$.

Indeed, for $\rho = -1$ the Hill formula (8.15) gives

$$\det(\mathbf{H}_{-1}) \det(I + P) > 0.$$

Corollary 8.5. *Suppose the iterate* $\mathbf{x}^2$ *of a periodic trajectory* $\mathbf{x}$ *has* $\operatorname{ind} \mathbf{H}(\mathbf{x}^2)$ *not of the same evenness as* $\operatorname{ind} \mathbf{H}(\mathbf{x})$. *Then* $\mathbf{x}$ *is unstable.*

Proof. Since $2n$-periodic vector fields along $\mathbf{x}^2$ are split into n-periodic and n-antiperiodic ones,

$$\operatorname{ind} \mathbf{H}(\mathbf{x}^2) = \operatorname{ind} \mathbf{H}(\mathbf{x}) + \operatorname{ind} \mathbf{H}_{-1}(\mathbf{x}).$$

If $\operatorname{ind} \mathbf{H}_{-1}(\mathbf{x})$ is odd, $\mathbf{x}$ is unstable by Corollary 8.4. Otherwise $\operatorname{ind} \mathbf{H}(\mathbf{x})$ and $\operatorname{ind} \mathbf{H}(\mathbf{x}^2)$ are even and odd simultaneously. □

In the case $m = 1$ there is a possibility to identify hyperbolicity or ellipticity of a periodic trajectory in terms of the index.

Corollary 8.6. *Suppose that* $m = 1$. *Then a nondegenerate periodic trajectory* $\mathbf{x}$ *is hyperbolic iff* $\operatorname{ind} \mathbf{H}(\mathbf{x}^2)$ *is even and elliptic iff* $\operatorname{ind} \mathbf{H}(\mathbf{x}^2)$ *is odd.*

For a straightforward application of the Hill formula consider the plane convex billiard system ($m = 1$). If $l(x, y)$ is the length of the line segment $[x, y]$, where x is a global cyclic coordinate on the billiard curve, then by (8.10) $B(x, y) < 0$. Corollaries 8.2–8.6 can be used if we take as a Lagrangian $L(x, y) = -l(x, y)$. We obtain the following two corollaries.

Corollary 8.7. *Suppose that* $\mathbf{x}$ *is a non-degenerate local maximum of the plane billiard length functional (i.e., a minimum of the action). Then* $\mathbf{x}$ *is hyperbolic by Corollary* 8.2.

Corollary 8.8. *Let* $\mathbf{x}$ *be a non-degenerate local minimum of the plane billiard length functional and let* n *be even. Then* $\mathbf{x}$ *is hyperbolic by Corollary* 8.3.

In particular, by the Birkhoff theorem [16] (see also [73], any convex billiard system has (at least) two periodic trajectories of period n with rotation number $k < n$, where one of them has a maximum length, and hence is generically hyperbolic. The other has index 1, and so has no positive multipliers > 1.

The requirement for n to be even in Corollary 8.8 at the first glance looks somewhat strange because the billiard trajectory minimizing $\mathscr{A}$ is naturally associated

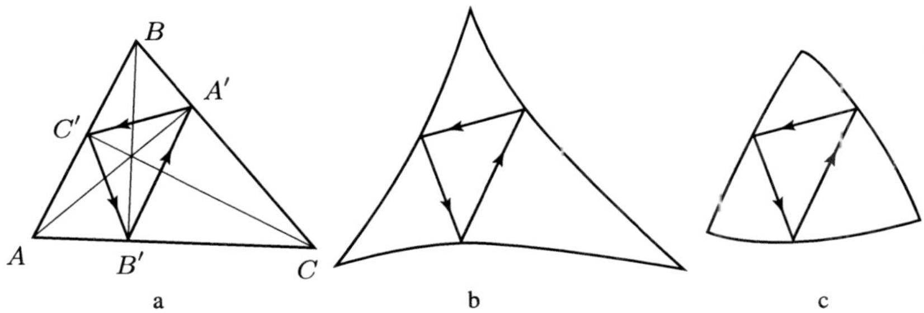

Fig. 8.2 Deformation of a parabolic periodic orbit.

with a locally shortest closed geodesic on a two-dimensional Riemannian manifold. Such geodesics due to Poincaré [105] are known to be hyperbolic. However one should keep in mind that this Poincaré's result is valid only for orientable geodesics (see details in Sect. 8.3) while a periodic billiard trajectory with an odd period is not orientable.

A simple example of an elliptic action minimizing billiard trajectory with odd period can be constructed as follows. Let the billiard curve be an acute-angled triangle ABC. Then by a well-known theorem from planimetry the projections A', B', C' of vertices to the opposite sides form a triangle (the orthotriangle) which presents a local non-degenerate minimum of the billiard action (Fig. 8.2a). The corresponding periodic trajectory is parabolic: its multipliers are $\rho_{1,2} = -1$.

A small deformation of the billiard curve does not destroy the periodic trajectory $A'B'C'$ and just slightly deforms it. If the boundary curve becomes concave, we obtain a Sinai billiard [27]. In this case the trajectory is hyperbolic (Fig. 8.2b). If the boundary curve becomes strictly convex (the curvature gets positive: (Fig. 8.2c)) then the trajectory becomes elliptic still having a locally minimal action provided the deformation is small.

8.3 Continuous Case

Let a smooth m-dimensional manifold M be the configuration space of a Lagrangian system. The Lagrangian $\mathscr{L}(x, \dot{x}, t)$, $x \in M$, $\dot{x} \in T_x M$, is defined on the extended phase space $TM \times \mathbb{R}$. We assume it to be smooth, strictly convex in the velocity v and τ-periodic in time. According to Hamilton's principle, τ-periodic trajectories of the Lagrangian system are critical points of the action functional

$$\mathscr{A}(\gamma) = \int_0^\tau \mathscr{L}(\gamma(t), \dot{\gamma}(t), t)\, dt$$

on the loop space Ω of smooth τ-periodic curves $\gamma : [0, \tau] \to M$ such that $\gamma(0) = \gamma(\tau)$.

The goal of this section is to prove an analog of Theorem 8.1 for continuous Lagrangian systems.

Remark 8.4. A continuous Lagrangian system generates a discrete Lagrangian system as follows. Let $\gamma(t)$, $0 \le t \le \tau$, be a solution joining the points x and y. If these points are not conjugate along γ, then $\gamma = \gamma_{x,y}$ is locally determined by x, y and it depends smoothly on them. This defines a function

$$L(x, y) = \mathscr{A}(\gamma_{x,y})$$

on an open set $U \subset M^2$. In general L is multivalued. It is easy to see that solutions $x(t)$, $0 \le t \le n\tau$, of the Lagrangian system correspond to trajectories of DLS defined by L.

Remark 8.5. Suppose that DLS is generated by a continuous Lagrangian system with an autonomous Lagrangian on TM. Define a vector field w on M^2 by $w(x, y) = (\dot{x}(0), \dot{x}(\tau))$, where $x : [0, \tau] \to M$ joins x, y, and let $h^s : M^2 \to M^2$ be the corresponding transformation group (the phase flow). Then $T = h^\tau$, and so $T \circ h^s = h^s \circ T$. In general the group h^s does not preserve the discrete Lagrangian. However, DLS has a first integral—the energy of the continuous Lagrangian system.

Let

$$B_t = D_{\dot{x}}^2 \mathscr{L}(x, \dot{x}, t)\big|_{x=\gamma(t), \dot{x}=\dot{\gamma}(t)}$$

be the Hessian of $\mathscr{L}$ with respect to the velocity. This is a positive definite quadratic form on $E_t = T_{\gamma(t)}M$ and it defines a scalar product $(\,,\,)$ on E_t by $(u, v) = \langle B_t u, v\rangle$.

The second variation of the functional A at the critical point γ is a bilinear form $h(\xi, \eta)$ on the set of smooth τ-periodic vector fields along γ. It has the form

$$h(\xi, \eta) = \int_0^\tau \big((\nabla\xi(t), \nabla\eta(t)) + (W(t)\xi(t), \nabla\eta(t)) + (V(t)\xi(t), \eta(t))\big)\, dt.$$

Here $\nabla\xi(t)$ is a covariant derivative of the vector field $\xi(t) \in E_t$, i.e. a linear differential operator on the set of smooth vector fields $\xi(t) \in E_t$ which is consistent with the metric:

$$\frac{d}{dt}(\xi(t), \eta(t)) = (\nabla\xi(t), \eta(t)) + (\xi(t), \nabla\eta(t)), \qquad \nabla(f(t)\xi(t)) = \dot{f}\xi + f\nabla\xi$$

for any smooth vector fields $\xi(t), \eta(t) \in E_t$ and a smooth function $f(t)$.[1] Here $V(t), W(t)$ are linear operators in E_t.

Without loss of generality it may be assumed that V is self-adjoint with respect to the metric $V(t) = V^*(t)$. For $\xi = \eta$, the integrand in h is the quadratic Lagrangian for the variational system. It is defined up to adding a full derivative.

By integration by parts, h can be represented in the form

[1] The covariant derivative is not uniquely defined: for an antisymmetric operator $A(t)$, $\tilde{\nabla} = \nabla + A(t)$ is also a covariant derivative.

$$h(\xi, \eta) = \int_0^\tau \big((D\xi(t), D\eta(t)) + (U(t)\xi(t), \eta(t))\big)\, dt,$$

where

$$D\xi = \nabla\xi + W - W^*, \quad U = V - W - W^*.$$

Here D is a modified covariant derivative:

$$(D\xi, \eta) + (\xi, D\eta) = \frac{d}{dt}(\xi, \eta).$$

Thus D is skew-symmetric relative to the L^2 scalar product

$$(\xi, \eta) = \int_0^\tau (\xi(t), \eta(t))\, dt.$$

Therefore

$$h(\xi, \eta) = ((-D^2 + U)\xi, \eta).$$

The variational system of the trajectory γ (linearized Lagrange system) has the form

$$-D^2\xi(t) + U(t)\xi(t) = 0. \tag{8.16}$$

We extend the bilinear form h to a Hermitian form $h(\xi, \bar{\eta})$ on the space X of complex vector fields $\xi(t)$ from the Sobolev space[2] $W^{1,2}$ such that $\xi(0) = \xi(\tau)$:

$$h(\xi, \bar{\eta}) = \int_0^\tau \big((D\xi(t), D\overline{\eta}(t)) + (U(t)\xi(t), \overline{\eta}(t))\big)\, dt. \tag{8.17}$$

Thus

$$h(\xi, \bar{\eta}) = (\mathbf{H}\xi, \eta), \qquad \mathbf{H} = -D^2 + U : X \to X^*.$$

We define on X the structure of a complex Hilbert space by setting

$$\langle\!\langle \xi, \bar{\eta} \rangle\!\rangle = (D\xi, D\overline{\eta}) + (\xi, \overline{\eta}) = ((-D^2 + I)\xi, \bar{\eta}) = (\mathbf{B}\xi, \bar{\eta}), \qquad \mathbf{B} = -D^2 + I.$$

Then

$$h(\xi, \bar{\eta}) = \langle\!\langle H\xi, \bar{\eta} \rangle\!\rangle,$$

where

$$H = \mathbf{B}^{-1}\mathbf{H} = (-D^2 + I)^{-1}(-D^2 + U) = I + (-D^2 + I)^{-1}(U - I)$$

is a self-adjoint bounded operator $H : X \to X$.

The operator $(-D^2 + I)^{-1}$ is compact, and the operator $U - I$ is bounded. Hence $H - I$ is compact, and it is easy to see that the trace of $H - I$ is finite. Thus, the operator $H - I$ is of trace class and hence the determinant $\det H$ exists (see e.g. [112]). We call it the Hill determinant of the trajectory γ.

[2] By definition $\xi(t) \in W^{1,2}$ iff $\int_0^\tau ((\nabla\xi, \nabla\xi) + (\xi, \xi))\, dt < \infty$.

Let $p = (\gamma(0), \dot\gamma(0))$ and let $P : T_p(TM) \to T_p(TM)$ be the linear Poincaré mapping of the periodic trajectory γ. Using the covariant derivative, we identify $T_p(TM)$ with $E_0 \oplus E_0$. Then P is the monodromy operator of the variational system

$$P(\xi(0), D\xi(0)) = \big(\xi(\tau), D\xi(\tau)\big).$$

Let $Q : E_0 \to E_0$ be the monodromy operator of the equation of parallel transport $D\xi(t) = 0$, i.e. $Q\xi(0) = \xi(\tau)$. Let $\sigma = \det Q$. Note that Q is orthogonal, and so $\det Q = \pm 1$. This is a purely topological quantity: $\sigma = \pm 1$ depending on whether the trajectory γ preserves or reverses the orientation. If M is orientable, then $\sigma = 1$ always.

Theorem 8.3. *For any τ-periodic trajectory γ*

$$\det H = \sigma(-1)^m \frac{e^{m\tau} \det(I - P)}{\det^2(e^\tau I - Q)}. \tag{8.18}$$

Since Q is an orthogonal operator, the denominator in (8.18) does not vanish.

Remark 8.6. The definition of the Hill determinant is similar to that given in the discrete case. Indeed, define a quadratic form b on X by

$$b(\xi, \bar\eta) = (D\xi, D\bar\eta) + (\xi, \bar\eta) = ((-D^2 + I)\xi, \bar\eta).$$

Then h and b both define operators $\mathbf{H}, \mathbf{B} : X \to X^*$, where $\mathbf{H} = -D^2 + U$ and $\mathbf{B} = -D^2 + I$. Then $H = \mathbf{B}^{-1}\mathbf{H}$, as in the discrete case. Note that the choice of b is very natural for DLS, but not for CLS. This is the reason for a strange denominator in (8.18).

The following generalization of the Hill determinant is essentially contained in Hill's work [61]. For a given $\rho \in \mathbb{C}$, $|\rho| = 1$, let X_ρ be the space of complex quasi-periodic $W^{1,2}$ vector fields $\xi(t) \in E_t$, $0 \le t \le \tau$, along γ such that $\xi(\tau) = \rho\xi(0)$. We define on X_ρ an index form [67] by the formula of second variation (8.17).

Let $\mu = \tau^{-1} \ln \rho$. Then $\mu \in i\mathbb{R}$ is defined up to addition of $\omega = 2\pi i/\tau$. For definiteness we choose μ so that $0 \le \mu\tau/i < 2\pi$. We identify X and X_ρ, assigning to a vector field $\xi \in X$ the vector field $e^{\mu t}\xi(t)$ in X_ρ. We obtain a Hermitian form h_ρ on X which is a generalization of the second variation and has the form

$$\begin{aligned} h_\rho(\xi, \bar\eta) &= h(e^{\mu t}\xi, \overline{e^{\mu t}\eta}) = ((D + \mu I)\xi, \overline{(D + \mu I)\eta}) + (U\xi, \bar\eta) \\ &= -((D + \mu I)^2\xi, \bar\eta) + (U\xi, \bar\eta) = (\mathbf{H}_\rho\xi, \eta), \end{aligned}$$

where $\mathbf{H}_\rho = -(D + \mu I)^2 + U$. We used the facts that $\bar\mu = -\mu$ and D is real and antisymmetric. Next define an operator H_ρ by

$$h_\rho(\xi, \bar\eta) = \langle\!\langle H_\rho\xi, \bar\eta \rangle\!\rangle.$$

Then

$$H_\rho = \mathbf{B}^{-1}\mathbf{H}_\rho = (-D^2 + I)^{-1}(-(D + \mu I)^2 + U)$$

is a bounded operator in X.

We henceforth assume that ρ may take any complex values. Note that H_ρ is self-adjoint for $|\rho| = 1$, but not in general. Although the operator H_ρ is not of trace class for $\rho \neq 1$, it is possible to define the Hill determinant $\det H_\rho$ as follows. Define $\det H_\rho$ by means of the finite-dimensional approximation

$$\det H_\rho = \lim_{N\to\infty} \det P_N H_\rho P_N^*, \tag{8.19}$$

where $P_N : X \to X_N$ is the orthogonal projection onto the finite-dimensional eigenspace of the operator D corresponding to the eigenvalues $\nu \in \Lambda$ such that $|\nu| \leq N$, and $P_N^* : X_N \to X$.

Theorem 8.4. *The determinant is well defined and*

$$\det H_\rho = \sigma(-1)^m \frac{e^{m\tau} \det(\rho I - P)}{\rho^m \det^2(e^\tau I - Q)}. \tag{8.20}$$

For $\rho = 1$ we obtain (8.18).

In the one-dimensional oriented case $M = \mathbb{R}$, we have $m = 1, \sigma = 1, Q = 1$, $\det(e^\tau I - Q) = e^\tau - 1$, and

$$\rho^{-1} \det(\rho I - P) = \rho + \rho^{-1} - 2 + \det(I - P).$$

Hence (8.20) gives

$$\det H_\rho = \det H - \frac{\rho + \rho^{-1} - 2}{e^\tau + e^{-\tau} - 2}. \tag{8.21}$$

Hence multipliers of the periodic orbit are roots of the equation

$$\frac{\rho + \rho^{-1} - 2}{e^\tau + e^{-\tau} - 2} = \det H.$$

This result was obtained by Hill [61].

Example (The Hill determinant for the Hill equation). Consider the Hill equation

$$\ddot{\xi} - a(t)\xi = 0, \qquad a(t) = \sum_{n\in\mathbb{Z}} a_n e^{int}.$$

Here $D\xi = \dot{\xi}$ and $U\xi = a(t)\xi$. Let us represent the operator H_ρ in the basis $\{e^{int}\}$. If

$$\xi(t) = \sum_{n\in\mathbb{Z}} \xi_n e^{int},$$

then

$$(-D^2+I)^{-1}(-(D+\mu I)^2+U)\,\xi(t) = \sum_{n\in\mathbb{Z}} \frac{\xi_n}{n^2+1}\left(-(in+\mu)^2 + \sum_{k\in\mathbb{Z}} a_\kappa e^{ikt}\right)e^{int},$$

and so H_ρ is represented by the infinite matrix $(h_{jk}(\rho))$, where

$$h_{jk}(\rho) = \frac{-(ik+\mu)^2\delta_{jk} + a_{k-j}}{k^2+1}$$

(compare with (8.1)). The determinant of this matrix is given by (8.21) and it is a variation of the determinant computed by Hill. The original Hill's determinant was defined in a slightly different way: instead of $(n^2+1)^{-1}$, Hill used regularizing multiplier $(n^2-u_0)^{-1}$. This regularization does not work well in many dimensions due to the possibility of a resonance.

Some applications. We mention a few easy corollaries of Theorem 8.4.

Suppose that the Lagrangian is quadratic in the velocity and does not depend explicitly on time. Then it defines a Riemannian metric $(\,,\,)$ on M, and the periodic orbit γ is a closed geodesic of this metric. A closed geodesic γ always has two unit multipliers, and so $\det H = \det(I-P) = 0$. Let $H_\rho^\perp$ be the restriction of H_ρ to the invariant subspace

$$Y = X^\perp = \{\xi \in X : (\xi(t), \dot\gamma(t)) \equiv 0\}$$

of vector fields orthogonal to γ. Let $P^\perp$ be the monodromy operator corresponding to the restriction of the variational equation to the set Y of vector fields orthogonal to γ. Then the initial condition satisfies $\xi(0) \perp \dot\gamma(0)$, $\dot\xi(0) \perp \dot\gamma(0)$ (here $\dot\xi$ is the covariant derivative, so that $(d/dt)(\xi,\dot\gamma) = (\dot\xi,\dot\gamma))$. Let $Q^\perp$ be the mapping of parallel transport along γ of vectors orthogonal to γ. Then we have the following corollary.

Corollary 8.9. *In the autonomous case*

$$\det H_\rho^\perp = \sigma(-1)^{m-1}\frac{e^{(m-1)\tau}\det(\rho I - P^\perp)}{\rho^{m-1}\det^2(e^\tau I - Q^\perp)}. \tag{8.22}$$

Indeed, $\det H_\rho$ can be represented in the form of the product of $\det H_\rho^\perp$ and the determinant Δ of the restriction of H_ρ to the set $Z = \{\xi \in X : \xi(t) = \lambda(t)\dot\gamma(t)\}$ of vector fields parallel to γ. We restrict the Lagrangian to the two-dimensional subspace $T\gamma \subset TM$ and apply Theorem 8.4 to the system obtained. On the left side of (8.20) in this case will be the determinant Δ. Thus by (8.21),

$$\Delta = -\frac{e^\tau(\rho-1)^2}{\rho(e^\tau-1)^2}.$$

But

$$\det(\rho I - P) = (\rho-1)^2\det(\rho I - P^\perp),$$
$$\det(e^\tau I - Q) = (e^\tau - 1)\det(e^\tau I - Q^\perp),$$

which was required to prove.

Remark 8.7. The investigation of any autonomous natural Lagrangian system in a potential force field reduces to the investigation of the problem of geodesics in the Jacobi metric. However, the period will not then be fixed, and the Morse index is not preserved in general. To eliminate nondegeneracy with fixed period, we need more work.

A periodic trajectory of a Lagrangian system (respectively, a closed geodesic on a Riemannian manifold) is called nondegenerate in the Poincaré sense if the spectrum of the matrix P (respectively, $P^{\perp}$) does not contain unity. A periodic trajectory (respectively, a closed geodesic) is called nondegenerate in the sense of Morse if the determinant $\det H$ (respectively, $\det H^{\perp}$) is nonzero. It follows easily from (8.20) that *nondegeneracy in the sense of Poincaré is equivalent to nondegeneracy in the sense of Morse.*

Let $|\rho| = 1$. Define the ρ-index $\mathrm{ind}_\rho\gamma$ of a periodic trajectory (a closed geodesic) as the index of the Hermitian form h_ρ. Then $\mathrm{ind}\,\gamma = \mathrm{ind}_1\gamma$ is the Morse index of γ. It is equal to the number of negative eigenvalues of the operator H (or $H^{\perp}$). Suppose that the trajectory γ is nondegenerate. Then

$$(-1)^{\mathrm{ind}_\rho \gamma} = \mathrm{sign}\det H_\rho = \sigma(-1)^m\,\mathrm{sign}(\rho^{-m}\det(\rho I - P)).$$

In the geodesic case we have

$$(-1)^{\mathrm{ind}_\rho \gamma} = \mathrm{sign}\ \det H_\rho^{\perp} = \sigma(-1)^{m-1}\,\mathrm{sign}(\rho^{1-m}\det(\rho I - P^{\perp})).$$

The argument of the sign function is real for $|\rho| = 1$, since the characteristic polynomial is reciprocal.

Suppose that the periodic trajectory (or closed geodesic) is nondegenerate and such that $\sigma(-1)^{m+\mathrm{ind}\,\gamma} < 0$ ($\sigma(-1)^{m+\mathrm{ind}\,\gamma} > 0$). Then $\det(I - P) < 0$ ($\det(I - P^{\perp}) < 0$), so that the characteristic polynomial $F(\rho) = \det(\lambda I - P)$ has a real root $\lambda > 1$. Therefore, the trajectory γ is unstable. In particular, *nondegenerate closed geodesics of locally minimal length on an even-dimensional orientable manifold are unstable.*

Suppose that $m = 1$ and the 2τ-periodic trajectory γ^2 is γ traversed twice. *If γ^2 is nondegenerate, then γ is of hyperbolic (elliptic) type if and only if* $\mathrm{ind}\,\gamma^2$ *is even (odd).*

Indeed, the multipliers $\lambda_1 = \lambda_2^{-1}$ of γ^2 are equal to the squares of the multipliers of γ. Hence, hyperbolicity of γ^2 is equivalent to the conditions that λ_1, λ_2 are real and positive or, equivalently, the condition $\mathrm{sign}(\det(I - P^2)) = (-1)^{1+\mathrm{ind}\,\gamma^2} = -1$ (γ^2 always preserves orientation). Similarly, ellipticity of γ^2 is equivalent to the condition $(-1)^{1+\mathrm{ind}\,\gamma^2} = 1$. It remains to use the fact that γ and γ^2 are simultaneously elliptic or hyperbolic.

The corresponding assertion for geodesics is as follows. *Let γ be a closed geodesic on a 2-dimensional Riemannian manifold. If γ^2 is non-degenerate, then γ has hyperbolic (elliptic) type if and only if* $\mathrm{ind}\,\gamma^2$ *is even (odd).*

Suppose now that $\rho = -1$. We have $(-1)^{\mathrm{ind}_{-1}\gamma} = \sigma\,\mathrm{sign}\,F(-1)$, where F is the characteristic polynomial. Thus if $\mathrm{ind}_{-1}\gamma$ is odd, there exists a real multiplier

$\rho < -1$. Note that, for $\rho = -1$, the space X_ρ corresponds to antiperiodic variations $\xi(\tau) = -\xi(0)$. Since 2τ-periodic vector fields are sums of τ-periodic and τ-antiperiodic, we obtain

$$\operatorname{ind}_{-1}\gamma = \operatorname{ind}\gamma^2 - \operatorname{ind}\gamma.$$

Proof (of Theorem 8.4). The proof follows the proof of Hill's result [61]. The real skew-Hermitian operator $D = \overline{D} = -D^*$ has compact resolvent $(D + \mu I)^{-1}$. Its spectrum $\Lambda \subset i\mathbb{R}$ coincides with the set of characteristic exponents of the equation $D\xi(t) = 0$, i.e., with the set of $\nu \in \mathbb{C}$ such that $\det(e^{\nu\tau}I - Q) = 0$. Thus

$$\Lambda = \{\nu_j + \omega\mathbb{Z},\ j = 1, \dots, m\}, \qquad \omega = 2\pi i/\tau.$$

If $\nu \in \Lambda$, then $-\nu$ and $\nu + \omega$ belong to Λ.

We have $H_\rho = ST$, where

$$S = -(-D^2 + I)^{-1}(D + \mu I)^2, \quad T = I - (D + \mu I)^{-2}U.$$

Suppose that $\mu \notin \Lambda = -\Lambda$. Since $P_N D = D P_N$, we have

$$\det H_\rho = \det S \det T.$$

The finite-dimensional approximation (8.19) of the determinant

$$f(\mu) = \det T = \lim_{N\to\infty} \det P_N T P_N^*$$

converges absolutely for $\mu \notin \Lambda$, since the operator $(D + \mu I)^{-2}U$ is of trace class. Thus, f is a holomorphic function on $\mathbb{C} \setminus \Lambda$ having at points of Λ poles of multiplicity no greater than double the multiplicity of the corresponding points of the spectrum of D [112].

The function f is periodic with period $\omega = 2\pi i/\tau$: $f(\mu + \omega) \equiv f(\mu)$. Indeed, if $\xi \in X$, then $e^{\omega t}\xi \in X$ and

$$(I - (D + \mu I)^{-2}U)e^{\omega t}\xi = e^{\omega t}(I - (D + (\mu + \omega)I)^{-2}U)\xi.$$

Thus we can write $f(\mu) = F(e^{\mu\tau})$, where F is a holomorphic function having poles $\rho = e^{\mu t}$ at the roots of $\det(\rho I - Q)$.

The poles of F are contained among the poles of $\det^{-2}(\rho I - Q)$. It is therefore possible to choose a polynomial $g(\rho)$ of degree no higher than $2m - 1$ such that the functions $F(\rho)$ and $g(\rho)\det^{-2}(\rho I - Q)$ have the same principal parts of the Laurent expansion at each pole. Since $F(\rho) \to 1$ as $|\rho| \to +\infty$, by Liouville's theorem,

$$F(\rho) = 1 + g(\rho)\det{}^{-2}(\rho I - Q). \tag{8.23}$$

The determinant $\det S$ converges conditionally, but it can be computed explicitly. By (8.19),

$$
\begin{aligned}
\det(-(-D^2+I)^{-1}(D+\mu I)^2) &= \lim_{N\to\infty} \prod_{\nu\in\Lambda,|\nu|\le N} \frac{(\nu+\mu)^2}{\nu^2-1} \\
&= \lim_{N\to\infty} (-\mu^2)^k \prod_{\nu\in\Lambda,|\nu|\le N, i\nu>0} \left(\frac{\nu^2-\mu^2}{\nu^2-1}\right)^2 \\
&= (-1)^k \prod_{\nu\in\Lambda} \frac{\nu^2-\mu^2}{\nu^2-1},
\end{aligned}
$$

where k is the multiplicity of zero in the spectrum of D. We have used the fact that $\Lambda = -\Lambda$.

From the form of the spectrum of the operator D it follows that the last product converges absolutely. Let us show that

$$
\prod_{\nu\in\Lambda} \frac{\nu^2-\mu^2}{\nu^2-1} = \frac{e^{m\tau}\det^2(\rho I - Q)}{\rho^m \det^2(\epsilon^\tau I - Q)}.
$$

We will use the formula (see [54])

$$
\prod_{n\in\mathbb{Z}} \left(1 - \frac{\mu^2}{(\nu+\omega n)^2}\right) = \frac{\cosh\mu\tau - \cosh\nu\tau}{1-\cosh\nu\tau}, \qquad \nu \notin \omega\mathbb{Z}.
$$

Let $\rho_1, \ldots, \rho_m$ be the roots of the characteristic polynomial $\det(\rho I - R)$. Then $\nu_j = \tau^{-1}\ln\rho_j$, $j = l, \ldots, m$. Suppose first that $\nu_j \notin \omega\mathbb{Z}$. We have

$$
\begin{aligned}
\prod_{\nu\in\Lambda} \frac{\nu^2-\mu^2}{\nu^2-1} &= \prod_{\nu\in\Lambda} \left(1-\frac{\mu^2}{\nu^2}\right)\left(1-\frac{1}{\nu^2}\right)^{-1} \\
&= \prod_{j=1}^m \prod_{n\in\mathbb{Z}} \left(1-\frac{\mu^2}{(\nu_j+\omega n)^2}\right)\left(1-\frac{1}{(\nu_j+\omega n)^2}\right)^{-1} \\
&= \prod_{j=1}^m \frac{\cosh\mu\tau - \cosh\nu_j\tau}{\cosh\tau - \cosh\nu_j\tau} = \prod_{j=1}^m \frac{\rho+\rho^{-1}-\rho_j-\rho_j^{-1}}{e^\tau + e^{-\tau} - \rho_j - \rho_j^{-1}} \\
&= \frac{e^{m\tau}\det^2(\rho I - Q)}{\rho^m\det^2(e^\tau I - Q)}.
\end{aligned}
$$

By continuity this holds also for $\nu_j \in \omega\mathbb{Z}$. Thus

$$
\det T = (-1)^k e^{m\tau} \rho^{-m} \frac{\det^2(\rho I - Q)}{\det^2(e^\tau I - Q)}.
$$

By (8.19) and (8.23),

$$\det H_\rho = (-1)^k \frac{e^{m\tau}(\det^2(\rho I - Q) + g(\rho))}{\rho^m \det^2(e^\tau I - Q)}.$$

Thus, $\rho^m \det H_\rho$ is a polynomial of degree $2m$ in ρ with leading coefficient equal to $(-1)^k e^{m\tau} \det^{-2}(e^\tau I - Q)$.

Let us show that, if the operator H_ρ is nonreversible, then $\det H_\rho = 0$. For $\rho = 1$, when H is of trace class, the determinant is uniformly convergent; this follows from the general properties of the determinant [112]. In general additional arguments are needed. Suppose first that $\mu \notin \Lambda$. Then $H_\rho = ST$, where S is reversible and T of trace class. If H_ρ is nonreversible, then so is T, and so $\det T = 0$. Then it follows that $\det H_\rho = 0$. If $\mu \in \Lambda$, we can repeat the same computation replacing S by $S = -(-D^2 + I)^{-1}((D + \mu I)^2 + \alpha I)$, and similarly for T. Then S is reversible for an appropriate choice of α, so the same argument works.

The kernel of the operator H_ρ consists of τ-periodic vector fields ξ such that $(-D^2 + U)e^{\mu t}\xi(t) = 0$. Therefore, the roots of the polynomial $\rho^m \det H_\rho$ and of the characteristic polynomial $\det(\rho I - P)$ of the variational equation coincide. Thus,

$$\rho^m \det H_\rho = (-1)^k \frac{e^{m\tau} \det(\rho I - P)}{\det^2(e^\tau I - Q)}.$$

We remark that k is the dimension of the subspace on which the orthogonal operator R is the identity, while $\sigma = (-1)^l$, where l is the dimension of the subspace on which R is a reflection. Since the dimension $m - k - l$ of the complementary subspace is even, formula (8.20) has been proved. □

Chapter 9
Appendix

9.1 Diophantine Frequencies

In this appendix we consider some problems related to the resonant, Diophantine and other arithmetic properties of frequency vectors. Here we deal with expressions of the form $\langle k, \nu\rangle$, where $k \in \mathbb{Z}^m$ and $\nu \in \mathbb{R}^m$ is a constant frequency vector. In the perturbation theory these expressions appear as small denominators. Below for any vector $v = (v_1, \dots, v_m)^T$ we put

$$\|v\| = \max_{1\le j\le m} |v_j|.$$

1. Only a few of the small denominators are really small. The majority of them are of order $\|k\|$. Nevertheless, for any vector ν the set of small denominators contains zero among limit points. Indeed, the following assertion holds.

Theorem 9.1 (Dirichlet). *Given $\nu \in \mathbb{R}^m$, for any $K \in \mathbb{N}$ there exists a nonzero vector $k \in \mathbb{Z}^m$ with $\|k\| \le 2K$, such that*

$$|\langle k, \nu\rangle| \le m\|\nu\|2^{-m}K^{-m+1}. \tag{9.1}$$

Corollary 9.1. *There exist infinitely many vectors $k \in \mathbb{Z}^m$ such that $|\langle k, \nu\rangle| \le m\|\nu\|\,\|k\|^{-m+1}$.*

Proof (of Theorem 9.1). For any $K \in \mathbb{N}$ we put $\mathscr{B}_K = \{k \in \mathbb{Z}^m : 0 \ne \|k\| \le K\}$. The set $\mathscr{B}_K$ contains $(2K+1)^m - 1$ elements. For any $k \in \mathscr{B}_K$ we have

$$|\langle k, \nu\rangle| \le m\,\|k\|\,\|\nu\| \le mK\|\nu\|.$$

Therefore, there exist two vectors $k', k'' \in \mathscr{B}_K$ such that

$$|\langle k', \nu\rangle - \langle k'', \nu\rangle| \le \frac{mK\|\nu\|}{(2K+1)^m - 2} \le \frac{m\|\nu\|}{2(2K)^{m-1}}.$$

The vector $k = k' - k'' \in \mathscr{B}_{2K}$ satisfies (9.1). □

D. Treschev, O. Zubelevich, *Introduction to the Perturbation Theory of Hamiltonian Systems*, Springer Monographs in Mathematics,
DOI 10.1007/978-3-642-03028-4_9, © Springer-Verlag Berlin Heidelberg 2010

2. We define the sets

$$D_m(c,\gamma) \subset \mathbb{R}^m, \qquad c > 0, \quad \gamma \geq 0,$$

as follows: $\nu \in D_m(c,\gamma)$ if for any nonzero $k \in \mathbb{Z}^m$ we have

$$|\langle k, \nu\rangle| \geq \frac{1}{c\,\|k\|^{\gamma}}. \tag{9.2}$$

Vectors lying in at least one set $D_m(c,\gamma)$, are called Diophantine.

Obviously, for $c' \leq c$ and $\gamma' \leq \gamma$

$$D_m(c',\gamma') \subset D_m(c,\gamma).$$

The Dirichlet theorem (see Corollary 9.1) implies that the sets $D_m(c,\gamma)$ are empty for $\gamma < m-1$.

Let us show that the sets $D_m(c, m-1)$ are not empty for sufficiently large $c > 0$. Recall that $\mathbb{Z}[x]$ denotes the space of polynomials

$$p(x) = a_l x^l + \cdots + a_1 x + a_0, \qquad l \in \mathbb{N},$$

with rational coefficients. If p does not vanish identically, the coefficient a_l is assumed to be nonzero. The quantity l is called the degree of the polynomial: $l = \deg p$. A polynomial with unit leading coefficient is called unitary. A polynomial $p(x) \in \mathbb{Z}[x]$ is called prime if it is not divisible by any $q \in \mathbb{Z}[x]$, $0 < \deg q < \deg p$. For example, any polynomial of degree two or three is prime over $\mathbb{Z}[x]$ if it has no rational roots.

Proposition 9.1. *Let α be a root of a prime polynomial $p \in \mathbb{Z}[x]$, $\deg p = m$. Then, for any $q \in \mathbb{Z}[x]$ such that $q(\alpha) = 0$, either q is divisible by p or $q \equiv 0$.*

Proof. Suppose that q is not divisible by p and $q \not\equiv 0$. Then the largest common divisor d of p and q lies in $\mathbb{Z}[x]$ and has the degree $\deg d \in \{1, \ldots, \deg p - 1\}$. This contradicts to the assumption that p is prime. □

Theorem 9.2 (Lagrange). *Let $\alpha_1, \ldots, \alpha_m$ be roots of a unitary*[1] *prime polynomial p of degree m with integer coefficients. Consider the vectors*

$$\nu_j = (1, \alpha_j, \alpha_j^2, \ldots, \alpha_j^{m-1})^T, \qquad j = 1, \ldots, m.$$

Then for any $j \in \{1, \ldots, m\}$

$$\nu_j \in D_m(c, m-1), \qquad c = \frac{m^{m-1}}{\|\nu_j\|} \prod_{j=1}^{m} \|\nu_j\|.$$

---

[1] The assumption that p is unitary is taken for simplicity. In [119] more general statements are contained. For example, the following one. Let the numbers $1, \alpha_1, \ldots, \alpha_m$ form a basis of a real number field of degree $m+1$. Then $(1, \alpha_1, \ldots, \alpha_m)^T \in D_{m+1}(c, m)$ for some $c > 0$.

Proof. Let us fix a nonzero vector $k \in \mathbb{Z}^m$.

Lemma 9.1. *The polynomial $f_k(x) = \prod_{s=1}^m (x - \langle k, \nu_s \rangle)$ is unitary. It has a nonzero constant term and integer coefficients.*

Proof (of Lemma 9.1). The unitarity of f_k is obvious. According to Proposition 9.1 the quantities $\langle k, \nu_s \rangle$ do not vanish. Therefore, the constant term of f_k is nonzero.

Since f_k depends on $\alpha_1, \ldots, \alpha_m$ polynomially, it is possible to write

$$f_k = F_k(x; \alpha_1, \ldots, \alpha_m),$$

where F_k is a polynomial of $m + 1$ variables with integer coefficients. Obviously, F is symmetric in $\alpha_1, \ldots, \alpha_m$. Hence, according to the theorem on symmetric polynomials, is can be represented in the form

$$F_k(x; \alpha_1, \ldots, \alpha_m) = G_k(x; \sigma_1, \ldots, \sigma_m),$$

where G_k is a polynomial of $m + 1$ variables with integer coefficients and

$$\sigma_1 = \sum_s \alpha_s, \quad \sigma_2 = \sum_{s_1 \neq s_2} \alpha_{s_1} \alpha_{s_2},$$
$$\sigma_3 = \sum_{s_1 \neq s_2,\, s_2 \neq s_3,\, s_3 \neq s_1} \alpha_{s_1} \alpha_{s_2} \alpha_{s_3}, \ldots, \qquad \sigma_m = \alpha_1 \cdots \alpha_m.$$

Recall that by the Vieta theorem

$$p(x) = x^m - \sigma_1 x^{m-1} + \sigma_2 x^{m-2} - \cdots + (-1)^m \sigma_m.$$

Therefore, the quantities $\sigma_1, \ldots, \sigma_m$ are integers. Hence,

$$G_k(x; \sigma_1, \ldots, \sigma_m) \in \mathbb{Z}[x]$$

is a polynomial with integer coefficients. The lemma is proved. □

The following corollary is a result of the application of the Vieta theorem to the polynomial f_k.

Corollary 9.2. *For any $k \neq 0$*

$$|\langle k, \nu_1 \rangle \cdots \langle k, \nu_m \rangle| = |\sigma_m| \geq 1. \tag{9.3}$$

Now we turn to the proof of Theorem 9.2. Since the inequality $|\langle k, \nu_s \rangle| \leq m \, \|k\| \times \|\nu_s\|$ holds for any $s \in \{1, \ldots, m\}$ then according to (9.3) we have

$$|\langle k, \nu_1 \rangle| \geq \frac{1}{|\langle k, \nu_2 \rangle| \cdots |\langle k, \nu_m \rangle|} \geq \frac{1}{m^{m-1} \, \|\nu_2\| \cdots \|\nu_m\| \, \|k\|^{m-1}}.$$

This is the required estimate. In the case $j \neq 1$ the proof is analogous. □

As an example consider the positive root α_g of the polynomial $x^2 + x - 1$. The number $\alpha_g = (\sqrt{5} - 1)/2$ is called the golden mean. Since

$$\alpha_g = (1 + \alpha_g)^{-1} = (1 + (1 + \alpha_g)^{-1})^{-1} = \cdots,$$

we have the continued fraction expansion

$$\alpha_g = \cfrac{1}{1 + \cfrac{1}{1 + \ddots}}.$$

The Lagrange theorem implies that $v_g = (1, \alpha_g) \in D_2(c, 1)$ for some positive c. Moreover, it is proved in the theory of continued fractions (see, for example, [66]) that for $k \in \mathbb{Z}^2 \setminus \{0\}$

$$|\langle k, v_g \rangle| \geq \frac{1}{\sqrt{5}\, \|k\|},$$

i.e., $v_g \in D_2(\sqrt{5}, 1)$. In a sense, the golden mean is the best irrational number from the viewpoint of the theory of rational approximations.

3. The measure of the set of Diophantine frequencies is estimated in the following theorem.

Theorem 9.3. *Let μ be the Lebesgue measure on $\mathbb{R}^m$, let $\gamma > m - 1$, and let $\mathscr{C}$ be the cube $\{v \in \mathbb{R}^m : \|v\| \leq 1\}$. Let S be the $(m-1)$-dimensional area of the maximal section of the cube $\mathscr{C}$ by a hyperplane. Then*

$$\mu(\mathscr{C} \setminus D_m(c, \gamma)) \leq \frac{4Sm3^{m-1}}{c}\left(1 + \frac{1}{\gamma - m + 1}\right).$$

Corollary 9.3. *For any $\gamma > m - 1$ the measure of the set $\mathscr{C} \setminus \bigcup_{c>0} D_m(c, \gamma)$ vanishes.*

Proof (of Theorem 9.3). For any nonzero vector $k \in \mathbb{Z}^m$ we define the set

$$\Pi_k = \left\{ v \in \mathbb{R}^m : |\langle v, k \rangle| < \frac{1}{c\, \|k\|^\gamma} \right\}.$$

Then the following estimate holds:

$$\mu(\mathscr{C} \setminus D_m(c, \gamma)) \leq \sum_{k \in \mathbb{Z}^m \setminus \{0\}} \mu(\mathscr{C} \cap \Pi_k).$$

Any set Π_k is a "strip", bounded by the two hyperplanes $\langle v, k \rangle = \pm \frac{1}{c\, \|k\|^\gamma}$. The distance ρ between these planes equals the distance between the points $\pm(c\, \|k\|^\gamma \times |k|^2)^{-1} k$. Therefore, $\rho = 2/(c\, \|k\|^\gamma |k|)$. Hence,

$$\mu(\mathscr{C} \cap \Pi_k) \leq \frac{2S}{c\, \|k\|^\gamma |k|} \leq \frac{2S}{c\, \|k\|^{\gamma+1}},$$

because $|k| \geq \|k\|$. We also note that the number of points in the set $\mathbb{Z}^m$ lying on the surface $\|k\| = l$ satisfies the inequality $\#\{k \in \mathbb{Z} : \|k\| = l\} \leq 2m(2l+1)^{m-1}$. As a result we obtain

$$\mu(\mathscr{C} \setminus D_m(c, \gamma)) \leq \sum_{k \in \mathbb{Z}^m \setminus \{0\}} \frac{2S}{c\, \|k\|^{\gamma+1}} = \sum_{l=1}^{\infty} \sum_{\|k\|=l} \frac{2S}{cl^{\gamma+1}}$$

$$\leq \sum_{l=1}^{\infty} \frac{2S\, 2m\, (2l+1)^{m-1}}{cl^{\gamma+1}} < \sum_{l=1}^{\infty} \frac{4Sm3^{m-1}}{cl^{\gamma-m+1}}.$$

The last sum can be estimated with the help of an integral:

$$\sum_{l=1}^{\infty} \frac{1}{l^{\gamma-m+2}} \leq \left(1 + \int_1^{\infty} \frac{dl}{l^{\gamma-m+2}}\right) = 1 + \frac{1}{\gamma - m + 1}.$$

This implies the required estimate. □

4. Now we discuss the Diophantine properties for resonance frequency vectors. Let $g_\nu \subset \mathbb{Z}^m$ be the set of resonances, corresponding to the vector $\nu \in \mathbb{R}^m$:

$$g_\nu = \{k \in \mathbb{Z}^m : \langle \nu, k \rangle = 0\}.$$

The set g_ν is a subgroup of the Abelian group $(\mathbb{Z}^m, +)$. We put

$$D_{m,l}(c, \gamma) = \{\nu \in \mathbb{R}^m : \text{rank}\, g_\nu = l \text{ and for any } k \in \mathbb{Z}^m \setminus g_\nu$$
$$\text{conditions (9.2) hold}\}.$$

Let K_0 be an integer unimodular ($\det K_0 = 1$) square ($m \times m$) matrix and let $k_1^*, \dots, k_n^*, k_1, \dots, k_l$, $l + n = m$ be its columns. We assume that the vectors $k_1, \dots, k_l$ generate g_ν. The existence of such a matrix follows from the theory of Abelian groups.[2] The last l components of the vector $K_0^T \nu$ vanish. Since the groups g_ν and $g_{K_0^T \nu}$ are isomorphic, the vector $\nu^* \in \mathbb{R}^n$, formed by the first n components of $K_0^T \nu$, is non-resonant.

Proposition 9.2. *The vector ν^* is Diophantine if and only if $\nu \in D_{m,l}(c, \gamma)$ for some $c, \gamma > 0$.*

Proof. (a) Let the vector ν^* be Diophantine. Putting $\nu' = K_0^T \nu = \binom{\nu^*}{0}$, we have $g_{\nu'} = \mathbb{Z}_0^l = \{j \in \mathbb{Z}^m : j_1 = \dots = j_n = 0\}$. Let us show that the inequality $|\langle \nu', j \rangle| \geq c\|j\|^{-\gamma}$ holds for any $j \in \mathbb{Z}^m \setminus \mathbb{Z}_0^l$. Indeed, let $j^* = (j_1, \dots, j_n)^T$. Then

---

[2] In the case $m = 2$ the matrix K_0 can be easily constructed directly. Indeed, in the case $l = 0$ or $l = 2$ it is possible to take as K_0 the identity matrix. In the case $l = 1$ we choose a vector $(b, d)^T \in g_\nu$ whose components are relatively prime. (It is impossible to do this only in two cases: either $(1, 0)^T \in g_\nu$, then we put $(b, d) = (1, 0)$, or $(0, 1)^T \in g_\nu$, then we put $(b, d) = (0, 1)$.) There exist integer numbers a, c such that $ad - bc = 1$. The matrix $K_0 = \left(\begin{smallmatrix} a & b \\ c & d \end{smallmatrix}\right)$ is the one we seek.

$$|\langle \nu', j\rangle| = |\langle \nu^*, j^*\rangle| \geq c\|j^*\|^{-\gamma} \geq c\|j\|^{-\gamma}.$$

For any $j \in \mathbb{Z}^m \setminus g_\nu$ we have $K_0^{-1} j \in \mathbb{Z}^m \setminus \mathbb{Z}_0^l$. Therefore

$$|\langle \nu, j\rangle| = |\langle \nu', K_0^{-1} j\rangle| \geq c\|K_0^{-1} j\|^{-\gamma} \geq c'\|j\|^{-\gamma}$$

for some constant c'.

(b) Suppose that for any $j \in \mathbb{Z}^m \setminus g_\nu$ we have $|\langle \nu, j\rangle| \geq c'\|j\|^{-\gamma}$. Then for any nonzero $j^* \in \mathbb{Z}^n$

$$|\langle \nu^*, j^*\rangle| = \left|\left\langle K_0^T \nu, \begin{pmatrix} j^* \\ 0 \end{pmatrix}\right\rangle\right| \geq c' \left\| K_0 \begin{pmatrix} j^* \\ 0 \end{pmatrix}\right\|^{-\gamma},$$

because $K_0\binom{j^*}{0} \in \mathbb{Z}^m \setminus g_\nu$. The estimate $|K_0\binom{j^*}{0})\| \leq c''\|j^*\|$ implies the Diophantine property for the vector ν^*. □

For the sets $D_{m,l}(c, \gamma)$ an analog of Theorem 9.3 holds. In particular, let

$$\mathscr{C}_l = \{\nu \in \mathbb{R}^m : \|\nu\| \leq 1, \operatorname{rank} g_\nu = l = m - n\}.$$

Then the n-dimensional measure of the set $\mathscr{C}_l \setminus \bigcup_{c>0} D_{m,l}(c, \gamma)$ vanishes for any $\gamma > n - 1$.

9.2 Closures of Asymptotic Curves

1. Theorems on closures of asymptotic curves. Let T be a smooth area-preserving diffeomorphism[3] of a two-dimensional manifold M and $\hat{z} \in M$ a hyperbolic fixed point of T. The point $\hat{z}$ generates 4 asymptotic curves (separatrix branches): the stable branches $\Gamma^s_{1,2}$ and the unstable ones $\Gamma^u_{1,2}$. By definition, the point $\hat{z}$ does not belong to $\Gamma^{s,u}_{1,2}$.

The separatrices generically form a complicated network. The dynamics in the vicinity of this network is highly unstable and irregular. Because of this it is accepted to call this vicinity the stochastic layer and to characterize the dynamics in the stochastic layer by the word "chaos".

The structure of such a chaos is weakly understood. It is well-known that in the stochastic layer there exists an invariant hyperbolic set on which T is isomorphic to the Bernoulli shift. However, the measure of this set vanishes and the question of what behavior is typical for trajectories in the stochastic layer, remains open.

We mention here two questions. Suppose that stable and unstable separatrices do not coincide: $\Gamma^s_1 \cup \Gamma^s_2 \neq \Gamma^u_1 \cup \Gamma^u_2$.

- Is the measure of the separatrix closure $\overline{\Gamma} = \overline{\Gamma}^s_1 \cup \overline{\Gamma}^u_1 \cup \overline{\Gamma}^s_2 \cup \overline{\Gamma}^u_2$ positive?

[3] It is sufficient to require C^1-smoothness.

- If the answer to the first question is positive, is the map T restricted to this set ergodic?

  For the hyperbolic automorphism of a torus

$$\mathbb{T}^2 \to \mathbb{T}^2, \quad \mathbb{T}^2 \ni z \mapsto Az \in \mathbb{T}^2, \qquad A = \begin{pmatrix} 2 & 1 \\ 1 & 1 \end{pmatrix},$$

we have $\overline{\Gamma} = \mathbb{T}^2$ and the answers to both questions are positive. The same is true for small deformations of this map. In the general situation the answers are unknown.[4]

Theorem 9.4. *Suppose that the following conditions hold.*

(1) *The set $\Gamma_1^s \cap \Gamma_1^u$ is not empty.*
(2) *The curves Γ_1^s and Γ_1^u lie in an invariant domain $D \subset M$, where the closure $\overline{D}$ is compact.*

Then the closure of the unstable branch $\overline{\Gamma}_1^u$ contains the stable one Γ_1^s.

Corollary 9.4. *Since conditions of the theorem are symmetric with respect to Γ_1^s and Γ_1^u, the inclusion $\Gamma_1^u \subset \overline{\Gamma}_1^s$ holds. Hence, the sets $\overline{\Gamma}_1^s$ and $\overline{\Gamma}_1^u$ coincide.*

Corollary 9.5. *If conditions of the theorem hold for two pairs*

$$\Gamma_j^s \quad \text{and} \quad \Gamma_j^u, \qquad j \in \{1, 2\},$$

the sets $\overline{\Gamma}_1^s$, $\overline{\Gamma}_2^s$, $\overline{\Gamma}_1^u$, $\overline{\Gamma}_2^u$ coincide.

Indeed, it is easy to show that if $\Gamma_j^s \cap \Gamma_j^u \neq \emptyset$, $j \in \{1, 2\}$, then $\Gamma_1^s \cap \Gamma_2^u \neq \emptyset$ and $\Gamma_2^s \cap \Gamma_1^u \neq \emptyset$. Hence, the conditions of Theorem 9.4 hold for any pair Γ_i^s and Γ_j^u, $i, j \in \{1, 2\}$.

Let p and q be hyperbolic periodic points of T and let i and j respectively be their periods. Since p and q are hyperbolic fixed points for the maps T^i and T^j, we can define the branches $\Gamma_{1,2}^{s,u}(p)$ and $\Gamma_{1,2}^{s,u}(q)$. We put

$$W_{1,2}^{s,u}(p) = \bigcup_{k=0}^{2i-1} T^k(\Gamma_{1,2}^{s,u}(p)), \quad W_{1,2}^{s,u}(q) = \bigcup_{k=0}^{2j-1} T^k(\Gamma_{1,2}^{s,u}(q)).$$

Theorem 9.5. *Suppose that the following conditions hold.*

(1′) *The set $W_1^s(p) \cap W_1^u(q)$ is not empty.*
(2′) *The set $W_1^s(q) \cap W_1^u(p)$ is not empty.*
(3′) *The asymptotic manifolds $W_1^s(p)$, $W_1^u(p)$, $W_1^s(q)$, $W_1^u(q)$ belong to an invariant domain $D \subset M$, where the closure $\overline{D}$ is compact.*

[4] These questions have the same nature as the question on the positiveness of the metric entropy of the map T.

Then the following inclusions hold

$$W_1^s(p) \subset \overline{W}_1^u(q) \cup \overline{W}_1^s(q), \tag{9.4}$$

$$W_1^s(p) \subset \overline{W}_1^u(q) \cup \overline{W}_1^u(p). \tag{9.5}$$

Remark 9.1. In the case $p = q$ conditions $(1')$ and $(2')$ coincide.

Corollary 9.6. *Interchanging q and p in Theorem 9.5 and (or) replacing T by T^{-1} we get the same conditions $(1')$–$(3')$. Instead of (9.4) we obtain the inclusions*

$$W_1^s(q) \subset \overline{W}_1^u(p) \cup \overline{W}_1^s(p),$$
$$W_1^u(p) \subset \overline{W}_1^u(q) \cup \overline{W}_1^s(q),$$
$$W_1^u(q) \subset \overline{W}_1^u(p) \cup \overline{W}_1^s(p).$$

In particular, $(1')$–$(3')$ imply that $\overline{W}_1^u(p) \cup \overline{W}_1^s(p) = \overline{W}_1^u(q) \cup \overline{W}_1^s(q)$.

Corollary 9.7. *If $p = q$ and conditions $(1')$–$(3')$ hold, according to (9.5) and to the corresponding symmetric inclusion, we have $\overline{W}_1^s(p) = \overline{W}_1^u(p)$.*

Recall that points of the sets

$$(W_1^s(p) \cup W_2^s(p)) \cap (W_1^u(q) \cup W_2^u(q)), \quad (W_1^s(q) \cup W_2^s(q)) \cap (W_1^u(p) \cup W_2^u(p))$$

are called homoclinic if p and q lie on the same periodic trajectory, and heteroclinic otherwise.

Takens proved [130] that, if M is compact, there exists a residual subset R in the topological space of area-preserving C^1-smooth diffeomorphisms of M such that, for any $T \in R$ and any hyperbolic periodic point p of T, the set of homoclinic points is dense on $W_{1,2}^{s,u}(p)$. Note that methods of [130] are essentially restricted to the C^1-topology.

The following conjecture was in fact formulated (in a weaker form) by Poincaré.

Conjecture 9.1. If $p = q$ and the conditions of Theorem 9.5 hold, the set of homoclinic points is dense on $W_1^s(p)$ and on $W_1^u(p)$.

Mather [87] proved that if M is compact, for a C^k-generic ($k \geq 4$) area-preserving map, any two branches of a hyperbolic periodic point have the same closure. The word "generic" is understood in the sense of Baire category.

Oliveira [100] obtained the following results related to the problems in question. Let T be a C^1 area-preserving diffeomorphism of a compact orientable surface. Assume that L and K are branches of a hyperbolic fixed point with either $L = K$ or $L \cap K = \emptyset$. If $K \cap \omega(L) \neq \emptyset$ then $K \subset \omega(L)$. (Here as usual, $\omega(L)$ is the ω-limit set of L.)

This result implies the following assertion [100]. Let M be a compact orientable surface and $1 \leq k \leq \infty$. Then $L \subset \omega(L)$ for any branch L of a C^k-generic area-preserving map.

Note that the manifold M in Theorems 9.4–9.5 can be neither compact nor orientable. It can even have a boundary. The further constructions depend on a certain integer r. The reader who is satisfied by the case when M is a sphere or a plane can put $r = 0$ and not pay any attention to geometric objects (homology groups, intersection indices, etc.) appearing below. The basic geometric fact we use in the case $r = 0$ is very simple: two closed curves transversal to one another on a sphere or on a plane cannot have exactly one common point.

In the general situation we consider the homology group $H_1(M, \partial M)$ of the manifold M with respect to the boundary ∂M with coefficients from $\mathbb{Z}_2$. Let G be the subgroup in $H_1(M, \partial M)$ generated by curves lying in D. We put $r = \operatorname{rank} G$. Since according to (2) or (3′) the closure $\overline{D}$ is compact, we have $0 \le r < \infty$.

Below we can assume that in the case of Theorem 9.4 the map T preserves each branch and, in the case of Theorem 9.5, T preserves each of the eight manifolds $W^{s,u}_{1,2}(p)$, $W^{s,u}_{1,2}(q)$. Indeed, if T does not satisfy this condition, we just change T to T^2.

2. Proof of Theorem 9.4. Let $U \subset D$ be any open set such that $U \cap \Gamma_1^s \neq \emptyset$. Suppose that

$$U \cap \Gamma_1^u = \emptyset. \tag{9.6}$$

Then Theorem 9.4 is proved if we are able to obtain a contradiction under its conditions. Considering if necessary instead of U a smaller domain (we will keep for it the same notation U), we can assume that $U = U^+ \cup U^0 \cup U^-$, where $U^\pm$ are open, connected, and $U^0 \subset \Gamma_1^s$ is a connected interval (see Fig. 9.1) Let $\hat{I}$ be the minimal connected piece of Γ_1^s such that $\hat{z}$ is its endpoint and $U^0 \subset \hat{I}$. We can assume that

$$U \cap T(\hat{I}) = \emptyset, \tag{9.7}$$

$$U \cap T^{-r-1}(\hat{I}) = U^0. \tag{9.8}$$

These equations hold for sufficiently small U.

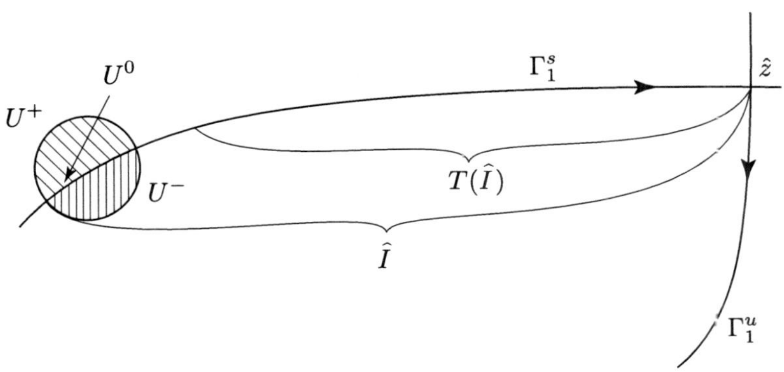

Fig. 9.1 The domain U and the intervals $\hat{I}$ and $T(\hat{I})$.

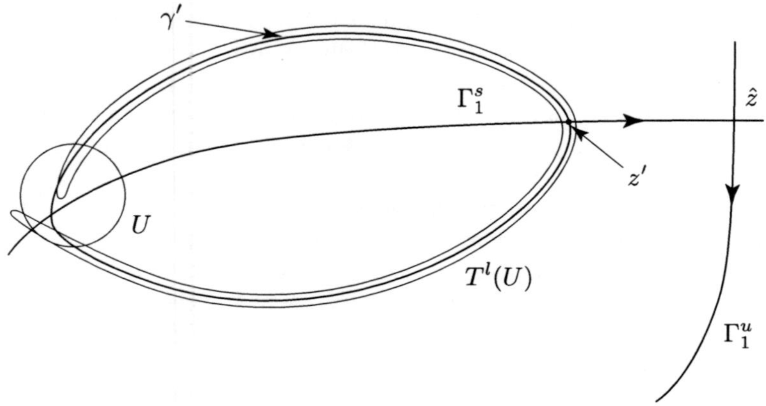

Fig. 9.2 The curve γ'.

Lemma 9.2. *There exists a natural number $l > r + 1$ such that*

$$U^+ \cap T^l(U^+) \neq \emptyset, \quad U^- \cap T^l(U^-) \neq \emptyset.$$

Proof (of Lemma 9.2). Consider the map

$$T \times T : M \times M \to M \times M, \quad T \times T(z_1, z_2) = (T(z_1), T(z_2)).$$

This map preserves the measure $\sigma \times \sigma$, where σ is the area on M. The set $U^+ \times U^-$ lies inside the compact invariant set $\overline{D} \times \overline{D}$. Hence, according to the Poincaré recurrence theorem,

$$(T \times T)^l(U^+ \times U^-) \cap (U^+ \times U^-) \neq \emptyset$$

for infinitely many $l \in \mathbb{N}$. The lemma is proved. □

There exists a smooth closed curve γ' satisfying the following properties.

(a) $\gamma' \subset U'$, $U' = U \cup T^l(U)$.
(b) The set $\gamma' \cap T^l(U^0)$ consists of a single point z' and the curves γ' and $T^l(U^0)$ intersect at z' transversely.

The curve γ' goes along the set $T^l(U^+)$ from the point z' to the set U^+. Then γ' passes through the interval U^0 to the set U^-, goes to the set $T^l(U^-)$ and finally returns to the point z' (see Fig. 9.2).

According to property (a) and equation (9.6) we have

$$\gamma' \cap \Gamma_1^u = \emptyset. \tag{9.9}$$

Lemma 9.3. *There exists an interval $I \subset \Gamma_1^s$ satisfying the following two properties.*

(A) *The endpoints of I are $\hat{z}$ and z_0, where z_0 is homoclinic.*
(B) $I \cap U' = T^{-r}(I) \cap U' = T^l(U^0)$.

Proof (of Lemma 9.3). The interval $I' = T^{l-1}(\hat{I}) \setminus T^l(\hat{I})$ contains a homoclinic point. Indeed, otherwise we have a contradiction to assumption (1) because of the equation $\Gamma_1^s = \bigcup_{k\in\mathbb{Z}} T^k(I')$. Let z_0 be this homoclinic point and let I be defined by (A). Then $I \subset T^{l-1}(\hat{I})$.

Equations (B) follow from the three inclusions

$$T^l(U^0) \subset (I \cap U') \subset (T^{-r}(I) \cap U') \subset T^l(U^0).$$

The first two are obvious. Let us check the third one. Since $I \subset T^{l-1}(\hat{I})$, we have: $T^{-r}(I) \subset T^{l-r-1}(\hat{I})$. This relation together with the inequality $l > r + 1$ (see Lemma 9.2) imply that

$$T^{-r}(I) \cap U \subset T^{l-r-1}(\hat{I}) \cap U \subset T(\hat{I}) \cap U = \emptyset$$

due to (9.7). Analogously,

$$T^{-r}(I) \cap T^l(U) \subset T^{l-r-1}(\hat{I}) \cap T^l(U) \subset T^l(T^{-r-1}(\hat{I}) \cap U) = T^l(U^0).$$

Here we have used assumption (9.8). Lemma 9.3 is proved. □

According to property (A) the points $\hat{z}$ and z_0 can be connected by an interval I^u of the unstable separatrix branch Γ_1^u. The curve $\sigma_0 = I \cup I^u$ is closed. According to the definition of γ' and property (B) the curves I and γ' have exactly one common point (the point z') and intersect at z' transversely. The curves I^u and γ' do not intersect because of (9.9). Hence, the intersection index (modulo 2) is

$$\mathrm{ind}_2(\sigma_0, \gamma') = 1. \tag{9.10}$$

Proposition 9.3. *In the case $r = 0$, the intersection index of any two closed curves on D vanishes.*

Since the equality (9.10) contradicts to Proposition 9.3, Theorem 9.4 is proved for $r = 0$. Proposition 9.3 is a simple corollary of the following lemma.

For any closed curve $a \subset M$ let $\{a\}$ be the corresponding element of the homology group $H_1(M, \partial M)$.

Lemma 9.4. *The intersection index of any two closed curves $a_1, a_2 \subset M$ is determined by the elements $\{a_1\}, \{a_2\}$:*

$$\mathrm{ind}_2(a_1, a_2) = \alpha(\{a_1\}, \{a_2\}), \qquad \alpha : H_1(M, \partial M) \times H_1(M, \partial M) \to \mathbb{Z}_2.$$

The function α is bilinear.

Lemma 9.4 is a standard geometric fact. The proof can be found in [38].

3. Continuation of the proof: the case $r > 0$. The map T is defined on the invariant domain $D \subset M$. Hence, the automorphism T_* of the homology group $H_1(M, \partial M)$

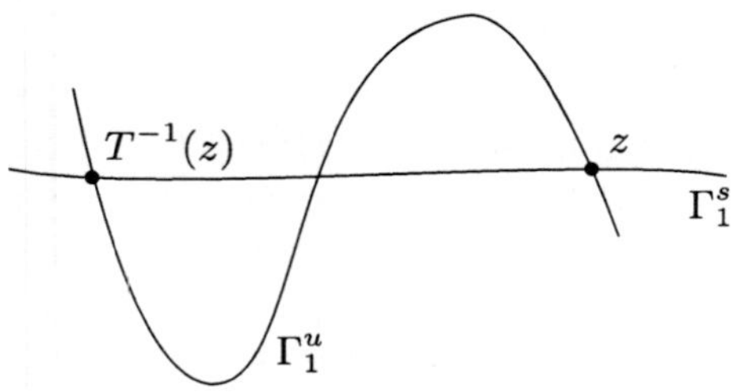

Fig. 9.3 The L-loop $\vartheta(z)$.

can be restricted to an automorphism of the invariant subgroup $G \subset H_1(M, \partial M)$. (Recall that the subgroup G is generated by closed curves lying in D.)

For any closed curves $u, v \subset M$

$$\alpha(\{u\}, \{v\}) = \alpha(T_*\{u\}, T_*\{v\}). \tag{9.11}$$

Lemma 9.5. *Let $\sigma_k = T^{-k}(\sigma_0)$, $0 \le k \le r$. Then $\alpha(\{\sigma_k\}, \{\gamma'\}) = 1$.*

Proof (of Lemma 9.5). Any curve σ_k is a union $\sigma_k^s \cup \sigma_k^u$, where $\sigma_k^s \subset \Gamma_1^s$ and $\sigma_k^u \subset \Gamma_1^u$. Let $0 \le k \le r$. Then $I \subset \sigma_k^s \subset T^{-r}(I)$. Hence, by Lemma 9.3, $\sigma_k^s \cap \gamma' = z'$ and the intersection is transversal. Because of (9.9) we have: $\sigma_k^u \cap \gamma' = \emptyset$. Lemma 9.5 is proved. □

For any homoclinic point $z \in \Gamma_1^s \cap \Gamma_1^u$ the closed curve formed by segments of Γ_1^s and Γ_1^u, confined by z and its preimage $T^{-1}(z)$ (see Fig. 9.3) will be called an L-loop $\vartheta(z)$. We put $\vartheta_0 = \vartheta(z_0)$ and $\vartheta_k = T^{-k}(\vartheta_0)$. Then for any natural number q

$$\{\sigma_q\} = \{\sigma_0\} + \sum_{k=0}^{q-1} \{\vartheta_k\}. \tag{9.12}$$

Lemma 9.6. *Let ϑ_k, $k \in \mathbb{Z}$, be an L-loop. Then $\alpha(\{\vartheta_k\}, \{\gamma'\}) = 0$.*

Proof (of Lemma 9.6). According to Lemma 9.5 and (9.12) we have

$$\alpha(\{\vartheta_k\}, \{\gamma'\}) = 0, \qquad 0 \le k \le r - 1.$$

Note that G can be regarded as a vector space over $\mathbb{Z}_2$ and $T_*|_G : G \to G$ as a linear operator. Consider the maximal m such that the vectors

$$\{\vartheta_0\}, \{\vartheta_1\}, \ldots, \{\vartheta_m\} \in G \tag{9.13}$$

are linearly independent. Obviously, $m \le r - 1$.

Let $\mathscr{L}$ be the linear hull of vectors (9.13). Then $\{\vartheta_k\} \in \mathscr{L}$ for any $k \in \mathbb{Z}$. For any vector $\{v\}$ from the vector space $\mathscr{L}$ we have $\alpha(\{v\}, \{\gamma'\}) = 0$. Lemma 9.6 is proved. □

According to identity (9.11) we have

$$\alpha(T_*^l\{\sigma_0\}, T_*^l\{\gamma'\}) = \alpha(\{\sigma_0\}, \{\gamma'\}) = 1.$$

Since $T_*^l\{\sigma_0\} - \{\sigma_0\}$ equals a sum of L-loops $\{\vartheta_k\}$,

$$\alpha(T_*^l\{\sigma_0\}, \{\gamma'\}) = 1.$$

Hence,

$$\alpha(T_*^l\{\sigma_0\}, \{\gamma'\} + T_*^l\{\gamma'\}) = 0. \tag{9.14}$$

The curves γ' and $T^l(\gamma')$ go through the same connected open set $T^l(U)$. Hence, inside the set $U'' = U \cup T^l(U) \cup T^{2l}(U)$ there exists a closed curve γ'' of the homology class $\{\gamma'\} + T_*^l\{\gamma'\}$ which differs from $\gamma' \cup T^l(\gamma')$ only inside the set $T^l(U)$. Since $U'' \cap \Gamma_1^u = \emptyset$, we have

$$\gamma'' \cap \Gamma_1^u = \emptyset. \tag{9.15}$$

Note that

$$T^l(\sigma_0) = T^l(I^u) \cup T^l(I), \qquad T^l(I^u) \subset \Gamma_1^u, \quad T^l(I) \subset \Gamma_1^s.$$

According to Lemma 9.3 and the obvious equation $U \cap T^l(I) = \emptyset$, we have $T^l(I) \cap U'' = T^{2l}(U^0)$. Hence, $T^l(I) \cap \gamma'' = T^l(z')$ and the intersection is transversal. On the other hand, $T^l(I^u) \cap \gamma'' = \emptyset$ because of (9.15). Thus, $\mathrm{ind}_2(T^l(\sigma_0), \gamma'') = 1$, which contradicts (9.14). Theorem 9.4 is proved.

4. Separatrices of periodic points. Now we prove Theorem 9.5. Let $U \subset D$ be an open set such that $U \cap W_1^s(p) \neq \emptyset$. Suppose that at least one of the equations

$$U \cap (W_1^u(q) \cup W_1^s(q)) = \emptyset, \tag{9.16}$$

$$U \cap (W_1^u(q) \cup W_1^u(p)) = \emptyset, \tag{9.17}$$

holds. Then Theorem 9.5 will be proved when we obtain a contradiction under its conditions.

Considering the map T^i instead of T, we define the sets $U^\pm$, U^0 and the interval $\hat{I}$. In particular, we again assume that U satisfies (9.7)–(9.8) with T^i instead of T.

We can obviously assume that $U \cap \Gamma_1^s(p) \neq \emptyset$. (Otherwise we take $T^k(U)$ with a proper $k \in \{1, \dots, i-1\}$ instead of U.) The closed curve $\gamma' \subset U'$ is defined in the same way as above. We can assume that γ' does not pass through points of the form $T^k(p)$ and $T^k(q)$, $k \in \mathbb{Z}$.

By using condition (1′) we get $\Gamma_1^s(p) \cap W_1^u(q) \neq \emptyset$. Hence, Lemma 9.3 remains true if in its statement $z_0 \in \Gamma_1^s(p) \cap W_1^u(q)$.

For some $j^u, j^s \in \{0, \dots, j-1\}$ we have

$$z_0 \in T^{j^u}(\Gamma_1^u(q)) \cap \Gamma_1^s(p), \quad \Gamma_1^u(p) \cap T^{j^s}(\Gamma_1^s(q)) \neq \emptyset.$$

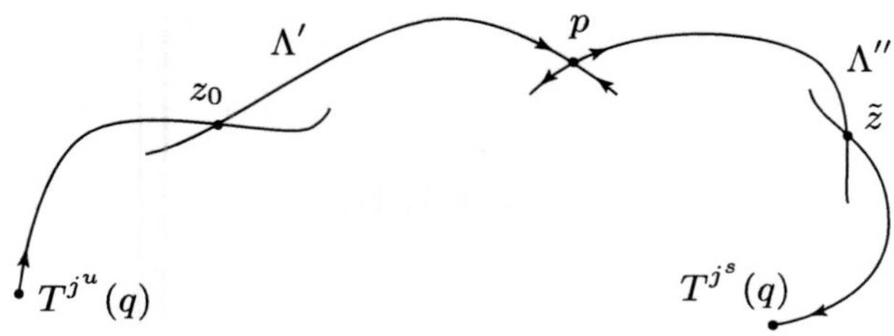

Fig. 9.4 The curves Λ' and Λ''.

Let Λ' be the curve that goes from $T^{j^u}(q)$ to p and is formed by the segment of the branch $T^{j^u}(\Gamma_1^u(q))$ from the point $T^{j^u}(q)$ to the point z_0 and by the segment of the branch $\Gamma_1^s(p)$ from z_0 to p. Analogously, let Λ'' be the curve that goes from p to $T^{j^s}(q)$ and is formed by the segment of the branch $\Gamma_1^u(p)$ from p to some point $\tilde{z} \in \Gamma_1^u(p) \cap T^{j_s}(\Gamma_1^s(q))$ and by the segment of the branch $T^{j_s}(\Gamma_1^s(q))$ from $\tilde{z}$ to $T^{j_s}(q)$ (see Fig. 9.4).

Let ν be the least common multiple of the numbers j and $\Delta = j^s - j^u$. If $\Delta = 0$, we put $\nu = 0$. The curve

$$\Lambda' \cup \Lambda'' \cup T^{\Delta}(\Lambda') \cup T^{\Delta}(\Lambda'') \cup T^{2\Delta}(\Lambda') \cup T^{2\Delta}(\Lambda'') \cup \cdots \cup T^{\nu-\Delta}(\Lambda') \cup T^{\nu-\Delta}(\Lambda'')$$

is closed. The same is true for the curve

$$\begin{aligned}\sigma_0 = \sigma_0(k', k'') = \Lambda' \cup T^{ijk''}(\Lambda'') \cup T^{ijk'+\Delta}(\Lambda') \cup T^{ijk''+\Delta}(\Lambda'') \cup \cdots \\ \cup T^{ijk'+\nu-\Delta}(\Lambda') \cup T^{ijk'+\nu-\Delta}(\Lambda''),\end{aligned}$$

where k', k'' are arbitrary integers. Obviously,

$$\sigma_0 \in W_1^s(p) \cup W_1^u(p) \cup W_1^s(q) \cup W_1^u(q).$$

We choose $k' > 0$ so large that

$$\gamma' \cap \sigma_0 \cap W_1^s(p) = \gamma' \cap I = z'. \tag{9.18}$$

(We can do this because the set $(\sigma(k', k'') \cap W_1^s(p)) \setminus I$ for large $k' > 0$ is situated in a small neighborhood of the trajectory generated by p and the curve γ' does not contain points of this trajectory.)

Suppose that equation (9.16) holds. Then choose $k'' < 0$ so large in absolute value that

$$\gamma' \cap \sigma_0 \cap W_1^u(p) = \emptyset. \tag{9.19}$$

(In the case (9.17) we take $k'' > 0$ such that $\gamma' \cap \sigma_0 \cap W_1^s(q) = \emptyset$ and use analogous arguments.)

According to (9.16), (9.18), and (9.19) we have $\sigma_0 \cap \gamma' = z'$, where the intersection is transversal. Hence, $\mathrm{ind}_2(\sigma_0, \gamma') = 1$. In the case $r = 0$ this index should vanish and we obtain the required contradiction. In the case $r > 0$ the arguments are the same as in **3**: a contradiction is obtained as a result of the evaluation of the intersection index for the curves $T^{il}(\sigma_0)$ and γ'', where $\{\gamma''\} = \{\gamma'\} + T_*^{il}\{\gamma'\}$.

9.3 Invariant Tori in a Neighborhood of a Resonance

In this section, following [142], we study the difference of frequency vectors on two invariant tori, which bound a resonance.

Two degrees of freedom. Consider the system with Hamiltonian

$$H(I, \varphi, \varepsilon) = H_0(I) + \varepsilon H_1(I, \varphi, \varepsilon). \tag{9.20}$$

Here $I \in \mathbb{R}^2$ and $\varphi = (\varphi_1, \varphi_2) \bmod 2\pi$ are canonically conjugated coordinates. The function H is assumed to be real-analytic in the phase variables and smooth in the parameter $\varepsilon \geq 0$.

For $\varepsilon = 0$ the phase space is foliated by two-dimensional tori $\{(I, \varphi) : I = \text{const}\}$, filled with quasi-periodic solutions:

$$\varphi = \nu(I)t + \varphi_0, \qquad \nu(I) = \frac{\partial H_0}{\partial I}(I).$$

Recall that an invariant torus of the unperturbed ($\varepsilon = 0$) system is called resonant if $\langle k, \nu \rangle = 0$ for some nonzero vector $k \in \mathbb{Z}^2$. A nonresonant torus is called Diophantine if there are $\alpha, \ \gamma > 0$ such that for any nonzero $k \in \mathbb{Z}^2$

$$|\langle k, \nu \rangle| \geq \frac{1}{\alpha |k|^\gamma}. \tag{9.21}$$

For small $\varepsilon > 0$ resonant tori are as a rule destroyed, and Diophantine ones just slightly deform.

Suppose that the unperturbed system is isoenergetically non-degenerate, i.e.,

$$\Delta = \det \begin{pmatrix} \frac{\partial^2 H_0}{\partial I^2}(I^0) & \nu(I^0) \\ \nu^T(I^0) & 0 \end{pmatrix} \neq 0,$$

and on the energy level h there is a Diophantine torus

$$\mathbb{T}^2_{\varkappa}(0, h) = \{(I, \varphi : I = I^0, \ H(I^0) = h, \ \nu_1(I^0)/\nu_2(I^0) = \varkappa\}.$$

Then (Theorem 2.3) for small $\varepsilon > 0$ on the same energy level

$$\Sigma(\varepsilon, h) = \{(I, \varphi) \in M : H(I, \varphi, \varepsilon) = h\}$$

there exists a KAM-torus $\mathbb{T}^2_{\varkappa}(\varepsilon, h)$ with the same frequency ratio. The family of tori $\mathbb{T}^2_{\varkappa}(\varepsilon, h) \subset \Sigma(\varepsilon, h)$ is smooth in ε.

Theorem 9.6. *Suppose that $H_0(I^0) = h$ and the following conditions hold.*

(1) $\nu(I^0) = \lambda(m, n)^T$, $\lambda \neq 0$, $n \neq 0$, *where m and n are either relatively prime integers, or $(m, n) = (0, 1)$.*
(2) *The function*[5]

$$\nu_{\mathrm{a}}(q) = \frac{1}{2\pi\Delta}\int_0^{2\pi} H_1\left(I^0, mt + \frac{q}{n}, nt, 0\right) dt \tag{9.22}$$

has a unique global minimum on the circle q mod 2π, and this minimum is non-degenerate.

Then for small $\varepsilon > 0$ there are two KAM-tori $\mathbb{T}^2_{\varkappa_j}(\varepsilon, h) \subset \Sigma(\varepsilon, h)$, $j = 1, 2$, such that

(a) *the action variable I, restricted to $\mathbb{T}^2_{\varkappa_j}$, satisfies the inequality $|I - I^0| \leq C\sqrt{\varepsilon}$,*
(b) $\varkappa_1 < m/n < \varkappa_2$,
(c) $|\varkappa_1 - \varkappa_2| < c\,\varepsilon$.

The positive constants c and C do not depend on ε. The function ν_{a} is 2π-periodic. The periodicity becomes obvious after the change $t \mapsto \tau = t + \frac{\gamma}{n}q$, where $\gamma \in \mathbb{Z}$ and $\delta n + \gamma m = 1$ for some $\delta \in \mathbb{Z}$.

The assumption on the analyticity of the Hamiltonian (9.20) is essential: in the case of a finite smoothness estimate (c) is much weaker:

$$|\varkappa_1 - \varkappa_2| < c\frac{\sqrt{\varepsilon}}{\log \varepsilon}.$$

One-and-a-half degrees of freedom. Consider the system with Hamiltonian

$$H(p, q, t, \varepsilon) = H_0(p) + \varepsilon H_1(p, q, t, \varepsilon), \tag{9.23}$$

where the scalar variables $p \in \mathbb{R}$ and $q = q$ mod 2π are canonically conjugated and $t = t$ mod 2π. The function H is assumed to be real-analytic on the extended phase space $M = D \times \mathbb{T}^2$ and smooth in ε.

For $\varepsilon = 0$ the phase space M is foliated by two-dimensional invariant tori $\{(p, q, t) : p = \text{const}\}$. The motion on the tori is quasi-periodic:

$$q = \varkappa(p)t + q_0, \qquad \varkappa(p) = \frac{\partial H_0}{\partial p}.$$

Theorem 9.7. *Suppose that for some $p = p^0$ the following conditions hold:*

(1) $\varkappa(p^0) = m/n$, $n \neq 0$, *where $m, n \in \mathbb{Z}$ are either relatively prime, or $m = 0$, $n = 1$.*

[5] The subscript "a" is from "autonomous".

(2) *The function*[6]

$$\nu_n(q) = \frac{1}{2\pi}\left(\frac{\partial^2 H_0}{\partial p^2}(p^0)\right)^{-1} \int_0^{2\pi} H_1\left(p^0, mt + \frac{q}{n}, nt, 0\right) dt \qquad (9.24)$$

has a unique global minimum on the circle $q \bmod 2\pi$, *and this minimum is non-degenerate.*

Then for small $\varepsilon > 0$ *there are two KAM-tori with frequencies* $\varkappa_1$ *and* $\varkappa_2$ *such that*

(a) *the variable* p *on these tori satisfies the inequality* $|p - p^0| < C\sqrt{\varepsilon}$,
(b) $\varkappa_1 < m/n < \varkappa_2$,
(c) $|\varkappa_1 - \varkappa_2| < c\varepsilon$.

Theorem 9.7 follows from Theorem 9.6. Indeed, let us put $p = I_1, q = \varphi_1$, $t = \varphi_2$. Then the projection $(I_1, I_2, \varphi_1, \varphi_2) \mapsto (I_1, \varphi_1, \varphi_2) = (v, q, t)$ maps trajectories of the system with Hamiltonian

$$\hat{H} = H_0(I_1) + I_2 + \varepsilon H_1(I_1, \varphi_1, \varphi_2, \varepsilon), \qquad (9.25)$$

to solutions of the system with Hamiltonian (9.23). The function (9.25) satisfies the conditions of Theorem 9.6.

Discrete case. Consider a family of symplectic self-maps P_ε of the cylinder

$$Z = \{(I, \varphi) : I \in \mathbb{R},\ \varphi \in \mathbb{T}^1\}.$$

The maps P_ε are determined by the generating function $S(J, \varphi, \varepsilon) = f(J) + \varepsilon W(J, \varphi, \varepsilon)$:

$$P_\varepsilon(I, \varphi) = (J, \psi), \qquad J = I - S_\varphi(J, \varphi, \varepsilon), \quad \psi = \varphi + S_J(J, \varphi, \varepsilon).$$

The function S is real-analytic in J and φ, 2π-periodic in φ, and smooth in ε. The map P_0 has the form:

$$J = I, \quad \psi = \varphi + f'(I).$$

The variable I enumerates circles (one-dimensional tori) invariant with respect to the action of P_0. We put

$$\mathbb{T}_\varkappa = \left\{(I, \varphi) : I = I^0,\ f'(I^0) = 2\pi\varkappa,\ \varphi \in \mathbb{T}\right\}.$$

Recall that in the discrete case the resonant and Diophantine conditions should be considered for the vector $(\varkappa,\ 1)^T$.

Theorem 9.8. *Suppose that for* $I = I^0$ *the following conditions hold:*

(1) $\varkappa(I^0) = \frac{m}{n}$, *where* m *and* n *are either relatively prime integers, or* $m = 0$, $n = 1$.

[6] The subscript "n" is from "nonautonomous".

(2) *The function*[7]

$$\nu_{\rm d}(q) = \frac{1}{f''(I^0)} \sum_{k=0}^{n-1} W\left(I^0, \frac{2\pi km + q}{n}, 0\right)$$

has a unique minimum on the circle $q \mod 2\pi$.

Then for small $\varepsilon > 0$ *there exist two invariant circles with frequencies* $\varkappa_1$ *and* $\varkappa_2$ *such that*

(1) *the variable* I *on these circles satisfies the inequality* $|I - I^0| < C\sqrt{\varepsilon}$, $C > 0$,
(2) $\varkappa_1 < \frac{m}{n} < \varkappa_2$,
(3) $|\varkappa_1 - \varkappa_2| < c\,\varepsilon$, $c > 0$.

Proof of Theorem 9.8 can be obtained in the following way. According to Theorem 1.10, P_ε can be regarded as a time-2π map in some non-autonomous Hamiltonian system with real-analytic Hamiltonian $H = H(J, \psi, t, \varepsilon)$, 2π-periodic in ψ and t. Moreover, we can assume that $H(J, \psi, t, 0) = \frac{1}{2\pi} f(J)$. To reduce Theorem 9.8 to Theorem 9.7, it is sufficient to check conditions 1) and 2) for the Hamiltonian H.

Let us represent H in the form

$$H(J, \psi, t, \varepsilon) = \frac{1}{2\pi} f(J) + \varepsilon H_1(J, \psi, t) + O(\varepsilon^2).$$

The time-2π map in this system is as follows:

$$\begin{aligned} \psi &= \varphi - f'(J) + O(\varepsilon), \\ J &= I - \varepsilon \int_0^{2\pi} \frac{\partial H_1}{\partial \psi}\left(I, \frac{f'(I)}{2\pi}\xi + \varphi, \xi\right) d\xi + O(\varepsilon^2). \end{aligned}$$

Here $O(\varepsilon)$ and $O(\varepsilon^2)$ are 2π-periodic in φ, real-analytic in φ and I, and smooth in ε. Comparing this map with the one generated by the function S:

$$\begin{aligned} \psi &= \varphi + f'(I) + O(\varepsilon), \\ J &= I - \varepsilon W_\varphi(I, \varphi, 0) + O(\varepsilon^2), \end{aligned}$$

we obtain $W(J, \varphi, 0) = \int_0^{2\pi} \partial H_1(I, \frac{f'(I)}{2\pi}\xi + \varphi, \xi)\, d\xi + C(J)$. Here $C(J)$ is a real-analytic function. Condition (1) of Theorem 9.7 obviously holds. Condition (2) follows from the equation $\nu_{\rm n}(q) = \frac{1}{2\pi n}\nu_{\rm d}(q)$.

Proof (of Theorem 9.6). Without loss of generality we assume that $H_0(I^0) = 0$. Let us perform the canonical change of the variables $(I, \varphi) \mapsto (J, \psi)$, $I = I^0 + \Lambda^T J$, $\varphi = \Lambda^{-1}\psi$, where

[7] No comment.

$$\Lambda = \begin{pmatrix} n & -m \\ \gamma & \delta \end{pmatrix}, \qquad \det \Lambda = n\delta + m\gamma = 1, \qquad \delta, \gamma \in \mathbb{Z}, \tag{9.26}$$

and reduce the order on the level of the energy integral

$$H(I^0 + \Lambda^T J, \Lambda^{-1}\psi, \varepsilon) = 0. \tag{9.27}$$

Solving equation (9.27) with respect to J_2, we get

$$-J_2 = \frac{\Delta}{2\lambda} J_1^2 + \frac{\varepsilon}{\lambda} H_1(I^0, \Lambda^{-1}\psi, 0) + f(J_1, \psi, \varepsilon). \tag{9.28}$$

The function f is real-analytic in J_1 and ψ and 2π-periodic in ψ. Moreover, $f = O(J_1^3, \varepsilon^2, \varepsilon J_1)$ as $J_1 \to 0$, $\varepsilon \to 0$. The function $-J_2 = -J_2(J_1, \psi_1, \psi_2, \varepsilon)$ is the Hamiltonian of the system on the zero energy level, where ψ_2 should be regarded as the new time, and the variables J_1 and ψ_1 are canonically conjugated. After the change of variables $J_1 = \sqrt{\varepsilon} p$, $\psi_1 = q$, $\psi_2 = t$, $J_2 = -\sqrt{\varepsilon}\hat{H}$, the Hamiltonian (9.28) takes the form

$$\begin{aligned} \hat{H}(p, q, t, \varepsilon) &= \frac{\Delta\sqrt{\varepsilon}}{\lambda}\left(\frac{p^2}{2} + V_1(q, t)\right) + \varepsilon V_2(p, q, t, \varepsilon), \\ V_1(q, t) &= \frac{1}{\Delta} H_1(I^0, \delta q + mt, nt - \gamma q, 0). \end{aligned} \tag{9.29}$$

The function V_2 is real-analytic and 2π-periodic in q and t, and smooth in $\sqrt{\varepsilon}$.

In the new variables the resonance we study takes the simplest form: the first frequency vanishes and the second one (corresponding to time) equals 1. Hence, it is now sufficient to find two invariant tori whose first frequencies are of order ε and have different signs.

The system with Hamiltonian (9.29) contains two slow variables p, q ($\dot{p}, \dot{q} \sim \sqrt{\varepsilon}$) and one fast variable t ($\dot{t} = 1$). Since $\hat{H}$ is real-analytic, we can apply Theorem 6.2 and Remark 6.3. As a result we obtain the following proposition.

Proposition 9.4. *There exists a canonical real-analytic, near-identical change of the coordinates $(p, q) \mapsto (\tilde{p}, \tilde{q})$ which is 2π-periodic in $\tilde{x}$ and t, such that in the new coordinates $(\tilde{p}, \tilde{q})$ the Hamiltonian* (9.29) *has the form:*

$$\begin{aligned} &\tilde{H}(\tilde{p}, \tilde{q}, t) = \sqrt{\varepsilon}\tilde{H}_0(\tilde{p}, \tilde{q}, \varepsilon) + \tilde{\psi}(\tilde{p}, \tilde{q}, \varepsilon, t), \\ &\tilde{H}_0 = \frac{\Delta}{\lambda}\left(\frac{\tilde{p}^2}{2} + v_a(\tilde{q})\right) + \sqrt{\varepsilon} U_2(\tilde{p}, \tilde{q}, \varepsilon), \quad |\tilde{\psi}| \le c \exp\left(-\frac{\alpha}{\sqrt{\varepsilon}}\right), \\ &c, \alpha > 0. \end{aligned} \tag{9.30}$$

The function $\tilde{H}$ is 2π-periodic in $\tilde{q}$ and t, and real-analytic in $\tilde{p}$, $\tilde{q}$.

If we put formally $\tilde{\psi} = 0$ then according to assumption (2) of Theorem 9.6 the remaining integrable part of the Hamiltonian (9.30) has for small $\varepsilon > 0$ a hyperbolic 2π-periodic solution $z(t) = (\hat{p}, \hat{q}, t)$. Here $(\hat{p}, \hat{q}) = (0, q_0) + O(\sqrt{\varepsilon})$, where q_0

is the global minimum of the function v_a on $\mathbb{T}$. Note that without loss of generality $\tilde{H}_0(\hat{p}, \hat{q}, \varepsilon) = 0$, i.e., the unperturbed Hamiltonian vanishes on the unperturbed separatrices.

The domain in the space $\{\tilde{p}, \tilde{q}, t\}$, bounded by surfaces asymptotic to $z(t)$, corresponds to the resonance. The solution $z(t)$ projects from the extended phase space $\{\tilde{p}, \tilde{q}, t\}$ to the space $\{\tilde{p}, \tilde{q}\}$ into a hyperbolic equilibrium, and the asymptotic surfaces to the corresponding separatrices. When the perturbation $\tilde{\psi}$ is taken into account, the solution $z(t)$ deforms slightly, and the asymptotic surfaces split.

Our further argument is shortly as follows. In the unperturbed ($\psi = 0$) system it is possible to introduce action-angle variables (y, x) so that $y \to 0$ when a point approaches the separatrices. Since the perturbation is exponentially small, from both sides of the resonance domain there exist invariant tori on which the action is exponentially small: $|y| < e^{-c/\sqrt{\varepsilon}}$. Unfortunately, this almost obvious fact is not a direct consequence of standard theorems from KAM theory. The reason is that frequencies on these tori are not quite usual: the frequency corresponding to the angle variable φ is small and tends to zero as $\varepsilon \searrow 0$. However, the situation can be reduced to standard by passing to the separatrix map. When it is proven that on some pair of invariant tori $|y| < e^{-c/\sqrt{\varepsilon}}$, it remains to use the fact that the frequency ratio on invariant tori close to separatrices is of order $-1/\log|y|$ (Lemma 9.7).

Consider the time-2π map in the system with Hamiltonian (9.30):

$$T_\varepsilon : N \to N, \qquad N = D_{\tilde{y}} \times \mathbb{T}_{\tilde{x}}, \quad \varepsilon > 0.$$

In Fig. 9.5 the fixed point $\hat{z}_\varepsilon = z_\varepsilon(0)$ and the corresponding separatrices of T_ε are represented. The map T_ε satisfies conditions (A)–(B) from Sect. 4.6 with $\delta_\varepsilon = \exp(-\alpha/\sqrt{\varepsilon})$. Hence by (4.39) the corresponding separatrix map S_ε satisfies the equations $S_\varepsilon : (y, x, \sigma) \mapsto (y', x', \sigma')$,

$$\begin{aligned} y' &= y + O(\delta_\varepsilon), \\ x' &= x + \omega(\varepsilon) + \frac{\log|y' + O(\delta_\varepsilon)|}{\log \mu_\varepsilon}, \\ \sigma' &= \sigma \ \mathrm{sign}(y'), \end{aligned} \tag{9.31}$$

where $y = y(\tilde{p}, \tilde{q}, \varepsilon)$ coincides with the "unperturbed" energy:

$$y = \tilde{H}_0 + O(\tilde{H}_0^2). \tag{9.32}$$

We look for invariant KAM curves of the separatrix map. To this end we perform in (9.31) the change

$$y = e^{-\alpha/(2\sqrt{\varepsilon})}(1 + u\lambda), \quad y' = e^{-\alpha/(2\sqrt{\varepsilon})}(1 + u'\lambda).$$

Here we assume that $u, u' \in [-1/2, 1/2]$. The map (9.31) takes the form

$$u' = u + a_1(x, u, \varepsilon), \quad x' = x + \varkappa(u, \varepsilon) + a_2(\varphi, u, \varepsilon), \quad \sigma' = \sigma = \sigma_0, \tag{9.33}$$

$$\varkappa = -\frac{\alpha}{2\lambda\sqrt{\varepsilon}} - \frac{1}{\lambda}\log(\alpha^2\lambda) + \frac{1}{\lambda}\log(1 + \lambda u). \tag{9.34}$$

The functions a_i are 1-periodic in φ, real-analytic in u, x, and satisfy the estimate $|a_1| + |a_2| = O(e^{-\alpha/(2\sqrt{\varepsilon})})$. Theorem 2.4 in this situation can be formulated as follows.

Theorem 9.9. *Suppose that for some* $u^0 \in [-1/2, 1/2]$ *the frequency vector* $(\varkappa^0, 1)^T$, $\varkappa^0 = \varkappa(u^0, \varepsilon)$ *is Diophantine. Then for small* $\varepsilon > 0$ *there exist real-analytic, 1-periodic functions* $P(s)$, $Q(s)$ *such that the curve*

$$\vartheta = \{(x, u) : (x, u) = (s + P(s), u^0 + Q(s)), \ s \in \mathbb{T}\}$$

is invariant with respect to the map (9.33), *and on* ϑ *the map is a rigid shift:* $s \mapsto s + \varkappa^0$. *Moreover,* $|P| + |Q| \le \delta e^{-\alpha/(2\sqrt{\varepsilon})}$, $\delta = \text{const} > 0$.

It is easy to show that for some $u_0 \in [-1/2, 1/2]$ the frequency (9.34) is Diophantine. The curve ϑ corresponds to a closed invariant curve of the map (9.31), where $|y| < c\,e^{-\alpha/(2\sqrt{\varepsilon})}$, and $\sigma = \sigma'$. This implies that the map T_ε has a closed invariant curve l, on which by (9.32)

$$|\tilde{H}_0(\tilde{p}, \tilde{q}, \varepsilon)| \le c\,e^{-\alpha/(2\sqrt{\varepsilon})}. \tag{9.35}$$

The equation $\sigma = \sigma'$ means that the curve l comes close to the hyperbolic fixed point $\hat{z}$ only once; in other words, it is outside the resonant domain.

Now it remains to estimate the frequency corresponding to l in terms of the variables $(\tilde{p}, \tilde{q})$. Putting in (9.30) formally $\tilde{\psi} = 0$, we introduce the action-angle variables (K, Θ). Then the full system (9.30) takes the form

$$\mathscr{H} = \sqrt{\varepsilon}\mathscr{H}_0(K, \varepsilon) + e^{-\alpha/\sqrt{\varepsilon}}\mathscr{H}_1(K, \Theta, \varepsilon, t), \qquad \mathscr{H}_0(K, \varepsilon) = \tilde{H}_0(\tilde{p}, \tilde{q}, \varepsilon). \tag{9.36}$$

The variables (K, Θ) are introduced in domains not containing separatrices of the system $(9.30)|_{\tilde{\psi}=0}$. We can assume that $K \to 0$ when a point approaches the separatrices. The curve l corresponds to an invariant torus in the system with Hamiltonian (9.36).

The function $K(\tilde{p}, \tilde{q}, \varepsilon)$ is a continuous first integral of this system provided $e^{-\alpha/\sqrt{\varepsilon}} = 0$. Outside the separatrices $\{\tilde{H}_0(\tilde{p}, \tilde{q}, \varepsilon) = 0\}$ it is real-analytic. According to Lemma 9.7 (see below), $K = O(K \log K)$ as $K \to 0$. Therefore, due to (9.35), the invariant torus is situated in the domain $|K| < K_0\varepsilon^{-1}e^{-\alpha/(2\sqrt{\varepsilon})}$. The frequency on this torus is $|\varkappa| = |\sqrt{\varepsilon}\frac{\partial H_0}{\partial K}|$. For small $\varepsilon > 0$ by Corollary 9.8, $|\varkappa| \le 4\alpha\varepsilon$. Theorem 9.6 is proved. □

Frequencies on invariant curves, close to separatrices. Consider an autonomous Hamiltonian system with one degree of freedom in the domain $D \subset \mathbb{R} \times \mathbb{T}$. The Hamiltonian $\tilde{H}_0 : D \to \mathbb{R}$, $\tilde{H}_0 = \tilde{H}_0(\tilde{p}, \tilde{q})$ is assumed to be real-analytic. Suppose that there is a hyperbolic equilibrium $\hat{z} \in D$, and the corresponding separatrices

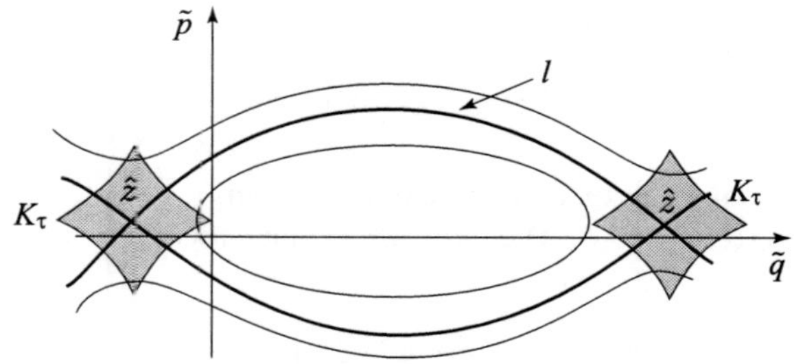

Fig. 9.5 Level lines of the function $\tilde{H}_0$, the curve l, and the square K_τ.

are doubled (Fig. 9.5). Three families of invariant closed curves have the separatrix loops of the point $\hat{z}$ as limit curves. For one of these families both loops form a limit curve, and for other two only one of the loops is limit. Consider one of the latter two families. Following the standard procedure, we assign to each curve a value of the action variable I, where $I \to 0$ as the curve approaches the separatrix.

Lemma 9.7. *There exists a function F, real-analytic at the point $(0, 0, 0) \in \mathbb{R}^3$ with $F(0, 0, 0) = 1$, such that*

$$\tilde{H}_0 = \mathscr{H}_0(I) = \frac{\lambda I}{\log|I|} F\left(\frac{1}{\log|I|}, \frac{\log|\log|I||}{\log|I|}, I\right). \tag{9.37}$$

Differentiating equation (9.37), we obtain the following corollary.

Corollary 9.8. *For I close to zero*

$$\frac{\partial \mathscr{H}_0}{\partial I} = \frac{\lambda}{\log|I|} R\left(\frac{1}{\log|I|}, \frac{\log|\log|I||}{\log|I|}, I\right),$$

where R is real-analytic at the point $(0, 0, 0)$, and $R(0, 0, 0) = 1$.

Corollary 9.9. *For I close to zero*

$$\frac{\partial \mathscr{H}_0}{\partial I} = \frac{\lambda}{\log|\mathscr{H}_0/\lambda|}\left(1 + O\left(\frac{\log\log \mathscr{H}_0}{\log \mathscr{H}_0}\right)\right).$$

Proof (of Lemma 9.7). Let (x, y) be real-analytic normal coordinates in a neighborhood of the hyperbolic point $\hat{z}$. The function $\tilde{H}$ has the form $\tilde{H} = \hat{H}(\xi) = \lambda\xi + O(\xi^2)$, $\xi = xy$. Solving the equation $h = \hat{H}(\xi)$ with respect to ξ, we have $\xi = \frac{1}{\lambda} h f(h)$, where f is real-analytic at the point $h = 0$, and $f(0) = 1$.

Consider the family of invariant curves, where $x > 0$, $y > 0$ (other cases are analogous). Let

$$l \subset \{(\tilde{p}, \tilde{q}) \in D : \tilde{H}_0(\tilde{p}, \tilde{q}) = h\}$$

be the limit curve for this family. Consider also the square $K_\tau = \{(x, y) \in \mathbb{R}^2 : x, y \in [0, \tau]\}$, where $\tau > 0$ is a small number.

The closed curve l can be broken into two parts: $l_1 = l \cap K_h$ and $l_2 = l \backslash K_h$.

The canonical coordinates $(\tilde{p}, \tilde{q})$ and (x, y) are related as follows: $\tilde{p}d\tilde{q} = ydx + dS$, where the function $S = S(x, y)$ is real-analytic in x, y at zero. Therefore,

$$2\pi I = \int_l \tilde{p}\, d\tilde{q} = \int_{l_1} y\, dx + \int_{l_1} dS + \int_{l_2} \tilde{p}\, d\tilde{q}.$$

We have:

$$\int_{l_1} y\, dx = \int_{\lambda^{-1}hf(h)/\tau}^{\tau} \frac{hf(h)}{\lambda x}\, dx = \frac{hf(h)}{\lambda}\left(\log\frac{\tau^2\lambda}{f(h)} - \log|h|\right).$$

The family of curves $l_2 = l_2(h)$ is real-analytic in h at zero. Therefore, $\int_{l_2} pdq$ is real-analytic at zero. Hence, to find the function $\mathscr{H}_0(I) = h$, we should solve the equation

$$I = -\frac{1}{\lambda} hf(h)\log|h| + hg(h),$$

with respect to h, where f and g are real-analytic at zero. We put $h = -\frac{\lambda I}{\log|I|}\chi$. The new unknown function $\chi(I)$ satisfies the equation

$$G(\chi, I, 1/\log|I|, \log|\log|I||/\log|I|) = 0,$$

where G is real-analytic at the point $(1, 0, 0, 0)$, $G(1, 0, 0, 0) = 0$, and $\frac{\partial G}{\partial \chi}(1, 0, 0, 0) \neq 0$. Now Lemma 9.7 follows from the implicit function theorem. □

9.4 Stability Islands in Separatrix Lobes

Let $T_\varepsilon : D \to D$ be a family of symplectic maps of a two-dimensional domain D onto itself. The map T_0 is assumed to be integrable, i.e., there is a smooth locally non-constant function $F: D \to \mathbb{R}$ such that $F \circ T_0 = F$. We assume that T_0 has a hyperbolic fixed point σ_0 generating two separatrix loops. The point σ_0 is critical for the first integral F. We assume that this critical point is non-degenerate.

The map T_ε has a hyperbolic fixed point σ_ε smoothly depending on ε. Generically for $\varepsilon \neq 0$ the separatrices of σ_ε split and their segments form two-angular domains, the lobes. The vicinity of the split separatrices (especially the union of the lobes) is a zone where the chaos is most visible. It is well-known in particular that the lobes are responsible for transport phenomena in the stochastic layer [114]. However chaos in lobes is not uniform. One piece of evidence for this is the existence of stability islands inside lobes.

Sinai [125] discovered that the elliptic periodic trajectories can visit separatrix lobes (Fig. 9.6). We call these trajectories EPL trajectories (elliptic, periodic, passing through lobes). In [125] the map is not near-integrable.

One can obtain many periodic solutions near the separatrices of the point σ_ε by the methods of symbolic dynamics, but all these solutions are unstable.

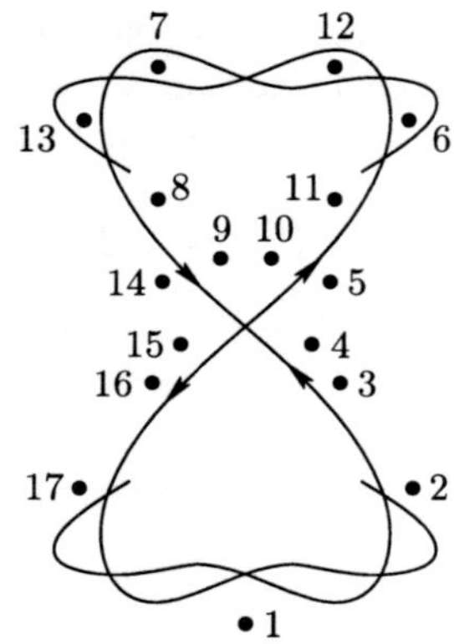

Fig. 9.6 The simplest type of periodic trajectory intersecting a separatrix lobe.

Kozlov showed that, if the Poincaré–Melnikov integral (3.4), Sect. 3.2, does not vanish identically, then the unperturbed resonance curves, located near the unperturbed separatrices, destruct. As a result, non-degenerate periodic solutions occur [71]. (For a multidimensional generalization of this observation, see [133].) All these solutions make exactly one full rotation inside the separatrix loop. Some of them are stable. Stable periodic trajectories obtained in this way can exist only outside the $\sqrt{|\varepsilon|}$-neighborhood of the separatrices [42], and hence cannot intersect the separatrix lobes, because the width of the latter is of order ε.

In this section, following [124], we show that EPL-trajectories typically appear in near-integrable systems.

To each periodic trajectory in a neighborhood of separatrices one can assign its type: the number of times it passes near the point σ_ε. In other words, a trajectory of type n is a periodic trajectory of period n for the separatrix map.

The simplest example of an EPL trajectory is shown in Fig. 9.6. The trajectory makes a full revolution inside the upper loop of the figure-eight separatrix and passes once along the whole figure-eight. According to our classification, the type $n = 3$. One can easily see that the type of an EPL trajectory is always at least three.

Let $\varepsilon_0 > 0$ be sufficiently small and let $J = J(T_\varepsilon) \subset (-\varepsilon_0, \varepsilon_0)$ be the set of values of the parameter ε, $|\varepsilon| < \varepsilon_0$, for which the map T_ε has an EPL trajectory. We assume that T_0 is fixed, and write $T' = dT_\varepsilon/d\varepsilon|_{\varepsilon=0}$. Then T' is a Hamiltonian vector field.

Let T_ε have smoothness[8] $C^{\mathbf{r}}$, $\mathbf{r} \in \{\mathbf{r}_0, \mathbf{r}_0 + 1, \dots, \infty, \omega\}$, and let $\mathscr{A}$ be the space of $C^{\mathbf{r}}$-smooth vector fields on N. We equip the space $\mathscr{A}$ with the C^2-topology.

Theorem 9.10. *There is an open set $\mathscr{S} \subset \mathscr{A}$ such that for any $T' \in \mathscr{S}$ the relative measure of the set J on the interval $(-\varepsilon, \varepsilon)$ exceeds a positive constant independent of $\varepsilon \in (0, \varepsilon_0)$.*

Conjecture 9.2. The set $\mathscr{S}$ can be chosen dense in $\mathscr{A}$.

[8] It is possible to take $\mathbf{r}_0 = 13$.

Theorem 9.10 follows from the corresponding result for reversible families T_ε. Let $I_\varepsilon : N \to N$ be a $C^{\mathbf{r}}$-smooth family of involutions ($I_\varepsilon^2 = \mathrm{id}$) such that $I_\varepsilon(\sigma_\varepsilon) = \sigma_\varepsilon$. The maps T_ε are said to be *reversible* with respect to I_ε (or I_ε-reversible) if

$$T_\varepsilon \circ I_\varepsilon = I_\varepsilon \circ T_\varepsilon^{-1}.$$

If T_0 is reversible, then I_0 maps the separatrix figure-eight onto itself. We assume that I_0 preserves the loops of the figure-eight, that is, it maps each loop onto itself. Informally speaking, this involution acts as a symmetry with respect to the vertical symmetry axis of the figure-eight.

Theorem 9.11. *Suppose that T_ε is reversible with respect to a family of loop-preserving involutions I_ε and at least one pair of separatrices of the point σ_ε splits in the first approximation with respect to ε. Then the relative measure of the set J on the interval $(-\varepsilon, \varepsilon)$ exceeds a positive constant independent of $\varepsilon \in (0, \varepsilon_0)$.*

As a rule, elliptic periodic trajectories generate stability islands. We assert that the area of such an island for a generic EPL trajectory is of the order of the area of the separatrix lobe. This means that, in principle, these islands can be observed on pictures obtained as a result of numerical simulation of the dynamics. However, naive attempts to observe stability islands in separatrix lobes usually fail. This happens for two reasons. First, the measure of the set J corresponding to 'not too small' islands is not large (although this depends strongly on the parameters of the map). If μ_0 stands for the leading multiplier of T_0 at the point σ_0, then rough numerical estimates show that this measure varies from 0.004 to 0.3 as μ_0 varies from 1.1 to 10. Second, the mean relative measure of a 'not too small' island in a lobe usually varies from 10^{-8} to 10^{-4} in the same interval of μ_0. For more detailed information about the numerical investigation of the problem, see [124].

Separatrix map in a reversible system. Consider a family of two-dimensional I_ε-reversible symplectic near-integrable maps T_ε.

Lemma 9.8. *Let the hyperbolic rotation*

$$\mathscr{T}(q, p) = (q\mathscr{M}(pq), p/\mathscr{M}(pq)), \qquad \mathscr{M}(0) = \mu > 1,$$

be reversible with respect to an involution $\mathscr{I}$ preserving the first and the third quadrants of $\mathbb{R}^2_{q,p}$. Then there are normal coordinates $(\widetilde{q}, \widetilde{p})$ in which

$$\mathscr{I}(\widetilde{q}, \widetilde{p}) = (\widetilde{p}(1 + O(\widetilde{q}\,\widetilde{p})), \widetilde{q}(1 + O(\widetilde{q}\,\widetilde{p}))). \tag{9.38}$$

Proof. Consider the function $\Phi(p, q) = qp$. Let us show first that

$$\Phi \circ \mathscr{I} = \Phi + O(\Phi^2). \tag{9.39}$$

Indeed, it follows from the definition of reversibility that

$$\mathscr{I} \circ \mathscr{T} = \mathscr{T}^{-1} \circ \mathscr{I}. \tag{9.40}$$

Since Φ is a first integral of $\mathscr{T}$, we have

$$\Phi \circ \mathscr{I} \circ \mathscr{T} = \Phi \circ \mathscr{T}^{-1} \circ \mathscr{I} = \Phi \circ \mathscr{I}.$$

This implies that $\Phi \circ \mathscr{I}$ is also a first integral of $\mathscr{T}$. It follows from Proposition 4.1 that $\Phi \circ \mathscr{I} = c_0 + c_1\Phi + O(\Phi^2)$. Moreover, $c_0 = 0$ because $\mathscr{I}(0) = 0$. Then

$$\Phi = \Phi \circ \mathscr{I} \circ \mathscr{I} = c_1\Phi \circ \mathscr{I} + O(\Phi \circ \mathscr{I})^2 = c_1^2\Phi + O(\Phi^2).$$

Hence, $c_1 = \pm 1$. In fact, $c_1 = 1$, because $\mathscr{I}$ preserves the first and the third quadrants of $\mathbb{R}^2_{q,p}$. This implies (9.39).

Let $\mathscr{I}^{(q)}$ and $\mathscr{I}^{(p)}$ be the first and second components of the image of $\mathscr{I}$. Since $\mathscr{I}$ interchanges the stable and unstable separatrices, we have

$$\mathscr{I}^{(q)}(q, 0) = 0, \quad \mathscr{I}^{(p)}(0, p) = 0.$$

Hence, we can put

$$\mathscr{I}^{(q)}(q, p) = p\,a(p, q), \quad \mathscr{I}^{(p)}(q, p) = q\,b(p, q), \tag{9.41}$$

where a and b are smooth functions. Using (9.40), we obtain

$$a(q\mu, p/\mu) = a(p, q) + O(qp), \quad b(q\mu, p/\mu) = b(p, q) + O(qp).$$

Therefore,

$$a(q, p) = a(0, 0) + O(qp), \quad b(q, p) = b(0, 0) + O(qp).$$

By (9.39) we have $a(0, 0)b(0, 0) = 1$. Since the involution $\mathscr{I}$ preserves the first and third quadrants, it follows that $a(0, 0) > 0$. Putting $p = \widetilde{p}\sqrt{a(0, 0)}$ and $q = \widetilde{q}/\sqrt{a(0, 0)}$, we obtain (9.38). □

Let the maps T_ε be I_ε-reversible, where I_ε is a family of involutions preserving the loops. Then it follows from Lemma 9.8 and Remark 4.1 that in a neighborhood of the hyperbolic fixed point σ_ε there are normal coordinates $(q, p) = (q_\varepsilon, p_\varepsilon)$ in which the involution I_ε satisfies (9.38).

Below it is convenient to replace the variable x defined in (4.13), Chap. 4, by $x - (\log \alpha_\pm)/\lambda$. In the new coordinates the separatrix map (4.1), Chap. 4, becomes

$$\mathscr{S}\mathscr{M}_\varepsilon \begin{pmatrix} x \\ y \\ s \end{pmatrix} = \begin{pmatrix} x^+ \\ y^+ \\ s^+ \end{pmatrix} = \begin{pmatrix} x + \frac{1+O(\varepsilon)}{\lambda}(\log \frac{\varepsilon}{\alpha_s^2\lambda} + \log|y^+|) \\ y + \lambda v_s(x - (\log \alpha_s)/\lambda)O(\varepsilon) \\ s \cdot \operatorname{sign}(y^+) \end{pmatrix}. \tag{9.42}$$

We also write out the formulas for the inverse map $\mathscr{S}\mathscr{M}_\varepsilon^{-1}$,

$$\mathscr{S}\mathscr{M}_\varepsilon^{-1}\begin{pmatrix}x\\y\\s\end{pmatrix} = \begin{pmatrix}x^-\\y^-\\s^-\end{pmatrix} = \begin{pmatrix}x - \frac{1+O(\varepsilon)}{\lambda}(\log\frac{\varepsilon}{\alpha_{s^-}^2\lambda} - \log|y|)\\ y - \lambda\nu_{s^-}(x^- - (\log\alpha_{s^-})/\lambda) + O(\varepsilon)\\ s\cdot\mathrm{sign}(y)\end{pmatrix}.$$

Proposition 9.5. *The reversibility of the map T_ε induces the reversibility of the separatrix map. The corresponding involution has the form*

$$I_\varepsilon\begin{pmatrix}x\\y\\s\end{pmatrix} = \begin{pmatrix}-x + 2(\log\alpha_s)/\lambda + O(\varepsilon)\\ y - \lambda\nu_s(-x + (\log\alpha_s)/\lambda) + O(\varepsilon)\\ s\end{pmatrix}. \tag{9.43}$$

The functions $\nu_\pm$ are odd: $\nu_\pm(x) = -\nu_\pm(-x)$.

Proof. The separatrix map is determined by equation (4.12) (Chap. 4). Therefore

$$\begin{aligned}\mathscr{S}\mathscr{M}_\varepsilon^{-1} &= \mathscr{G}_\varepsilon^{-1}\circ\mathscr{T}_\varepsilon^{-n_r} = \mathscr{G}_\varepsilon^{-1}\circ\mathscr{I}_\varepsilon\circ\mathscr{T}_\varepsilon^{n_r}\circ\mathscr{I}_\varepsilon\\ &= \mathscr{G}_\varepsilon^{-1}\circ\mathscr{I}_\varepsilon\circ\mathscr{T}_\varepsilon^{n_r}\circ\mathscr{G}_\varepsilon\circ\mathscr{G}_\varepsilon^{-1}\circ\mathscr{I}_\varepsilon\\ &= \mathscr{G}_\varepsilon^{-1}\circ\mathscr{I}_\varepsilon\circ\mathscr{S}\mathscr{M}_\varepsilon\circ\mathscr{G}_\varepsilon^{-1}\circ\mathscr{I}_\varepsilon.\end{aligned}$$

Since $\mathscr{I}_\varepsilon\circ\mathscr{G}_\varepsilon^{-1} = \mathscr{G}_\varepsilon\circ\mathscr{I}_\varepsilon$, we see that in the normal coordinates the map $\mathscr{S}\mathscr{M}_\varepsilon$ is reversible with respect to the involution $\mathscr{I}_\varepsilon\circ\mathscr{G}_\varepsilon^{-1}$.

Using (4.11) (Chap. 4) and (9.38), we get

$$(\mathscr{G}_\varepsilon)^{-1}\circ\mathscr{I}_\varepsilon\begin{pmatrix}q\\p\end{pmatrix} = \begin{pmatrix}\alpha_s^2/q + O(\varepsilon, p)\\ pq^2\alpha_s^{-2} - \varepsilon q\nu_s(\log|q|/\lambda) + O_2(\varepsilon, p)\end{pmatrix}. \tag{9.44}$$

Writing this system in the coordinates (x mod 1, y, s), we obtain (9.43). The fact that the functions $\nu_\pm(x)$ are odd follows from the identity $I_\varepsilon^2 = \mathrm{id}$. □

Symmetric trajectories of period three. To use the reversibility, we seek EPL trajectories symmetric with respect to the involution I_ε. Let $z_0 = (x_0, y_0, -1)$ be the initial condition for a periodic trajectory of period 3 of the map $\mathscr{S}\mathscr{M}_\varepsilon$. Suppose that z_0 belongs to the symmetry axis, that is, $I_\varepsilon(z_0) = z_0$. Then by (9.43) we have $x_0 = \lambda^{-1}\log\alpha_- + O(\varepsilon)$.

We write the periodicity condition in the form $\mathscr{S}\mathscr{M}_\varepsilon^{-1}(z_0) = \mathscr{S}\mathscr{M}_\varepsilon^2(z_0)$:

$$\begin{cases}\frac{\log\alpha_+}{\lambda} - \gamma + O(\varepsilon) = \gamma + \frac{1}{\lambda}(\log\frac{\varepsilon}{\alpha_+\lambda} + \log(y_0 + \lambda\nu_+(\gamma))) + O(\varepsilon) - m,\\ y_0 - \lambda\nu_+(-\gamma) + O(\varepsilon) = y_0 + \lambda\nu_+(\gamma) + O(\varepsilon),\\ 0 < -\mathrm{sign}(y_0) = -\mathrm{sign}(y_0)\,\mathrm{sign}(y_0 + \alpha_+^2\nu_+(\gamma)).\end{cases}$$

Here $m \in \mathbb{N}$ is the period of the corresponding trajectory of the map T_ε, and

$$\gamma = \frac{1}{\lambda}\log\frac{\varepsilon|y_0|}{\alpha_-\alpha_+\lambda}.$$

The reversibility of the map and the symmetry of the trajectory imply that the first two equations are dependent. They are equivalent to the equation

$$1 + \mu^{\kappa - m - 3\chi} = \mu^r \nu_+(\chi) + O(\varepsilon), \tag{9.45}$$

where

$$y_0 = -\lambda\mu^{-r}, \quad \varepsilon = \alpha_+^2 \mu^{\chi - r + \kappa}, \quad \alpha_+/\alpha_- = \mu^\kappa. \tag{9.46}$$

Stability conditions. The ellipticity condition has the form $|\operatorname{tr} M| < 2$, where the monodromy matrix M is the differential of the map $\mathcal{S}\mathcal{M}_\varepsilon^3$ at the point z_0. Neglecting terms of the form $O(\varepsilon)$, we put

$$z_1 = \mathcal{S}\mathcal{M}_\varepsilon(z_0) = \begin{pmatrix} \frac{1}{\lambda}\log\frac{\varepsilon|y_0|}{\alpha_-\lambda} \\ y_0 \\ 1 \end{pmatrix}, \tag{9.47}$$

$$z_2 = \mathcal{S}\mathcal{M}_\varepsilon^2(z_0) = \mathcal{S}\mathcal{M}_\varepsilon^{-1}(z_0) = \begin{pmatrix} -\frac{1}{\lambda}\log\frac{\varepsilon|y_0|}{\alpha_-\alpha_+^2\lambda} \\ y_0 + \lambda\nu_+(\gamma) \\ 1 \end{pmatrix}. \tag{9.48}$$

The differential of the map $\mathcal{S}\mathcal{M}_\varepsilon$ is

$$D\mathcal{S}\mathcal{M}_\varepsilon(z) = \begin{pmatrix} 1 + \lambda\nu_s'(x - \log\alpha_s/\lambda)/y^+ & 1/y^+ \\ \lambda\nu_s'(x - \log\alpha_s/\lambda) & 1 \end{pmatrix}, \tag{9.49}$$

where $z = (x, y, s)$ and $y^+ = y + \lambda\nu_s(x - \log\alpha_s/\lambda)$.

By (9.45)–(9.49), the equation $\operatorname{tr} M = \operatorname{tr}(D\mathcal{S}\mathcal{M}_\varepsilon(z_2)\, D\mathcal{S}\mathcal{M}_\varepsilon(z_1) \times D\mathcal{S}\mathcal{M}_\varepsilon(z_0))$ can be written in the form

$$\begin{aligned} \operatorname{tr} M &= -2 + \delta\Lambda + O(\varepsilon), \\ \Lambda &= \mu^{r-\kappa+m+3\chi}\nu_-'(0) + (2 - \mu^r\nu_-'(0))(2 + \mu^{r-\kappa+m+3\chi}\nu_+'(\chi)), \\ \delta &= 1 - \mu^r\nu_+'(\chi). \end{aligned}$$

EPL trajectories exist. The existence condition for a symmetric EPL trajectory of period 3 is

$$1 + \mu^{\kappa - m - 3\chi} = \mu^r \nu_+(\chi) + O(\varepsilon), \tag{9.50}$$

$$|-2 + \delta\Lambda + O(\varepsilon)| < 2. \tag{9.51}$$

Proposition 9.6. *Suppose that the variables (m, r, χ), $m \in \mathbb{N}$, satisfy (9.50)–(9.51) with the terms $O(\varepsilon)$ omitted. Then $(m + 3, r, \chi - 1)$ also satisfy this system.*

Corollary 9.10. *The set of $\varepsilon > 0$ for which there is an EPL trajectory given by (9.50)–(9.51) is asymptotically invariant under the contraction $\varepsilon \mapsto \varepsilon/\mu$.*

Corollary 9.11. *To prove Theorem* 9.11, *it suffices to establish the existence of at least one solution of the system* (9.50)–(9.51)$|_{O(\varepsilon)=0}$.

Corollary 9.10 is obvious. To prove Corollary 9.11, we assume that the system (9.50)–(9.51)$|_{O(\varepsilon)=0}$ has a solution (m_0, r_0, χ_0). Since (9.51) is an inequality, it follows that for $m = m_0$ the system (9.50)–(9.51)$|_{O(\varepsilon)=0}$ has a curvilinear interval of solutions on the (r, χ)-plane. In particular, there is an interval $(\varepsilon_1, \varepsilon_2)$ lying in J. According to Corollary 9.10, the set J contains a large portion (in the sense of measure) of the points in the set $\bigcup_{n=0}^{\infty}(\mu^{-n}\varepsilon_1, \mu^{-n}\varepsilon_2)$. This implies Theorem 9.11.

Since the separatrices of T_ε split in the first approximation in ε, at least one of the functions $\nu_\pm$ does not vanish identically. We assume that $\nu_+ \not\equiv 0$. (Otherwise we interchange the symbols $+$ and $-$ in the above formulas.)

Proposition 9.7. *For any* $\kappa \in \mathbb{R}$ *the system* (9.50)–(9.51)$|_{O(\varepsilon)=0}$ *has a solution.*

Proof. We fix κ and put $O(\varepsilon) = 0$ in (9.50)–(9.51). We recall that the periodic functions $\nu_\pm$ have zero mean value.

For any $m \in \mathbb{Z}$ there is a smooth curve $\Phi_m \subset \{(\chi, r) \in \mathbb{R}^2 : \chi \in [0, 2\pi]\}$, a solution of (9.50). The curves Φ_m can be assumed to be oriented. Let ∂ be the operator of differentiation along Φ_m. One can prove in an elementary way that

(a) on each of the curves Φ_m there is a point (χ_m, r_m) at which $\delta = 0$,
(b) the equation $\Lambda(\chi, r) = 0$ holds at most at finitely many points (χ_m, r_m),
(c) the equation $\partial\delta(\chi, r) = 0$ also holds at most at finitely many points (χ_m, r_m).

Let $(\chi_m, r_m) \in \Phi_m$ be a point at which $\Lambda \neq 0$ and $\partial\delta \neq 0$. Then by (9.51)

$$\operatorname{tr} M(\chi_m, r_m) = -2, \qquad \partial(\operatorname{tr} M)(\chi_m, r_m) \neq 0.$$

Hence, in a small neighborhood of the point (χ_m, r_m) on Φ_m there are points at which $|\operatorname{tr} M| < 2$. □

Proof (of Theorem 9.10). Every integrable map T_0 is reversible with respect to some loop-preserving involution. Let us include the map T_0 in a family T_ε which is reversible with respect to an involution preserving the loops. Consider a family of symplectic maps $\hat{T}_\varepsilon$ such that

$$T_0 = \hat{T}_0, \qquad \left| dT_\varepsilon/d\varepsilon|_{\varepsilon=0} - d\hat{T}_\varepsilon/d\varepsilon|_{\varepsilon=0} \right| = \Delta$$

for some small $\Delta > 0$. The separatrix map corresponding to $\hat{T}_\varepsilon$ also satisfies (9.42), but the periodic functions $\hat{\nu}_\pm$ playing the role of the functions $\nu_\pm$ are no longer odd in general. However, they are $O(\Delta)$-close to odd functions $\nu_\pm$:

$$|\hat{\nu}_+ - \nu_+| \sim \Delta, \quad |\hat{\nu}_- - \nu_-| \sim \Delta.$$

If Δ is small, then any EPL trajectory of T_ε can be extended to an EPL trajectory of the map $\hat{T}_\varepsilon$. Since the set J for the maps T_ε consists of intervals, there is an interval $(\hat{\varepsilon}_1, \hat{\varepsilon}_2)$ such that the map $\hat{T}_\varepsilon$ has an EPL trajectory for any $\varepsilon \in (\hat{\varepsilon}_1, \hat{\varepsilon}_2)$. Theorem 9.10 is now a consequence of the following simple assertion.

Suppose that the map T_ε has an EPL trajectory for some small number ε. Then the map $T_{\varepsilon/\mu}$ also has an EPL trajectory (with a larger period).

This assertion is similar to Corollary 9.10 and follows from equations (9.42). □

The area of the stability island around an EPL trajectory has the same order of smallness as the area of the separatrix lobe, because both areas turn out to be of order 1 in the variables (x, y).

9.5 Elements of Analysis on Scales

The continuous averaging method produces initial value problems for countable systems of partial differential equations. These systems are the generalization of the standard Cauchy–Kovalevskaya problem to the case when the vector of the unknown functions is infinite dimensional.

The Nirenberg–Nishida theorem (see below) guarantees that the solution to such a system exists provided the evolution variable belongs to a small interval. But in the applications of the continuous averaging method it is important to know good estimates for the length of this interval from below. Such estimates follow neither from the Nirenberg–Nishida theorem nor from other general results, and they rest heavily on the precise nature of the equations.

In this section we develop a technique required for the analysis of these problems. This technique is based on the theory of locally convex spaces [113]. The main cause for such a choice is the properties of the differentiation operation on the set of analytic functions. We provide the set of analytic functions with the topology of compact convergence and obtain a locally convex space. The differentiation is not a bounded operator in any reasonable Banach space but in this locally convex space it is. Fortunately most fundamental theorems for Banach spaces have generalizations for locally convex ones.

9.5.1 Locally Convex Spaces

A linear topological space is called a locally convex space if it has a basis that consists only of convex neighborhoods of the origin. In this section we study a special class of locally convex spaces.

Let $\mathscr{S}$ be an arbitrary set. Let E stand for a linear space. Suppose the space E is endowed with a collection of seminorms $\{\|\cdot\|_s\}_{s\in\mathscr{S}}$, where for any non-zero element $u \in E$ there exists a seminorm $\|\cdot\|_s$ such that $\|u\|_s \neq 0$.

Taking the sets

$$U_{\varepsilon,s_1,\dots,s_n} = \left\{u \in E : \max_{s\in\{s_1,\dots,s_n\}} \|u\|_s < \varepsilon\right\} \tag{9.52}$$

for a basis of neighborhoods of the origin, we provide the space E with topology. Here ε is a positive number and $\{s_1, \dots, s_n\} \subset \mathscr{S}$ is an arbitrary finite set.

The topology of any Hausdorff locally convex space can be described in such a way. But we will study locally convex spaces with an extra hypothesis: suppose that the topology of the space E can be defined equivalently if in (9.52) we take indices $\{s_1, \dots, s_n\}$ not from the whole set $\mathscr{S}$ but only from a countable subset $\mathscr{S}' \subseteq \mathscr{S}$. If E satisfies this hypothesis we say that E is *countably normed*.

For example, if $\mathscr{S} = (0, S)$ is an interval and the seminorms of the space E satisfy the inequality

$$\| \cdot \|_s \le \| \cdot \|_{s'}, \qquad 0 < s < s' < S,$$

then we can take $\mathscr{S}' = (0, S) \cap \mathbb{Q}$ and the space E turns out to be countably normed.

Below all topological spaces are countably normed.

Definition 9.1. We say that a sequence $\{u_j\}_{j\in\mathbb{N}} \subset E$ converges to $u \in E$ if for any $s \in \mathscr{S}$

$$\lim_{j\to\infty} \|u_j - u\|_s = 0.$$

All the topological notions on a countably normed space can be described in terms of sequences.

Theorem 9.12 ([113]). *A set $K \subset E$ is compact iff any sequence $\{u_k\}_{\kappa\in\mathbb{N}} \subset K$ contains a subsequence that is convergent to an element from K.*

A set $K \subseteq E$ is called relatively compact if the closure $\overline{K}$ is compact.

Definition 9.2. A set $U \subset E$ is said to be bounded if $\sup_{u\in U} \|u\|_s\} < \infty$ for all $s \in \mathscr{S}$.

Now consider the space $C([a, b], E)$ of continuous functions on $[a, b]$ with values in E. The space is locally convex with respect to a collection of seminorms

$$\|u(\cdot)\|_s^C = \max_{t\in[a,b]} \|u(t)\|_s, \qquad s \in \mathscr{S}.$$

Definition 9.3. A set $G \subseteq C([a, b], E)$ is said to be equicontinuous if for any $s \in \mathscr{S}$ and for any $\varepsilon > 0$ there exists a constant $\delta > 0$ such that $\sup_{u\in G} \|u(t_1) - u(t_2)\|_s < \varepsilon$ for all $t_1, t_2 \in [a, b]$, $|t_1 - t_2| < \delta$.

Theorem 9.13 (Ascoli [120]). *Suppose that a set $W \subset C([a, b], E)$ is equicontinuous and the set*

$$W_t = \{u(t) \mid u(t) \in W\} \subset E$$

is compact for any $t \in [a, b]$. Then W is compact in $C([a, b], E)$.

Now let us formulate a locally convex space generalization of the Schauder fixed point theorem [56] (see also [26]).

Theorem 9.14 (Schauder–Tikhonov Theorem).[9] *Let $K \subset E$ be convex and compact. Then any continuous mapping $F : K \to K$ has a fixed point $\hat{u} = F(\hat{u}) \in K$.*

9.5.2 Scales of Banach Spaces

Many problems of partial differential equations can not be described satisfactory in a single Banach space. These problems require the consideration of some specially organized sets of Banach spaces, the so called scales.

Definition 9.4. A set of Banach spaces $\{(E_s, \|\cdot\|_s)\}_{0<s<1}$ is called a scale of Banach spaces if the following conditions hold

$$E_{s'} \subseteq E_s, \quad \|\cdot\|_s \le \|\cdot\|_{s'}, \qquad 0 < s < s' < 1.$$

As an example we put

$$\|w\| = \max_j |w_j|, \qquad w = (w_1, \dots, w_m) \in \mathbb{C}^m.$$

Let $D \subset \mathbb{R}^m$ be a bounded domain and let

$$D_s = \{z = x + w \mid x \in D,\ w \in \mathbb{C}^m,\ \|w\| < sR\}, \qquad 0 < s < 1 \tag{9.53}$$

be a complex neighborhood of this domain. Let H_s be a set of functions $f : D_s \to \mathbb{C}$ analytic in D_s and continuous in the closure $\overline{D}_s$. Being endowed with the norm

$$\|f\|_s^H = \sup_{z \in D_s} |f(z)|$$

the set H_s becomes a Banach space. The spaces H_s, $0 < s < 1$, form a scale of Banach spaces.

Consider the differentiation operator

$$\frac{\partial}{\partial x_k} : H_{s'} \to H_s, \qquad 0 < s < s' < 1, \quad k = 1, \dots, m.$$

By the Cauchy estimate (see Lemma 2.1)

$$\left\| \frac{\partial}{\partial x_k} \right\|_{H_{s'} \to H_s} \le \frac{C}{s' - s}. \tag{9.54}$$

Here the constant C does not depend on s and s'.

Now let us return to the general construction. Let us associate with a scale $\{(E_s, \|\cdot\|_s)\}_{0<s<1}$ a locally convex space

[9] Theorems 9.13 and 9.14 remain valid not only for the case of a countably normed space E but also if E is an arbitrary locally convex space.

$$E = \bigcap_{0<s<1} E_s$$

with the collection of norms $\|\cdot\|_s$, $s \in (0, 1)$.

Theorem 9.15. *Let L be a closed subspace of E and let L_s be the closure of L in E_s. Then*

$$L = \bigcap_{0<s<1} L_s, \quad L_{s'} \subset L_s, \quad \text{for all } s' > s.$$

This theorem shows that, if a locally convex space is associated with a scale of Banach spaces, then its closed subspace is also associated with a scale of Banach spaces.

Proof. The inclusions $L \subset \bigcap_{0<s<1} L_s$, $L_{s'} \subset L_s$, are obvious. To verify that $\bigcap_{0<s<1} L_s \subseteq L$, we assume the converse: there exists a point $x \in E$ such that $x \in \bigcap_{0<s<1} L_s$, $x \notin L$. Then there are sequences $\{x_k^s\}_{k\in\mathbb{N}} \subset L$ such that $\|x_k^s - x\|_s \to 0$ as $k \to \infty$.

Take an increasing sequence $s_j \to 1$ as $j \to \infty$ and construct the following neighborhoods of the point x in E:

$$U_j = \left\{ y \in E : \|y - x\|_{s_j} < \frac{1}{j} \right\}.$$

Now pick from the sequence $\{x_k^{s_j}\}_{k\in\mathbb{N}}$ an element x_j which belongs to U_j, $j \in \mathbb{N}$. Then the sequence $\{x_j\} \subset L$ converges to the element x in E. Indeed, take any $\varepsilon > 0$ and $s \in (0, 1)$. Choose a number N such that $1/N \le \varepsilon$ and $s_N \ge s$. Then for all x_k, $k \ge N$, we have

$$\|x_k - x\|_s \le \|x_k - x\|_{s_k} < \frac{1}{k} \le \varepsilon.$$

Thus $x \in L$. The contradiction proves the theorem. □

Introduce a locally convex space $H = \bigcap_{0<s<1} H_s$ with the collection of norms $\{\|\cdot\|_s^H\}$. This space is associated with the scale $\{H_s\}$ but it can be described directly. Consider an open neighborhood

$$D(R) = \{z = x + w \mid x \in D, \ w \in \mathbb{C}^m, \ \|w\| < R\}$$

of the domain D. Then H is the space of holomorphic functions $f : D(R) \to \mathbb{C}$.

Recall some notation. If $z = (z_1, \ldots, z_m) \in D(R)$ and $j = (j_1, \ldots, j_m) \in \mathbb{Z}_+^m$ then

$$z^j = z_1^{j_1} \cdots z_m^{j_m}, \qquad |j| = j_1 + \cdots + j_m, \quad \partial z^j = \partial z_1^{j_1} \cdots \partial z_m^{j_m}.$$

By (9.54) the n-th order differential operator

$$D^n = \sum_{|j|\le n} a_j(z)\frac{\partial^{|j|}}{\partial z^j} : H \to H, \qquad a_j \in H$$

is continuous. One consequence of this fact is as follows. The set $\ker D^n$ is a closed subspace of H, and thus by Theorem 9.15 the space $\ker D^n$ is associated with the corresponding scale.

Let T_1, T_2 be locally convex spaces. Recall that the embedding $T_1 \subseteq T_2$ is completely continuous if any bounded set of T_1 is relatively compact in T_2.

Theorem 9.16 (Montel).[10] *The embedding $H_{s'} \subset H_s$, $s < s'$, is completely continuous.*

Theorem 9.17. *Let $\{(E_s, \|\cdot\|_s^E)\}_{0<s<1}$ and $\{(V_s, \|\cdot\|_s^V)\}_{0<s<1}$ be two scales of Banach spaces. Assume that for any s there exists s' such that the embedding $E_{s'} \subseteq V_s$ is completely continuous. Then the embedding*

$$E \subseteq V = \bigcap_{0<s<1} V_s$$

is also completely continuous.

Proof. Enumerate a countable set $Q = (0, 1) \cap \mathbb{Q}$ in the following way:

$$Q = \{s_j\}_{j\in\mathbb{N}}.$$

Take a bounded set $U \subset E$ and let a number s be such that the embedding $E_s \subset V_{s_1}$ is completely continuous. Then by Definition 9.2 U is bounded with respect to the norm $\|\cdot\|_s^E$. Thus U contains a sequence $\{u_k\}$ convergent with respect to the norm $\|\cdot\|_{s_1}^V$. Then pick $\tilde{s}$ such that the embedding $E_{\tilde{s}} \subset V_{s_2}$ is completely continuous. The sequence $\{u_k\}$ is bounded with respect to the norm $\|\cdot\|_{\tilde{s}}^E$. Then this sequence contains a subsequence, say $\{u_{k_l}\} \subset \{u_k\}$, that is convergent with respect to the norm $\|\cdot\|_{s_2}^V$, and so on. The diagonal sequence is convergent in all the norms $\|\cdot\|_{s_j}^V$, $j \in \mathbb{N}$. This implies that this sequence is convergent in all the norms $\|\cdot\|_s^V$, $s \in (0, 1)$.

By Theorem 9.12, Theorem 9.17 is proved. □

Corollary 9.12. *Every bounded subset of the space H is relatively compact.*

Indeed, by the Montel theorem the embedding $H_{s'} \subset H_s$, $s < s'$, is completely continuous. Therefore by Theorem 9.17 the embedding $H \subseteq H$ is also completely continuous.

9.5.3 Other Examples of Scales

Now we introduce two scales of analytic functions and study their properties.

[10] The general version of the Montel theorem is contained in [120].

The first scale consists of Banach spaces $(G_s, \|\cdot\|_s^G)$. The space G_s is the space of the sequences $u = \{u_k(x)\}_{k\in\mathbb{Z}}$, $u_k \in H_s$, with the norm

$$\|u\|_s^G = \sup_{k\in\mathbb{Z}}\{(1+|k|^2)\|u_k\|_s^H\}. \tag{9.55}$$

For any $s < s'$ and $j = 1, \dots, m$, we have the operation

$$\frac{\partial}{\partial x_j} : G_{s'} \to G_s, \qquad \frac{\partial}{\partial x_j}u = \left\{\frac{\partial}{\partial x_j}u_k(x)\right\}_{k\in\mathbb{Z}}.$$

Now let us introduce a scale $(F_s, \|\cdot\|_s^F)$. The Banach spaces F_s consist of the sequences $u = \{u_k(x)\}_{k\in\mathbb{Z}}$, $u_k \in H_s$, but the norm is

$$\|u\|_s^F = \sqrt{\sum_{k\in\mathbb{Z}}(1+|k|^2)(\|u_k\|_s^H)^2}. \tag{9.56}$$

As a direct consequence of formula (9.54) we have the inequality

$$\left\|\frac{\partial}{\partial x_j}u\right\|_s^F \le \frac{C}{s'-s}\|u\|_{s'}^F, \qquad 0 < s < s' < 1. \tag{9.57}$$

The constant $C > 0$ does not depend on u, s, s'.

Proposition 9.8. *The embeddings $G_{s'} \subset F_s$, $0 < s < s' < 1$, are completely continuous.*

Proof. Consider the projector $P_n u = \{u_k\}_{|k|\le n}$, $u \in F_s$.

Let a set $W \subset G_{s'}$ be bounded. That is, there exists $M > 0$ such that

$$\|w\|_{s'}^G \le M \quad \text{for any } w = \{w_k\}_{k\in\mathbb{Z}} \in W.$$

This implies that

$$\|w_k\|_{s'}^H \le \frac{M}{1+|k|^2}. \tag{9.58}$$

The set $P_n(W)$ is an ε-net for W in F_s. Indeed, take an element $w \in W$. Then by (9.58)

$$(\|w - P_n w\|_s^F)^2 = \sum_{|k|>n}(\|w_k\|_s^H)^2(1+|k|^2) \le M^2\sum_{|k|>n}\frac{1}{1+|k|^2}.$$

The last expression of this formula is arbitrarily small if n is big enough.

By the Montel Theorem the set $P_n(W)$ is relatively compact in F_s. Therefore W is relatively compact in F_s. □

Theorem 9.17 implies the following corollary.

Corollary 9.13. *The embedding*

$$G \subset F, \quad G = \bigcap_{0<s<1} G_s, \quad F = \bigcap_{0<s<1} F_s,$$

is completely continuous.

Now we endow the space F_s with an algebraic structure. Let $u = \{u_k(x)\}_{k\in\mathbb{Z}} \in F_s$ and $v = \{v_k(x)\}_{k\in\mathbb{Z}} \in F_s$.

Proposition 9.9. *If a set of constants $\{b_{m,n}\}_{m,n\in\mathbb{Z}} \subset \mathbb{C}$ is bounded:*

$$\sup_{m,n\in\mathbb{Z}} |b_{m,n}| \le c_1 < \infty$$

then the space F_s is a Banach algebra with respect to the multiplication

$$uv = \left\{ \sum_{m+n=k} b_{m,n} u_n v_m \right\}_{k\in\mathbb{Z}},$$

so that

$$\|uv\|_s^F \le c\, \|u\|_s^F \, \|v\|_s^F,$$

where the constant $c > 0$ depends only on c_1.

Remark 9.2. Note that this multiplication is distributive but in general it is not commutative nor associative. Such algebraic properties depend on the constants $b_{m,n}$. But in our applications the constants $b_{m,n}$ are such that this multiplication is commutative and associative.

Proof. To prove the proposition, we introduce two 2π-periodic functions

$$U(t) = \sum_{k\in\mathbb{Z}} \|u_k\|_s^H e^{ikt}, \quad V(t) = \sum_{k\in\mathbb{Z}} \|v_k\|_s^H e^{ikt}.$$

These functions belong to the Sobolev space[11] $H^1(\mathbb{T})$ and

$$\|u\|_s^F = \|U\|_{H^1(\mathbb{T})}, \quad \|v\|_s^F = \|V\|_{H^1(\mathbb{T})}.$$

Since the space $H^1(\mathbb{T})$ is a Banach algebra with respect to the standard multiplication [2] we have:

[11] The Sobolev space $H^1(\mathbb{T})$ is the Banach space of the Fourier series $u(x) = \sum_{k\in\mathbb{Z}} u_k e^{ikx}$ with finite norm $\|u\|^2_{H^1(\mathbb{T})} = \sum_{k\in\mathbb{Z}} (1 + |k|^2)|u_k|^2$ [128].

$$
\begin{aligned}
(\|uv\|_s^F)^2 &= \sum_{k\in\mathbb{Z}}(1+|k|^2)\left(\left\|\sum_{m+n=k} b_{m,n}u_n v_m\right\|_s^H\right)^2 \\
&\le \sum_{k\in\mathbb{Z}}(1+|k|^2)\left(\sum_{m+n=k} |b_{m,n}|\|u_n\|_s^H\|v_m\|_s^H\right)^2 \\
&\le c_1^2\,\|UV\|^2_{H^1(\mathbb{T})} \le c^2\,\|U\|^2_{H^1(\mathbb{T})}\|V\|^2_{H^1(\mathbb{T})} \\
&= c^2\,(\|u\|_s^F)^2(\|v\|_s^F)^2. \quad \square
\end{aligned}
\tag{9.59}
$$

Corollary 9.14. *For any* $0 < s < s' < 1$ *and* $u, v, u', v' \in F_s$

$$
\left\|u\frac{\partial v}{\partial x_j} - u'\frac{\partial v'}{\partial x_j}\right\|_s^F \le \frac{c\max\{\|v\|_{s'}^F, \|u'\|_{s'}^F\}}{s'-s}(\|u-u'\|_{s'}^F + \|v-v'|_{s'}^F).
$$

The positive constant c *depends only on* c_1.

Indeed, this estimate easily follows from the identity

$$
u\frac{\partial v}{\partial x_j} - u'\frac{\partial v'}{\partial x_j} = (u-u')\frac{\partial v}{\partial x_j} + u'\left(\frac{\partial v}{\partial x_j} - \frac{\partial v'}{\partial x_j}\right)
$$

and formula (9.57).

To conclude this section let us make an important remark on the spaces G_s and F_s.

Remark 9.3. We mainly use sequences $w(z) = \{w_j(z)\}_{j\in\mathbb{Z}}$ which consist not of the scalar functions $w_j(z)$ but of the vector functions $w_j(z) = (w_{j,1}, \ldots, w_{j,q})(z)$.

The spaces of such sequences are also denoted by G_s and F_s. We also use formulas (9.55) and (9.56) with $\|w_j\|_s^H = \sup_{z\in D_s} |w_j(z)\|$, where $\|\cdot\|$ is any norm in $\mathbb{C}^q$.

Another way to consider the vector-valued sequence $w = \{w_j\}_{j\in\mathbb{Z}}$ as an element of G_s or F_s is to stretch it into the scalar sequence

$$
(\ldots, w_{-1,1}, \ldots, w_{-1,q}, w_{0,1}, \ldots, w_{0,q}, w_{1,1}, \ldots, w_{1,q}, \ldots).
$$

In both cases all the theorems about the spaces G_s and F_s remain valid.

9.5.4 Majorant Functions

Here we state two functional analytic facts on the relation "$\ll$". (See Sect. 6.2.)

Suppose formally that the domain D consists of a single point $0 \in \mathbb{R}^m$ and put

$$
D_s = \{z \in \mathbb{C}^m : \|z\| < sR\}.
$$

Thus the set D_s becomes a polycircle of radius sR.

Proposition 9.10. *Suppose that* $u, U \in H_s$, *where* $u \ll U$. *Then* $\|u\|_s^H \le \|U\|_s^H$.

Proof. Let $u(z) = \sum_{j\in\mathbb{Z}_+^m} u_j z^j$, $U(z) = \sum_{j\in\mathbb{Z}_+^m} U_j z^j$. Then $|u_j| \le U_j$. There exists a point $\hat{z} \in \overline{D_s}$ such that

$$\|u\|_s^H = |u(\hat{z})| \le \sum_{j\in\mathbb{Z}_+^m} |u_j|\,|\hat{z}^j| \le \sum_{j\in\mathbb{Z}_+^m} U_j (sR)^{|j|} = U(\tilde{z}), \qquad \tilde{z} = sR\cdot\mathbf{1}. \quad \square$$

Proposition 9.11. *The set* $V = \{u \mid u \ll U \in H_s\}$ *is closed in* H_s.

Proof. Let $\{u_k\} \subset V$ and $u_k = \sum_{j\in\mathbb{Z}_+^m} u_{k,j} z^j \to u = \sum_{j\in\mathbb{Z}_+^m} u_j z^j$ in H_s as $k \to \infty$. Evidently, it is sufficient to check that $u_{k,j} \to u_j$. But this follows from the Cauchy estimate

$$\left| \frac{\partial^{|j|}(u_k - u)}{\partial z_1^{j_1} \cdots \partial z_m^{j_m}}(0) \right| \le c_{|j|,s} \|u_k - u\|_s^H.$$

Here $c_{|j|,s} > 0$ is a constant independent of u, u_k. $\square$

The properties of the spaces H_s and particularly formula (9.54) prompt the idea of generalizing the classical Cauchy–Kovalevskaya theorem from the space of analytic functions to general scales of Banach spaces. This generalization in its final form has been accomplished by Nirenberg [97] and Nishida [98]. The corresponding theorem is formulated in Sect. 6.2.

References

1. Abraham, R. and Marsden, J. E., Foundations of mechanics, Benjamin/ Cummings, Redwood City, 1978.
2. Adams, R., Sobolev spaces, Academic Press, New York, 1975.
3. Ahn, T., Kim, G. and Kim, S., Analysis of the separatrix map in Hamiltonian systems, *Physica D* **89** (1996), 315–328.
4. Alekseev, V. M., Quasi-random dynamical systems I, II, III, *Mat. Sb.* **76** (1968), 72–134, **77** (1968), 545–601, **78** (1969), 3–50 (Russian). Engl. transl.: *Math. USSR-Sb.* **5** (1968), 73–128, **6** (1968), **7** (1969), 1–43.
5. Arnold, V. I., Small denominators I. On self-maps of a circle, *Izv. Russ. Akad. Sci. Ser. Mat.* **25**:1 (1961), 21–86.
6. Arnold, V. I., Proof of a theorem of A. N. Kolmogorov on the persistence of quasi-periodic motions under small perturbations of the Hamiltonian, *Usp. Mat. Nauk* **18**:5 (1963), 13–40. English transl.: *Russ. Math. Surv.* **18**:5 (1963).
7. Arnold, V. I., Small denominators and problems of stability of motion in classical and celestial mechanics, *Usp. Mat. Nauk* **18**:6 (1963), 85–191. English transl.: *Russ. Math. Surv.*
8. Arnold, V. I., Instability of dynamical systems with several degrees of freedom, *Dokl. Akad. Nauk SSSR* **156**:1 (1964), 9–12. English transl.: *Sov. Mat. Dokl.* **5** (1964), 581–585.
9. Arnold, V. I., Mathematical methods in classical mechanics, Springer, Berlin, 1982.
10. Arnold, V. I. and Givental, A. B., Symplectic geometry, Encyclopedia of Math. Sciences, vol. 4, Springer, Berlin, 1989.
11. Arnold, V. I., Kozlov, V. V. and Neishtadt, A. I., Mathematical aspects of classical and celestial mechanics, Encyclopedia of Math. Sciences, vol. 3, Springer, Berlin, 1989.
12. Aubry, S. and Abramovici, G., Chaotic trajectories in the standard map: the concept of anti-integrability, *Physica D* **43** (1990), 199–219.
13. Banyaga, A., de la Llave, R. and Wayne, C. E., Cohomology equations near hyperbolic points and geometric versions of Sternberg linearization theorem, *J. Geom. Anal.* **6**:4 (1996), 613–649.

D. Treschev, O. Zubelevich, *Introduction to the Perturbation Theory of Hamiltonian Systems*, Springer Monographs in Mathematics,
DOI 10.1007/978-3-642-03028-4,

14. Bessi, U., An approach to Arnold's diffusion through the calculus of variations, *Nonlinear Anal.* **20** (1993), 1303–1318.
15. Birkhoff, G. D., Surface transformations and their dynamical applications, *Acta Math.* **43** (1920), 1–119.
16. Birkhoff, G. D., Dynamical systems, Amer. Math. Soc. Colloquium Publications, vol. 9, Am. Math. Soc., Providence, 1927.
17. Bogolyubov, N. N. and Mitropolsky, Yu. A., Asymptotic methods in the theory of nonlinear oscillations, Nauka, Moscow, 1974.
18. Bolotin, S. V., Condition for Liouville integrability of Hamiltonian systems, *Vestn. Mosk. Univ. Ser. I Math. Mech.* **3** (1986), 58–64.
19. Bolotin, S. V., On the Hill determinant of a periodic orbit, *Vestn. Mosk. Univ. Ser. I Mat. Meh.* **3** (1988), 30–34.
20. Bolotin, S. V., Doubly asymptotic to invariant tori motions in the perturbation theory of Hamiltonian systems, *Prikl. Mat. Meh.* **54**:3 (1990), 497–501.
21. Bolotin, S. V., Homoclinic orbits to invariant tori of Hamiltonian systems, *Am. Math. Soc. Transl. (2)* **168** (1995), 21–89.
22. Bolotin, S. V. and MacKay, R., Multibump orbits near the anti-integrable limit for Lagrangian systems, *Nonlinearity* **10** (1997), 1015–1029.
23. Bolotin, S. and Treschev, D., Unbounded growth of energy in nonautonomous Hamiltonian systems, *Nonlinearity* **12** (1999), 365–387.
24. Bolotin, S. and Treschev, D., Remarks on the definition of hyperbolic tori of Hamiltonian systems, *Regul. Chaotic Dyn.* **5**:4 (2000), 401–412.
25. Broer, H. W., Huitema, G. B. and Sevryuk, M. B., Quasi-periodic motions in families of dynamical systems: order amidst chaos, Lect. Notes in Math., vol. 1645, Springer, Berlin, 1996.
26. Browder, F. E., A new generalization of the Schauder fixed point theorem, *Math. Ann.* **174** (1967), 285–290.
27. Bunimovich, L. A., Sinai, Ya. G. and Chernov, N. I., Statistical properties of two-dimensional hyperbolic billiards, *Usp. Mat. Nauk* **46**:4(280) (1991), 43–92 (Russian). Transl.: *Russ. Math. Surv.* **46**:4 (1991), 47–106.
28. Cheng, C.-Q. and Wang, S., The surviving of lower-dimensional tori from a resonant torus of Hamiltonian systems, *J. Differ. Equ.* **155** (1999), 311–326.
29. Cheng, C.-Q. and Yan, J., Existence of diffusion orbits in a priori unstable Hamiltonian systems, *J. Differ. Geom.* **67**:3 (2004), 457–517.
30. Chierchia, L. and Falcolini, C., A direct proof of a theorem by Kolmogorov in Hamiltonian systems. *Preprint*, Scuola Normale Superiore, Pisa (1994). Serie IV, Vol. XXI, Fasc. 4.
31. Chierchia, L. and Gallavotti, G., Drift and diffusion in phase space, *Ann. Inst. Henri Poincaré, Phys. Theor.* **60**:1 (1994), 1–144.
32. Chirikov, B. V., A universal instability of many-dimensional oscillation systems, *Phys. Rep.* **52**:5 (1979), 263–379.
33. de la Llave, R., A tutorial on KAM theory, in Smooth ergodic theory and its applications (Seattle, WA, 1999), Proc. Sympos. Pure Math., vol. 69, pp. 175–292, Am. Math. Soc., Providence, 2001.

34. Delshams, A. and Gutierrez, P., Splitting potential and the Poincaré-Melnikov method for whiskered tori in Hamiltonian systems, *J. Nonlinear Sci.* **10**:4 (2000), 433–476.
35. Delshams, A., Gelfreich, V., Jorba, A. and Seara, T., Exponentially small splitting of separatrices under fast quasiperiodic forcing, *Commun. Math. Phys.* **189** (1997), 35–71.
36. Delshams, A., de la Llave, R. and Seara, T. M., A geometric approach to the existence of orbits with unbounded energy in generic periodic perturbations by a potential of generic geodesic flows of $\mathbb{T}^2$, *Commun. Math. Phys.* **209** (2000), 353–392.
37. Delshams, A., de la Llave, R. and Seara, T., A geometric mechanism for diffusion in Hamiltonian systems overcoming the large gap problem: heuristic and rigorous verification on a model, *Mem. Am. Math. Soc.* **179**:844 (2006), 1–141.
38. Dold, A., Lectures on algebraic topology, Springer, Berlin, 1972.
39. Douady, R., Une démonstration directe de l'écvivalence des théoremes de tores invariants pour difféomorphismes et champs de vecteures, *C. R. Acad. Sci. Paris A* **295** (1982), 201–204.
40. Douady, R., Applications du théoreme des tores invariantes. *Thesis*, Université Paris VII (1982).
41. Douady, R., Stabilité ou instabilité des points fixes elliptiques, *Ann. Sci. Ec. Norm. Super. IV Ser.* **21** (1988), 1–46.
42. Dovbysh, S. A., The splitting of separatrices and the generation of isolated periodic solutions in Hamiltonian systems with one and a half degrees of freedom, *Russ. Math. Surv.* **44**:2 (1989), 281–282.
43. Dovbysh, S. A., The structure of the Kolmogorov set near separatrices of a plane map, *Mat. Zametki* **46**:4 (1989), 112–114 (in Russian).
44. Dovbysh, S. A., Transversal intersection of separatrices and branching of solutions as obstructions to the existence of an analytic integral in many-dimensional systems I. Basic results: separatrices of hyperbolic periodic points, *Collect. Math.* **50**:2 (1999), 119–197.
45. Duarte, P., Plenty of elliptic islands for the standard family of area preserving maps, *Ann. Inst. Henri Poincaré Anal. Non Lineaire* **11**:4 (1994), 359–409.
46. Elliasson, L. H., Absolutely convergent series expansions for quasi-periodic motions, *Univ. of Stockholm, Sweden* **2** (1988), 1–31.
47. Elliasson, L. H., Perturbations of stable invariant tori, *Ann. Sc. Norm. Super. Pisa, Cl. Sci. Ser. IV* **15** (1988), 115–147.
48. Eliasson, L. H., Biasymptotic solutions of perturbed integrable Hamiltonian systems, *Bol. Soc. Bras. Mat.* **25** (1994), 57–76.
49. Filonenko, N. N., Sagdeev, R. Z. and Zaslavsky, G. M., Destruction of magnetic surfaces by magnetic field irregularities, *Nucl. Fus.* **7** (1967), 253–266.
50. Gallavotti, G., Twistless KAM tori, *Commun. Math. Phys.* **164** (1994), 145–156.
51. Gelfreich, V., Reference systems for splitting of separatrices, *Nonlinearity* **10**:1 (1997), 175–193.

52. Gelfreich, V. and Lazutkin, V., Splitting of separatrices: perturbation theory and exponential smallness, *Russ. Math. Surv.* **56**:3 (2001), 499–558.
53. Gelfreich, V. G., Lazutkin, V. F. and Svanidze, N. V., Refined formula to separatrix splitting for the standard map, *Physica D* **71** (1994), 101.
54. Gradshtein, I. S. and Ryzhik, I. M., Tables of integrals, series, and products, Academic Press, San Diego, 1980.
55. Graff, S. M., On the conservation of hyperbolic tori for Hamiltonian systems, *J. Differ. Equ.* **15**:1 (1974), 1–69.
56. Granas, A. and Dugundji, J., Fixed point theory, Springer, New York, 2003.
57. Greene, J. M., A method for determining a stochastic transition, *J. Math. Phys.* **20** (1979), 1183–1201.
58. Herman, M. R., Sur les courbes invariantes par les difféomorphismes de l'anneau, *Astérisque* (1983), 103–104.
59. Herman, M., Inégalités "a priori" pour des tores Lagrangiens invariants par des difféomorphismes symplectiques, *Inst. Hautes Études Sci. Publ. Math.* **70** (1990), 47–101.
60. Herman, M., Examples of compact hypersurfaces in $R^{2p}, 2p \geq 6$, with no periodic orbits, in Hamiltonian systems with three or more degrees of freedom (S'Agaro, 1995), NATO Adv. Sci. Inst. Ser. C Math. Phys. Sci., vol. 533, p. 126, Kluwer Academic, Dordrecht, 1999.
61. Hill, G. W., On the part of the motion of the lunar perigee which is a function of the mean motions of the sun and moon, *Acta Math.* **VIII**:1 (1886), 1–36.
62. Holmes, P. J. and Marsden, J. E., Melnikov's method and Arnold diffusion for perturbations of Hamiltonian systems, *J. Math. Phys.* **23** (1982), 669–675.
63. Kaloshin, V., Geometric proofs of Mather's connecting and accelerating theorems, in Proc. (Katsively), Cambridge University Press, Cambridge, 2003.
64. Kappeler, T. and Poschel, J., KdV & KAM, Ergebnisse der Mathematik und ihrer Grenzgebiete, Folge. A Series of Modern Surveys in Mathematics, vol. 45, Springer, Berlin, 2003 (Results in Mathematics and Related Areas. 3rd Series. A Series of Modern Surveys in Mathematics).
65. Khanin, K. and Sinai, Ya., Renormalization group method and Kolmogorov-Arnold-Moser theory, in Sagdeev, R. Z. (ed.), Nonlinear phenomena in plasma physics and hydrodynamics, pp. 31–64, Mir, Moscow, 1986.
66. Khinchin, A. Ja , Chain fractions, Nauka, Moscow, 1978 (in Russian).
67. Klingenberg, W., Lectures on closed geodesics, Springer, Berlin, 1978.
68. Kolmogorov, A. N., On conservation of conditionally periodic motions for a change in Hamiltonian's function, *Dokl. Akad. Nauk SSSR* **98**:4 (1954), 527–530.
69. Kosygin, D., Multidimensional KAM theory for the renormalization group viewpoint, in Dynamical systems and statistical mechanics, Adv. Sov. Math., vol. 3, pp. 99–129, Am. Math. Soc., Providence, 1991.
70. Kozlov, V. V., Methods of qualitative analysis in the dynamics of a rigid body, Moskov. Gos. Univ., Moscow, 1980.
71. Kozlov, V. V., The splitting of separatrices and the generation of isolated periodic solutions in Hamiltonian systems with one-and-a-half degrees of freedom, *Russ. Math. Surv.* **41**:5 (1986), 149–150.

72. Kozlov, V. V., Symmetries, topology and resonances in Hamiltonian mechanics, Springer, Berlin, 1996.
73. Kozlov, V. and Treschev, D., Billiards: a genetic introduction to the dynamics of systems with impacts, Translations of Mathematical Monographs, vol. 89, Am. Math. Soc., Providence, 1991.
74. Kuksin, S., On the inclusion of an almost integrable analytic symplectomorphism into a Hamiltonian flow, *Russian J. Math. Phys.* **1**:2 (1993), 191–207.
75. Kuksin, S., Analysis of Hamiltonian PDEs, Oxford Univ. Press, London, 2000.
76. Kuksin, S. and Pöschel, J., On the inclusion of analytic symplectic maps in analytic Hamiltonian flows and its applications, in Kuksin, S., Lazutkin, V. F. and Pöschel, L. (eds.), Seminar on dynamical systems, Euler international mathematical institute, St. Petersburg 1991, Progress in Nonlinear Differential Equations and Their Applications, vol. 12, pp. 96–116, Birkhäuser, Basel, 1994.
77. Lazutkin, V. F., To the Moser's theorem on invariant curves, in Vopr. dinamich. teorii rasprostr. seismich., vol. 14, pp. 109–120, Nauka, Leningrad, 1974 (in Russian).
78. Lazutkin, V. F., Splitting of separatrices for standard Chirikov's mapping, VINITI no. 6372-84, 24 Sept. 1984 (in Russian).
79. Lazutkin, V. F., On the width of a stability zone near separatrices of the standard map, *Dokl. Akad. Nauk SSSR* **313**:2 (1990), 268–272 (in Russian). English transl.: *Sov. Math. Dokl.* **42**:1 (1991), 5–9.
80. Lochak, P., Canonical perturbation theory via simultaneous approximation, *Usp. Mat. Nauk* **47**:6 (1992), 59–140 (in Russian). English transl.: *Russ. Math. Surv.* **47**:6 (1992), 57–133.
81. MacKay, R. S., A renormalization approach to invariant circles in area-preserving maps, *Physica D* **7** (1983), 283–300.
82. MacKay, R. S. and Meiss, J. D., Linear stability of periodic orbits in Lagrangian systems, *Phys. Lett. A* **98** (1983), 92–94.
83. MacKay, R. S. and Meiss, J. D., Cantori for symplectic maps near the anti-integrable limit, *Nonlinearity* **5**:149 (1992), 1–12.
84. MacKay, R. S. and Percival, I. C., Converse KAM: theory and practice, *Commun. Math. Phys.* **98** (1985), 469–512.
85. Marco, J.-P. and Sauzin, D., Stability and instability for Gevrey quasi-convex near-integrable Hamiltonian systems, *Publ. Math.* **96** (2002), 199–275.
86. Markeev, A. P., Libration points in celestial mechanics and cosmodynamics, Nauka, Moscow, 1978 (in Russian).
87. Mather, J., Invariant subsets for area-preserving homeomorphisms of surfaces, in Mathematical Analysis and Applications, Part B. Advances in Mathematics Supplementary Studies, vol. 7B, pp. 531–562, Academic Press, San Diego, 1981.
88. Mather, J., Non-existence of invariant circles, *Ergod. Theory Dyn. Syst.* **4** (1984), 301–311.
89. Mather, J., Talk given at the conference in dynamical systems, Oberwolfach (1997)

90. Melnikov, V. K., On the stability of the center for time-periodic perturbations, *Tr. Mosk. Mat. Obs.* **12** (1963), 3–52 (Russian). English transl.: *Trans. Mosc. Math. Soc.* **1963** (1965), 1–56.
91. Morse, M., A fundamental class of geodesics in any closed surface of genus greater than one, *Trans. Am. Math. Soc.* **26** (1924), 25–61.
92. Moser, J., The analytical invariants of an area-preserving mapping near hyperbolic point, *Commun. Math. Phys.* **9**:4 (1956), 673–692.
93. Moser, J., Convergent series expansions for quasiperiodic motions, *Math. Ann.* **169** (1967), 136–176.
94. Neishtadt, A. I., Estimates in the Kolmogorov theorem on conservation of conditionally periodic motions, *Prikl. Mat. Meh.* **45**:6 (1981), 1016–1025 (in Russian). English transl.: *J. Appl. Math. Mech.* **45**:6 (1981), 766–772.
95. Neishtadt, A. I., The separation of motions in systems with rapidly rotating phase, *Prikl. Mat. Meh.* **48**:2 (1984), 197–204 (in Russian). English transl.: *J. Appl. Math. Mech.* **48**:2 (1984), 133–139.
96. Nekhoroshev, N. N., An exponential estimate of the time of stability in Hamiltonian systems close to integrable, *Usp. Mat. Nauk* **32**:6 (1977), 5–66 (in Russian). English transl.: *Russ. Math. Surv.* **32**:6 (1977), 1–65.
97. Nirenberg, L., An abstract form of the nonlinear Cauchy-Kowalewski theorem, *J. Differ. Geom.* **6** (1972), 561–576.
98. Nishida, T., *J. Differ. Geom.* **12**:4 (1977), 629–633.
99. Nitecki, Z., Differentiable dynamics, MIT Press, Cambridge, 1971.
100. Oliveira, F., On the generic existence of homoclinic points, *Ergod. Theory Dyn. Syst.* **7** (1987), 567–595.
101. Olvera, A. and Simó, C., An obstruction method for the destruction of invariant curves, *Physica D* **26** (1987), 181–192.
102. Piftankin, G. N., Diffusion speed in the Mather problem, *Nonlinearity* **19** (2006), 2617–2644.
103. Piftankin, G. N. and Treschev, D. V., Separatrix map in Hamiltonian systems, *Usp. Mat. Nauk* **62**:2(374) (2007), 3–108 (Russian). Transl.: *Russ. Math. Surv.* **62**:2 (2007), 219–322.
104. Plotnikov, P. I., Morse theory for conditionally periodic solutions of Hamiltonian systems, *Sib. Math. J.* **35** (1994), 590–604.
105. Poincaré, H., Les métodes nouvelles de la mécanique céleste, vols. 1–3, Gauthier–Villars, Paris, 1892, 1893, 1899.
106. Pöschel, J., Über invariante Tori in differenzierbaren Hamiltonschen Systemen, *Bonner Math. Schr.* **37** (1980), 302–310.
107. Pöschel, J., Integrability of Hamiltonian systems on Cantor sets, *Commun. Pure Appl. Math.* **35**:1 (1982), 653–695.
108. Pöschel, J., On elliptic lower-dimensional tori in Hamiltonian systems, *Math. Z.* **202** (1989), 559–608.
109. Pöschel, J., Nekhoroshev estimates for quasi-convex Hamiltonian systems, *Math. Z.* **213**:2 (1993), 187–216.
110. Pronin, A. and Treschev, D., On the inclusion of analytic maps into analytic flows, *Regul. Chaotic Dyn.* **2**:2 (1997), 14–24.

111. Ramis, J. P. and Schäfke, R., Gevrey separation of fast and slow variables, *Nonlinearity* **9** (1996), 353–384.
112. Reed, M. and Simon, B., Methods of modern mathematical physics, Academic Press, San Diego, 1978.
113. Robertson, A. P. and Robertson, W. J., Topological vector spaces, Cambridge Univ. Press, Cambridge, 1964.
114. Rom-Kedar, V. and Wiggins, S., Transport in two-dimensional maps, *Arch. Ration. Mech. Anal.* **109**:3 (1990), 239–298.
115. Rüssmann, H., On the frequencies of quasi-periodic solutions of analytic nearly-integrable Hamiltonian systems, in Kuksin, S., Lazutkin, V. F. and Pöschel, L. (eds.), Seminar on dynamical systems, Euler international mathematical institute, St. Petersburg 1991, Progress in Nonlinear Differential Equations and Their Applications, vol. 12, pp. 160–183, Birkhäuser, Basel, 1994.
116. Samoilenko, A. M., Elements of mathematical theory of multi-frequency oscillations. Invariant tori, Nauka, Moscow, 1987 (in Russian).
117. Samoilenko, A. M. and Kulik, V. L., Exponential dichotomy of an invariant torus of dynamical systems, *Differ. Uravn.* **15** (1979), 1434–1444 (in Russian).
118. Sauzin, D., Caractère Gevrey des solutions formelles dún problème de moyennisation, *C. R. Acad. Sci. Paris, Ser. I* **315** (1992), 991–995.
119. Schmidt, W. M., Diophantine approximations, Lect. Notes in Math , vol. 785, Springer, Berlin, 1988.
120. Schwartz, L., Analyse mathematique, Hermann, Paris, 1967.
121. Sevryuk, M. B., KAM-stable Hamiltonians. *J. Dyn. Control Syst.* **1**:1 (1995), 351–366.
122. Sevryuk, M. B., Invariant tori of Hamiltonian systems nondegenerate in the sense of Rüssmann, *Dokl. Russ. Akad. Nauk* **346**:5 (1996), 590–593. English transl.: *Russian Acad Sci. Dokl. Math.*
123. Shilnikov, L. P., A case of the existence of a denumerable set of periodic motions, *Sov. Math. Dokl.* **6** (1965), 163–166.
124. Simó, C. and Treschev, D., Stability islands in the vicinity of separatrices of near-integrable symplectic maps, *J. Discrete Contin. Dyn. Syst. Ser. B* **10**:2–3 (2008), 681–698.
125. Sinai, Ya. G., Several exact results on decay of correlations, in Zaslavskii, G. M. (ed.), Statistical irreversibility in nonlinear systems, Current Problems in Physics, pp. 124–139, Nauka, Moscow, 1970.
126. Stepin, A. M., Integrable Hamiltonian systems, in Qualitative methods in the studying of nonlinear differential equations and of linear oscillations, pp. 116–143, Math. Inst. of Akad. Nauk USSR, Kiev. 1981.
127. Svanidze, N. V., Small perturbations of a dynamical system with an integral invariant on a torus, in Boundary problems of mathematical physics, 10, *Tr. Mat. Inst. Akad. Nauk SSSR* **147** (1980), 124–146
128. Taylor, M. E., Partial differential equations, vol. 1, Springer, New York, 1996.
129. Takens, F., A C^1 counterexample to Moser's twist theorem, *Indag. Math.* **33**:4 (1971), 379–386.

130. Takens, F., Homoclinic points of conservative systems, *Inv. Math.* **18** (1972), 267–292.
131. Treschev, D., On the question of stability of periodic trajectories of the Birkhoff billiard, *Vestn. Mosk. Univ. Ser. I Mat-Meh.* **2** (1988), 44–50.
132. Treschev, D., A mechanism for the destruction of resonance tori in Hamiltonian systems, *Mat. Sb.* **180**:10 (1989), 1325–1346 (Russian). English transl.: *Math. USSR-Sb.* **68**:1 (1991), 181–203.
133. Treschev, D., Hyperbolic tori and asymptotic surfaces in Hamiltonian systems, *Russ. J. Math. Phys.* **2** (1994), 93–110.
134. Treschev, D., An averaging method for Hamiltonian systems, exponentially close to integrable ones, *Chaos* **6**:1 (1996), 6–14.
135. Treschev, D., Separatrix splitting from the point of view of symplectic geometry, *Mat. Zametki* **61**:6 (1997), 890–906 (in Russian). Transl.: *Math. Notes* **61**:5–6 (1997), 744–757.
136. Treschev, D., The continuous averaging method in the problem of separation of fast and slow motions, *Regul. Chaotic Dyn.* **2**:3/4 (1997), 9–20 (in Russian).
137. Treschev, D., Separatrix splitting for a pendulum with rapidly oscillating suspension point, *Russ. J. Math. Phys.* **5**:1 (1997), 63–98.
138. Treschev, D., Width of stochastic layers in near-integrable two-dimensional symplectic maps, *Physica D* **116**:1–2 (1998), 21–43.
139. Treschev, D., Multidimensional separatrix maps, *J. Nonlinear Sci.* **12**:1 (2002), 27–58.
140. Treschev, D., Trajectories in a neighborhood of asymptotic surfaces of a priori unstable Hamiltonian systems, *Nonlinearity* **15** (2002), 2033–2052.
141. Treschev, D., Evolution of slow variables in a priori unstable Hamiltonian systems, *Nonlinearity* **17** (2004), 1803–1841.
142. Treschev, D. and Zubelevich, O., Invariant tori in Hamiltonian systems with two degrees of freedom in a neighborhood of a resonance, *Regul. Chaotic Dyn.* **3**:3 (1998), 73–81.
143. Trifonov, S. I., Analytic diffeomorphisms as monodromy maps of analytic differential equations, *Vestn. Mosk. Univ., Ser. Mat.* **5** (1986), 70–72 (in Russian). English transl.: *Moscow Univ. Math. Bull.* **41**:5 (1986), 63–65.
144. Veselov, A. P., Integrable mappings, *Russ. Math. Surv.* **46**:5 (1991), 1–51.
145. Wiggins, S., Global bifurcations and chaos, Springer, New York, 1988.
146. Zaslavsky, G. M. and Filonenko, N. N., Stochastic instability of trapped particles and the conditions of applicability of the quasilinear approximation, *Zh. Eksp. Teor. Fiz.* **54** (1968), 1590–1602 (in Russian).
147. Zehnder, E., Generalized implicit function theorem with applications to some small divisor problems I, II, *Commun. Pure Appl. Math.* **28**:1 (1975), 91–140, **29**:1 (1976), 49–111.
148. Ziglin, S. L., Branching of solutions and non-existence of first integrals in Hamiltonian mechanics I, II, *Funkc. Anal. Ego Prilozh.* **16**:3 (1982), 30–41, **17**:1 (1983), 8–23 (in Russian).
149. Zubelevich, O., On the majorant method for the Cauchy–Kovalevskaya problem, *Math. Notes* **69**:3–4 (2001), 329–339.

Index

D. Treschev, O. Zubelevich, *Introduction to the Perturbation Theory of Hamiltonian Systems*, Springer Monographs in Mathematics,
DOI 10.1007/978-3-642-03028-4, © Springer-Verlag Berlin Heidelberg 2010

Lightning Source UK Ltd.
Milton Keynes UK
14 March 2010

151358UK00007B/160/P